COMPUTER TECHNOLOGY IN
NEUROSCIENCE

COMPUTER TECHNOLOGY IN NEUROSCIENCE

EDITED BY

PAUL B. BROWN

West Virginia University Medical Center
Morgantown

HEMISPHERE PUBLISHING CORPORATION
Washington   London

A HALSTED PRESS BOOK

JOHN WILEY & SONS
New York   London   Sydney   Toronto

Copyright © 1976 by Hemisphere Publishing Corporation. All rights reserved. No part of this book may be reproduced in any form, by photostat, microform, retrieval system, or any other means, without the prior written permission of the publisher.

Hemisphere Publishing Corporation
1025 Vermont Ave., N. W., Washington, D. C. 20005

Distributed solely by Halsted Press, a Division of John Wiley & Sons, Inc., New York.

1 2 3 4 5 6 7 8 9 0   P U P U   7 8 4 3 2 1 0 9 8 7 6

Library of Congress Cataloging in Publication Data

Main entry under title:

Computer technology in neuroscience.

    1. Neurobiology–Data processing–Congresses.
I. Brown, Paul Burton, 1942–
QP357.5.C65   612'.8' 02854   76-6884
ISBN 0-470-15061-0

Printed in the United States of America

CONTENTS

Contributors   xi

Preface   xv

chapter 1
DESIGN AND DEVELOPMENT OF A MULTILABORATORY COMPUTER COMPLEX FOR BIOMEDICAL RESEARCH
J. Macy, Jr., and M. P. White   1

chapter 2
DESIGN AND DEVELOPMENT OF A REAL-TIME PROGRAMMING LANGUAGE: RPL
M. P. White, J. Macy, Jr., and J. D. Gerard   17

chapter 3
DESIGN CHARACTERISTICS OF BIOMEDICAL RESEARCH INFORMATION SYSTEMS
B. J. Ransil   29

chapter 4
SNAX: A LANGUAGE FOR INTERACTIVE NEURONAL MODELING AND DATA PROCESSING
D. K. Hartline   41

chapter 5
SIMULATION OF NERVE CELL KINETICS USING INTERACTIVE SIMULATION LANGUAGE
R. D. Benham and D. K. Hartline   67

chapter 6
A CELL KINETICS SIMULATION SYSTEM
C. E. Donaghey   85

chapter 7
RECORDING AND ANALYSIS OF 3-D INFORMATION FROM SERIAL
SECTION MICROGRAPHS: THE CARTOS SYSTEM
  E. R. Macagno, C. Levinthal, C. Tountas, R. Bornholdt,
  and R. Abba   97

chapter 8
A MINICOMPUTER-BASED IMAGE ANALYSIS SYSTEM
  M. J. Shantz   113

chapter 9
AN ALGORITHM FOR THE ALIGNMENT OF SERIAL SECTIONS
  M. L. Dierker   131

chapter 10
COUNTING HIGH CONTRAST CLOSED OBJECTS IN BIOLOGICAL
IMAGES USING A 525-LINE RASTER SCAN TELEVISION
CAMERA AND A MINICOMPUTER
  D. F. Wann   135

chapter 11
AN ALGORITHM FOR THE DISPLAY AND MANIPULATION OF
LINES IN THREE DIMENSIONS
  M. L. Dierker   139

chapter 12
APPLICATIONS OF PLATO IN COMPUTER ASSISTED RESEARCH
  D. Walter and R. McKown   153

chapter 13
REAL TIME CLASSIFICATION AND DISPLAY OF POSTSYNAPTIC
POTENTIAL AMPLITUDES AND INTERVALS
  T. O. Clarke and J. H. Peacock   169

chapter 14
TWO TIME-VARYING AUTOREGRESSIVE PREDICTORS FOR THE
ELECTROENCEPHALOGRAM: A PRELIMINARY STUDY
  C. R. Parisot and D. O. Walter   177

chapter 15
THE ACQUISITION AND GRAPHICAL DISPLAY OF NEUROPHYSIOLOGY
DATA WITH AN ONLINE MULTIPROGRAMMING COMPUTER SYSTEM
  W. I. Wood   211

chapter 16
MULTI-CHANNEL NERVE IMPULSE SEPARATION TECHNIQUES
  W. M. Roberts and D. K. Hartline   221

chapter 17
THE SPIKE PROGRAM: A COMPUTER SYSTEM FOR ANALYSIS
OF NEUROPHYSIOLOGICAL ACTION POTENTIALS
    J. J. Capowski   237

chapter 18
NEURAL UNIT DATA ANALYSIS SYSTEM
    W. S. Rhode and V. Soni   253

chapter 19
SEQUENTIAL INTERVAL HISTOGRAM ANALYSIS OF NON-
STATIONARY SPIKE TRAIN DATA
    A. C. Sanderson and B. Kobler   271

chapter 20
A GENERAL-PURPOSE ALGORITHM FOR HISTOGRAM
GENERATION
    P. B. Brown, J. Froelich, C. Roby, and J. Marler   293

chapter 21
A CONTOUR MAPPING ALGORITHM SUITABLE FOR SMALL
COMPUTERS
    L. S. Davidow and P. B. Brown   321

chapter 22
NEURON POPULATION ANALYSIS WITH THE NVAR SYSTEM
    G. W. Harding, A. L. Towe, and T. H. Kehl   337

chapter 23
AN ON-LINE MEAN-SQUARE-ERROR ANALYSIS TECHNIQUE
USING WHITE NOISE INPUTS
    D. P. O'Leary, C. Wall, III, and L. Traini   371

chapter 24
THE TABLE AS A BASIS FOR DATA ANALYSIS
    B. J. Ransil   385

chapter 25
A NUMERICAL APPROACH TO DIFFERENTIAL EQUATION
MODELS IN NEUROBIOLOGY
    H. C. Howland   395

chapter 26
COMPUTER SYSTEM ARCHITECTURE AT THE UCLA
BRAIN-COMPUTER INTERFACE LABORATORY
    J. Vidal and R. H. Olch   411

chapter 27
RETRIEVAL OF DATA BY ATTRIBUTE USING PARAMETER-FLAGGED DATA STORAGE
   D. A. Ronken and D. H. Eldredge   439

chapter 28
A SAMPLING ALGORITHM FOR BANDWIDTH LIMITATION OF ACTION POTENTIAL TRAINS
   A. S. French   447

chapter 29
SOFTWARE FOR SPECTRAL ANALYSIS OF NEUROPHYSIOLOGICAL DATA
   A. S. French   459

chapter 30
REAL TIME PROGRAMMING CONTROL OF NEUROPHYSIOLOGICAL BEHAVIORAL EXPERIMENTS
   T. Medlin   475

chapter 31
DADTA IV: A COMPUTER BASED VIDEO DISPLAY CONTROL AND DATA COLLECTION SYSTEM FOR BEHAVIORAL TESTING
   K. J. Drake and K. H. Pribram   509

chapter 32
AUTOMATED SYSTEMS FOR BEHAVIORAL NEUROPHYSIOLOGY—THREE STRATEGIES FOR INTERACTION WITH NEUROPHYSIOLOGICAL EXPERIMENTS
   S. L. Moise, Jr.   529

chapter 33
A DIGITAL SYSTEM FOR AUDITORY NEUROPHYSIOLOGICAL RESEARCH
   W. S. Rhode   543

chapter 34
THREE COMPUTER INTERFACES FOR NEUROPHYSIOLOGISTS
   P. B. Brown, D. Duffy, and T. W. McIntyre   569

chapter 35
USES OF THE LM^2 IN NEUROBIOLOGY
   T. H. Kehl and L. Dunkel   591

chapter 36
ON UNIFORMITY OF DIGITAL COMPUTER INTERFACE DESIGN
   T. H. Kehl   601

chapter 37
HIGH LEVEL LANGUAGE COMPUTERS: POTENTIALS AND
LIMITATIONS IN NEUROBIOLOGY
   T. H. Kehl   613

chapter 38
LABORATORY PROGRAMMING: HOW CAN WE REALLY
GET IT DONE?
   H. Moraff   623

chapter 39
PANEL DISCUSSION   637

Index   649

CONTRIBUTORS

R. Abba   Department of Biological Sciences, Columbia University, New York, New York 10027

R. D. Benham   Interactive Mini-Systems, Inc., Kennewick, Washington 99336

R. Bornholdt   Department of Biological Sciences, Columbia University, New York, New York 10027

J. J. Capowski   School of Medicine–Department of Physiology, University of North Carolina at Chapel Hill, Chapel Hill, North Carolina 94305

T. O. Clarke   Department of Neurology, Stanford University Medical Center, Stanford, California 94305

L. S. Davidow   Department of Biology, Massachusetts Institute of Technology, Cambridge, Massachusetts 02139

M. L. Dierker   Department of Anatomy, Washington University, St. Louis, Missouri 63130

C. E. Donaghey   Department of Industrial Electrical Engineering, University of Houston, Houston, Texas 77004

K. Drake   Poughkeepsie, New York 12603

D. J. Duffy   Hospital Systems, West Virginia University Medical Center, Morgantown, West Virginia 26506

L. Dunkel   Department of Physiology and Biophysics, University of Washington, Seattle, Washington 98195

D. H. Eldredge   Central Institute for the Deaf, St. Louis, Missouri 63110

A. S. French   Physiology Department, University of Alberta, Edmonton 7, Alberta, Canada

J. W. Froelich   Department of Physiology and Biophysics, West Virginia University Medical Center, Morgantown, West Virginia 26506

J. D. Gerard   Division of Biophysical Science, University of Alabama, Birmingham, Alabama 35294

G. W. Harding   Department of Physiology and Biophysics, University of Washington, Seattle, Washington 98195

D. K. Hartline   Biology Department, University of California at San Diego, La Jolla, California 92037

H. Howland Department of Neurobiology and Behavior, Cornell University, Ithaca, New York 14850

T. H. Kehl Department of Physiology and Biophysics, University of Washington, Seattle, Washington 98195

B. Kobler Biotechnology Program, Carnegie-Mellon University, Pittsburgh, Pennsylvania 15213

C. Levinthal Department of Biological Sciences, Columbia University, New York, New York 10027

E. R. Macagno Department of Biological Sciences, Columbia University, New York, New York 10027

J. Macy, Jr. Division of Biological Sciences, University of Alabama, Birmingham, Alabama 35294

J. Marler Department of Physiology and Biophysics, West Virginia University Medical Center, Morgantown, West Virginia 26506

T. W. McIntyre Department of Physiology and Biophysics, West Virginia University Medical Center, Morgantown, West Virginia 26506

R. McKown Department of Physiology and Biophysics, University of Illinois at Urbana-Champaign, Urbana, Illinois 61801

T. Medlin National Institute of Dental Research, Bethesda, Maryland 20014

S. L. Moise, Jr. Center for Health Sciences, University of California, Los Angeles, California 90021

H. Moraff New York Veterinary College Computer Facility, Cornell University, Ithaca, New York 14850

R. H. Olch Computer Science Department, University of California, Los Angeles, California 90024

D. O'Leary Department of Otolaryngology, University of Pittsburgh School of Medicine, Pittsburgh, Pennsylvania 15213

C. Parisot Brain Research Institute, University of California, Los Angeles, California 90024

J. H. Peacock Department of Neurology, Stanford University Medical Center, Stanford, California 94305

K. H. Pribram Psychology Department, Stanford University, Stanford, California 94305

B. J. Ransil Clinical Research Center, Beth Israel Hospital, Boston, Massachusetts 02215

W. S. Rhode Laboratory Computer Facility, University of Wisconsin, Madison, Wisconsin 53706

W. Roberts Department of Biology, University of California at San Diego, La Jolla, California 92037

C. Roby Department of Physiology and Biophysics, West Virginia University Medical Center, Morgantown, West Virginia 26506

D. A. Ronken Central Institute for the Deaf, St. Louis, Missouri 63110

A. C. Sanderson Biotechnology Program, Carnegie Mellon University, Pittsburgh, Pennsylvania 15213

M. J. Shantz Information Science Department, California Institute of Technology, Pasadena, California 91125

V. Soni Laboratory Computer Facility, University of Wisconsin, Madison, Wisconsin 53706

L. Traini Department of Otolaryngology, University of Pittsburgh School of Medicine, Pittsburgh, Pennsylvania 15213

C. Tountas Department of Biological Sciences, Columbia University, New York, New York 10027

A. L. Towe Department of Physiology and Biophysics, University of Washington, Seattle, Washington 98195

J. Vidal Computer Science Department, University of California, Los Angeles, California 90024

C. Wall Department of Otolaryngology, University of Pittsburgh School of Medicine, Pittsburgh, Pennsylvania 15213

D. O. Walter Brain Research Institute, University of California, Los Angeles, California 90024

D. Walter Department of Physiology and Biophysics, University of Illinois at Urbana-Champaign, Urbana, Illinois 61801

D. F. Wann Department of Electrical Engineering, Washington University, St. Louis, Missouri 63130

M. P. White Division of Biological Sciences, University of Alabama, Birmingham, Alabama 35294

W. I. Wood Department of Biochemistry and Molecular Biology, Harvard University, Cambridge, Massachusetts 02183

PREFACE

This volume presents, insofar as possible, computer technology currently being applied to neuroscience research. The chapters represent independent development, in a number of laboratories, of workable computer systems to be applied. None of the systems presented can achieve all the objectives of all the laboratories, and most of the systems utilize a hardware environment that is more than the minimum needed. Each system required substantial investment of time and money for development (usually with little use made of similar systems developed elsewhere).

This work presents the first part of a two-part project aimed at reducing unnecessary inefficiencies involved in the independent implementation of computer-aided research from scratch in hundreds of laboratories around the country. Its objectives are: to specify a minimum hardware environment needed to perform each research task, with options to expand or adapt to existing hardware systems; to specify a minimum software package to perform each task or multi-task project; and to design these systems for easy modification and implementation. Part two will involve the implementation of these specifications by developing all the specified software. When this is accomplished, the package will be made available to the research community. Further expansions and refinements will be made available as need for them develops and as advances in computer science are made.

The need for this project is clear. To those who offer support for neuroscience research, duplication of effort in the development of computer research facilities is an unacceptable luxury. If the development and maintenance costs for software can be minimized or even eliminated, granting agencies may come to recognize well-developed general-purpose research computer systems as routine items of equipment. The development of inexpensive processors and peripherals, all of which can be assembled in the laboratory by technicians, should encourage such a trend. The possibility of a computer system that can be used by neuroscientists who have a minimum of computer science background should increase the use of such systems and open up new avenues of research not previously available to most neuroscientists.

This volume will alert researchers to the types of research made possible with computers applied to neurobiology. In some cases, the specific requirements for useful computer research systems are indicated. In others, specific software procedures and information are presented, which readers can apply in their own laboratories. Every effort was made to accurately represent the great breadth of computer applications in modern neurobiology.

I wish to thank the many neuroscientists from around the country who have taken time from their research programs to share their expertise. Most of them are not computer scientists, but biologists, whose major activity is biological research.

This volume is an outgrowth of a symposium held at West Virginia University in April, 1975. It was the first of three workshops funded by The National Science Foundation (grant number DCR74-24765) for the specification of a general-purpose computer language to be used in neurobiology research. Preparation of the camera-ready copy was accomplished using the university's system 360 and the Department of Physiology and Biophysics' PDP-12. Cheryl Seim, Dr. T. W. McIntyre, and Darrell Duffy provided immeasurable assistance in making these systems work.

Paul B. Brown

COMPUTER TECHNOLOGY IN NEUROSCIENCE

chapter 1

DESIGN AND DEVELOPMENT OF A MULTILABORATORY COMPUTER COMPLEX FOR BIOMEDICAL RESEARCH

J. MACY, JR. AND M. P. WHITE

Division of Biological Sciences
University of Alabama
Birmingham, Alabama

The Multiple Laboratory Computer Center (MLCC) is a multiple-processor, distributed logic system designed to support simultaneously the on-line computing needs of many independent laboratories. It was designed to handle a wide range of data-rate requirements from these laboratories, and to provide reasonable efficiency by shared use of central resources.

This system is a natural outgrowth of previous experience with on-line hybrid computing in neurophysiology, neurology, and physiology laboratories during the 1960's. (1) This experience was the basis for the development of methods for on-line and real-time analysis, and closed loop experimental control. It established the methods needed to handle data rates (sampling rates) required for these experiments. These rates ranged from 10 samples per second to more than 50,000 samples per second. Sampling rates were chosen to preserve the bandwidth and avoid aliasing, for complex waveforms, or to preserve a needed time resolution in the detection and characterization of specific neural events.

The work described in this chapter was supported in part by the U. S. Public Health Service, National Institute of Health Grant RR00145.

This experience led to the development of rather specific forms of "real-time" computing, and established the special needs of such computing. A brief discussion of the nature of "real-time" computing will point out these distinctions. When this term is used, it appears to mean many different things to different people. Experience with one form of on-line computing really gives very little idea of what any other kinds of real-time computing actually involve. Probably the simplest and most common form of real-time computing is used by an airline reservation system, or any other system which provides terminals into which a human can put information and can later make inquiries to get some part of the information back. The terminal may be a teletype or some combination of cathode ray screen and keyboard. This "real-time system" can be defined as a system in which the acceptable time for the computer performance is determined by the user and not by the program. The computer, in other words, must keep up with the demands of the real world. Most timeshared multiple-user computing systems have access times on the order of seconds, or some fraction of a second and, as the number of users and the usage of each terminal increases, the system reacts to the increased load by decreasing its performance. If the load is very heavy on a conventional time-shared system, there is a longer wait for response from the terminal. This is probably the only real-time system in which such a wait is acceptable. If a human at the terminal has to wait three seconds instead of one second he may become a bit annoyed but no data is lost or mangled because of the wait.

The next order of magnitude of speed for real-time systems is the world of the process-control system. These are the systems used in many factories, steel mills, and oil refineries to control machinery and equipment. This is, in effect, the kind of system and time scale also used in patient monitoring in the intensive care units and for delivering patient care directly by computer. The response times are generally in milliseconds. In monitoring an EKG, for such things as arrhythmias, for example, the time resolution needed is milliseconds.

The order of magnitude of real-time, on-line computing for which the MLCC was designed originated from the demands of neurophysiology laboratories, and depends on microsecond response and additional special features.

Some examples from neurophysiological work will illustrate this. At the time the system was first conceived, much of our work was based on multiple-electrode, single-cell measurements, because we were particularly interested in the coding schemes of different areas of the nervous system and in the time series performance of some of these cells. We were attempting to make measurements for which it was necessary to convert the data continuously amd accumulate statistics over periods of many minutes. We had runs which would go continuously for as long as forty-five minutes, and we were generally working with two to five channels. To get the time resolution required for most of the measurements implied a sampling rate for each channel of at least 5,000 samples per second, sometimes as many as 10,000 or 20,000. The time between samples for each channel was less than 200 microseconds, sometimes as little as 20 microseconds. This is very different from the situation in an inquiry terminal or in intensive care monitoring. "Real real-time" could be considered as a context in which the time between the arrival of samples is not more than eight or ten times the basic machine cycle of the computer being used. The computer is no longer very much faster than the real world with which it is trying to cope but operates at about the same speed as the world being computed. Much of our work in the 1960's used a computer with a basic machine cycle of 6.5 microseconds, and an average execution time for commands of 15 microseconds. The interval between samples in this research was between 50 and 200 microseconds. In the worst case there was time between samples to execute three instructions. Furthermore, programming to take in 50, 100 or 5,000 samples and store them to be computed later was not possible, because in many cases it was necessary to keep up on a continuous basis wih a continuous process. Data points came in at a rate of 10,000 to 50,000 a second for about half an hour. The only possible technique was to compute fast enough to keep up, starting to build the answer and throwing away as much as possible of the raw samples. If the samples arrive on many channels at such a speed there must be some provision to dispose of irrelevant or combinable data, because no machine has enough storage to bring in and store such a flood.

For some processes it is possible to compress the data. "Outboard" analog or hybrid hardware will recognize in one way or another the particular phenomenon of interest and start sampling when it occurs. This happens frequently in monitoring where there is no real interest in analyzing every heartbeat for a twenty-four hour period but only in checking and accumulating derived measurements at stated intervals. Intermittently a series of heartbeats will be sampled, the complete analysis performed and the raw data discarded.

The MLCC was designed to meet the requirements of laboratories with a legitimate need for sampling rates up to 50,000 samples per second per channel. Some laboratories have requirements well below 1,000 samples per second per channel, so a range of time scales is needed to match the full range of requirements.

In many cases, there is an additional requirement for local control of the experiment in response to the already acquired data so that the progress of the experiment is actually dependent upon the results just processed: a closed-loop situation.

We started our multilaboratory design on the assumption that no computer can keep up with the flow of all the data from ten or more laboratories with 50 kc sampling rates per channel. It would be necessary to install additional preprocessing logic at the laboratory to cut down the actual rate of arrival of data at the central computer. In such a system, the logic would be spread and distributed so that the computing power with high precision, floating point, versatile commands and other refinements would be located at the center of the system. In general, the operations to be performed on the samples as they first come in, at the initial high rate of speed, are fairly simple and the conversions will be, at most, twelve bit conversions, so that extremely high precision is not essential. When the preprocessing operations have been performed at a very high rate of speed, the resulting sampling rate may be one-tenth of the original rate (perhaps 100 per second) and the machine can operate as a more conventional time-sharing system with the effectively reduced time scale. The burden of continuously computing and keeping up has been shifted out so that a small amount of relatively inexpensive logic can be dedicated to preprocessing the data full-time, unshared by other users.

The design philosophy of our system is based on four considerations. The first is the necessity for the tight control of time. In terms of Central Processing Unit

(CPU) cycles per millisecond, or per hundred milliseconds, used by the executive system and by each of the users, time must be controlled and the system must be able to guarantee to each user the minimum necessary number of CPU cycles and the rate at which they will be delivered. Without this control the users will be unable to use programs which keep up continuously with a flow of data from the outside world at some fixed rate.

The second consideration is the principle of "distributed logic"; a concentric array or nested series of processing units with the smallest at the periphery of the system, in the laboratory. These are backed by somewhat more sophisticated CPUs and controlled from the center by a single relatively large and powerful CPU. Each CPU is dedicated, as much as possible, to those operations it is uniquely qualified to perform; as much preprocessing as possible is done by fast but simple processors as close as possible to the source of the data. The high data rate laboratories have some laboratory processing capability which communicates directly with a central memory under the control of an input-output processor which can proceed independently of the central CPU. The low-data-rate laboratories send analog signals directly to a processor which shares its preprocessing capability among several laboratories and stores the partially processed results for final computation by the central system. The central or major CPU is freed of as many computationally trivial jobs and internal "housekeeping" details as possible by providing independent processors to supervise all of the input-output operations.

A third design consideration was a desire to provide as much program and data storage integrity as possible to each of the user laboratories. We planned a system in which no laboratory could interfere with the stored data or the running program of any other laboratory. In the event of peripheral hardware failure or applications programming errors, the system must be able to fail in pieces without disturbing the essential operation of the remaining laboratories. These considerations imply a system in which hardware and software devices protect the core memory and disk areas of each lab from access by any other lab, and which preserves the integrity of computer time for each laboratory separately. A fourth consideration was the development of a system of programming and languages which would make use by any one laboratory as convenient and flexible as possible. In the initial stages, programming for the multiple processors involved in this system was rather complicated and difficult. Our first priority was to develop a hardware system and the executive programs to satisfy the design criteria mentioned above. The executive has been developed with the language and programming improvements clearly in mind. These laboratory-oriented languages and programming systems are not an exact duplicate of any of the common existing languages. Most of the general purpose programming languages are not well adapted to the close control of time and most of their compilers do not allow the close control over program loop execution time, which is a frequent necessity in a sustained real-time programming context. The major general purpose language RPL is discussed in Chapter 2.

Commercially available components have been used. Although it is not likely that other institutions will want an exact copy of the total system, our design provides for many subsets and supersets of the hardware system and programming system which are viable on their own.

This project represents one part of a continuing effort to bring mathematics, computer science, and modern measurement and data processing techniques into the research program of the University Medical Center. We expect it to encourage the adoption of many of these methods by the laboratories associated with the project and to allow the growth of laboratory research efforts which would have been impossible without the system. Although no array of hardware can solve the problems of developing new research methodologies, this computer system is one part of an effort to strengthen medical research through the collaboration of the basic sciences, especially mathematics, information sciences, statistics and engineering.

In the system design, laboratories were divided into two classes, "fast" and "slow"; a "fast" laboratory is one whose data rates or other requirements imply the need for a satellite preprocessor, a "slow" laboratory is one whose preprocessing needs and data rates can be handled by a shared intermediate preprocessor. The dividing line is generally a rate of 1,000 samples per second; at this rate or less, for a reasonable number of channels, a lab can transmit and receive its data in analog form, and is a "slow" lab. If the rate per channel is greater than

1,000 samples per second, or the number of channels large enough to create a high aggregate rate, the laboratory is classed as a "fast" laboratory and uses a local satellite preprocessor. Communication between this processor and the more central equipment is digital, at high rates. Figure 1 illustrates the general system configuration, and shows the different styles of laboratory connection.

The central or major processor is a Sigma 7 (Xerox Data Systems). This CPU is equipped with nine sets of sixteen registers for temporary storage and computation; it has thirty-two levels of interrupt, each with separate and program adjustable priority and response address. It runs the operating executive system and the major

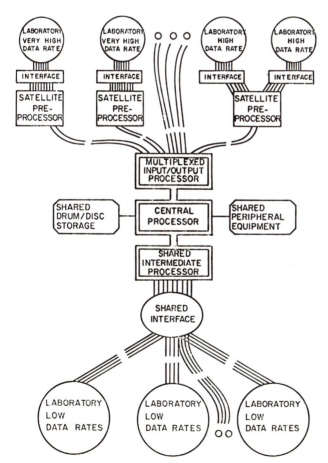

Figure 1. Overall structure of the Multiple Laboratory Computer Center.

portions of the applications programs, and controls the multiplexing input-output processors (MIOP) which handle the peripheral epuipment.

Auxiliary to this Sigma 7 is a Sigma 2 shared intermediate processor, which shares some, but not all, memory with the Sigma 7. This smaller processor is used to preprocess analog inputs and operate the central hybrid interface.

Memory of the central system is composed of four 16k 32-bit units. The first is the operating memory for the Sigma 7. It provides running space for the executive and for the active applications programs. The second and third are reserved as a data and program buffer for the remote satellites. They are the only memories accessed by the special MIOP, which handles communication with the satellites. The fourth memory is primarily a data storage area for signals which come from low-rate labs via analog lines, and are preprocessed by the Sigma 2 rather than by remote satellites. This fourth unit is used by the executive and is also accessible to the Sigma 2 CPU. The Sigma 2 CPU also has its own 8k (16-bit words) of private memory, for programs and its resident part of the system executive.

Peripheral devices on the first Sigma 7 MIOP include two digital tapes, printer, teletypewriter, communications controller, paper tape punch and reader, a CALCOMP digital plotter, a 1.5 million byte rapid access drum (RAD), two CALCOMP disc drives, and a card reader. A second MIOP handles four fast, large capacity drums of 24 million bytes. Average access time is 17 msec. These drums are used for data storage, and for scratch storage during real-time computation. The small drum of 1.5 million bytes and 88,000 bytes/second transfer rate is used for system programs.

On the Sigma 2, peripheral devices include a teletype, the paper tape punch and reader (via switch), and a large and versatile hybrid interface, with digital and analog input, output and linked multiplexers. These devices are controlled by a separate IOP internal to the Sigma 2, or by the Sigma 2 itself. Figure 2 shows the configuration of the central part of the system.

This central system receives Analog signals from many laboratories at once, via a multiple patch-bay and the Sigma 2. It can also connect to as many as eight satellite computers in high-data-rate laboratories. Satellites are DEC PDP-8 with 4k or more of (12-bit)

MULTILABORATORY COMPUTER COMPLEX

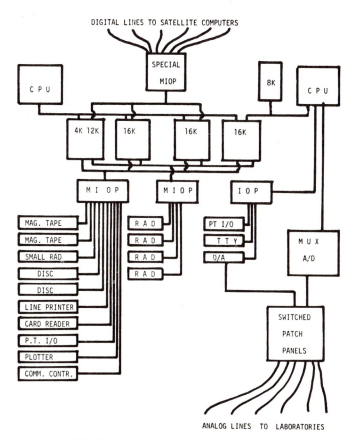

Figure 2. Detailed structure of the shared processors and memories for the Multiple Laboratory Computer Center.

memory, or PDP-8-Is with the AC-08 interface, or PDP-12. Any of the DEC "family of 8" are plug-in compatible as satellites.

The typical satellite station consists of computer, the laboratory (analog) interface and teletype printer-keyboard with paper tape reader and punch. This is the necessary minimum to allow program compatibility and interchange. Special equipment for any particular laboratory is added where needed. Satellites are connected to the laboratory via the "data-break". Data transfer from the PDP-8 can be preset by the central computer program, or controlled by means of instructions from the PDP-8. These include interrupt signals to notify the central computer to initiate transfers.

Transfers are made at rates up to 167,000 12-bit words per second; transfer of the entire 4k memory of the PDP-8 takes as little as 24 milliseconds. Normal block transfers of 1,000 words take 6 to 10 milliseconds. The rate of transfer from a PDP-8 can be limited by centrally programmed command, matching the rate to the timing requirements of a particular program.

The satellite laboratory interface permits flexible choice of input format, conversion accuracy, and conversion rates up to 50 kc. Any input bits not used for conversion can contain digital inputs in a packed format or can be filled with the sign bit. Digital input can come from response of detection logic to specific analog events in the laboratory or can signal the occurence of stimuli or some changed condition. Digital output can be used to generate stimuli or control laboratory equipment. The digital-to-analog channels can be used to create displays on laboratory oscilloscopes or pen recorders.

The printer-keyboard allows the experimenter to communicate with his satellite, or with the central computer via his satellite. This means that the satellite can be used as a "stand-alone" computer, as an input-output device with laboratory equipment and data sources, as a preprocessing facility for data-reduction and as a remote console for "conversation mode" interaction with the central computer.

The most critical part of the real-time multiprogramming system is the real-time executive. This executive has several functions, which correspond roughly to the categories of programming tasks it will have to supervise. The categories are: low-data-rate real-time programs servicing one laboratory, with one or more channels, at a fixed sampling rate; transfer programs designed to service requests for data transfer from the satellite stations; satellite computing programs, designed to perform computations, during or at the end of a run, on data preprocessed and stored by the satellite stations; input-output programs, which operate standard peripheral devices such as the RADs and magnetic tape; and background programs, which can be used during periods of light load to fill in background time. The feature which makes this executive different from most operating systems is its ability to handle the mixing, choosing, and supervision of several programs, each of which runs independently, in real time, with fixed but different sampling rates, and which must perform computations of different complexities. The executive must be able to decide at each new request whether it can accommodate the program requested. It must be able to insert the real-time parts of this program into the running collection with due attention to its sampling requirements, and without disturbing the sampling requirements of the existing programs; and it must always save enough time for its own operations and for the intermittent requests for block transfer of data or computation which may arrive from the satellite stations.

This executive imposes a load of 20% or less of the CPU time for its own operation. It functions as the major control element in this time-sharing environment and coordinates all other software programs, program sequences, data management, and input-output for the system. It is constructed on a modular basis, so that a change in the application will not necessitate a corresponding change in the executive monitor.

The most essential job the executive performs is the allocation of Sigma 7 system resources to the demands of applications programs. There is no such thing as permissible system degradation; each applications program must be guaranteed the CPU time, I/O facilities and core space required for the duration of its run before its laboratory is allowed to come on line. Core allocation is straightforward and I/O facilities are assigned according to the volume of data per unit time specified in the request by the applications program. For high speed scratch data requirements, keyed disc records are utilized, with indexes core-resident until the termination of the job. Allocation of CPU time is made on the basis of the specific amount required during a defined time cycle, such as 50 microseconds per 10 milliseconds. The duration of the cycle and the volume of immediate real-time response from the Sigma 7 necessary to keep the process running are factors which are quantified to determine if the resources are available for the program to run. The job queue is stepped through on the basis of clock interrupts set according to the parametric cycle durations of the applications programs comprising it. The time remaining in the cycle is used to perform the less critical functions, to supervise I/O operations, and to process any experimental data which has been queued and does not necessarily need to be processed while the associated job is on-line.

The programs and preprocessing programs for the low-data-rate laboratories vary greatly in nature and complexity of the computations. Typical examples are the on-line determination of cardiac output, cardiac work and stroke volume; the computation of mean and variance of evoked responses in EEG; the compilation of latency distributions from reaction time experiments, and the generation of stimulus sequences based on immediate past performance. In general, for preprocessing, computing times of 20 to 50 microseconds per sample point and sampling rates of 50 to 100 samples per second are sufficient. The programs are often written to react to each sample, examining it and performing the necessary computation within a fixed time limit. This method avoids the need to store blocks of input, and reduces storage requirements. The preprocessing capability of the Sigma 2 materially decreases the time needed at the Sigma 7 to process any source of data.

The low-data-rate labs usually have a minumum of equipment in the lab for communication with the computer. Coaxial analog cable, driver amplifiers, a few simple indicators and push-buttons and an intercom station are the usual elements. Data is transmitted to and from these labs in analog form, and displays are on the laboratory's own analog or digital recording equipment. The satellite stations have four different classes of programs, ("stand-alone" programs, data-acquisition programs, output and display programs, and conversation-mode programs) to allow different uses of the satellite.

The "stand-alone" programs can run on the satellite without the central system. The satellite has an assembly program, FORTRAN compiler, standard subroutine packages, and all the PDP-8 users group programs available to it. A real-time data acquisition and preprocessing language, called ODAC, has been written for use in these satellites, and allows programming in a "semi-higher-level" language. ODAC is generally a macro-based language with limited syntax. Each satellite has a flexible interface with its laboratory, and is capable of a wide range of operations as an independent computer.

The data-acquisition programs perform the sampling and preprocessing that allow the laboratory to work on-line with the central facility. These programs not only preprocess the data, but also extract from it preliminary results which can be used to control stimuli or other experimental parameters directly. In some such cases back-up from the central computer permits both anaysis and control functions which would not often be possible for the satellite alone. At the end of an experimental run, the central computer can finish the analysis process and produce the results. At this point a new program is transferred to the satellite, enabling it to act as an output device and display buffer for monitoring the results in the laboratory.

The central computer's ability to locate and transmit a 3,000 word program from drum to satellite in less than 50 milliseconds gives the satellite rapid access to a much wider range of programs than can be stored in its limited memory. The rapid program-swap technique means that only the program in use at the moment needs to be stored in the satellite memory. This "expansion" of the satellite capability is more convenient and economical than buying additional memory for each satellite.

The distances to the periphery of this system are within a maximum of 2,000 feet from the center at present. We have reason to believe we could go about twice that far, without any additional arrangements for data transmission. So far all of our links are directly wired. We have not attempted to use the telephone system, microwave linkages, or any alternative to direct wiring for transmitting data at high rates over longer distances, largely because there has been no demand for such arrangements.

The development of real-time, user-oriented languages is discussed in the second chapter. Such real-time languages provide further convenience in the use of the system. The system has been operational for more than three years, and is providing daily shared on-line service to an increasing number of laboratories, plus associated batch computing for participating laboratories.

One of the things we had hoped to do after the development period ended was to determine the bottleneck or limiting factor. What becomes saturated first? How much could we continue to expand the system by adding more memory, more multiplexer points, or more peripheral equipment? We have not been able to get any useful measures of these saturation effects, since the system configuration as described has carried the actual needs of all the laboratories now using it. As the number of laboratories using the system continues to increase, we hope to develop a better measure of the maximum carrying capacity.

The MLCC system is an example of one way in which the requirements of the different kinds of real-time (including what I have earlier called "real real-time") needs can be met. At this point and in the foreseeable future, it is not possible to meet such requirements by sharing all the varied time scales into one central machine.

DISCUSSION

MCINTYRE: What do you do in a situation where a job written by a user in one of the labs says he wants the CPU 100% of the time? Does he sit there and keep asking for it until he gets it?

MACY: Such a user could do **one** of two things: get his request up and just keep putting it in until sooner or later the system didn't have any users and he got it, or he could arrange to come sneaking in at five in the morning and be the first one to get it. We have not had to introduce that kind of political protection. It would presumably be easy enough to do if we had to.

MCINTYRE: Are the programs that run out of the user labs subject to some supervision from your central development facility?

MACY: Only in the following sense: The program that runs in the laboratory satellite -- and these are generally DEC family of 8, 12 and so forth -- with regard to these programs, pretty much anything goes. Now, we do have some standard programs and a standard language, which we will not talk about in this session, but there's no reason not to do anything you want in your satellite. The programs that run in the central system are supervised to some extent; that is, there are only certain modes of machine usage that you are allowed. You cannot use I/O instructions, they're reserved for the executive. You can't get out of your assigned data area; you can't get at the disk space. But within those limits, no, you could put a program in that had some weird bug in it and hopefully we would be protected in such a way that all you could do would be to bomb yourself.

DAVID WALTER: Could you give us a rough cost analysis per fancy neurophysiological user?

MACY: That would depend a good deal on the nature of the user. It is probably quite inefficient at the two ends of the spectrum: that is, if you have a need that requires, say, 85% of the total system, it becomes inefficient because in effect you have now pretty well taken all of it. You would then be better off with a dedicated system of your own. If on the other hand you have a need for five words of memory and one sample every five minutes, you'd be paying far too much. The system is designed for the medium user. What is our current rate per slice second, Mike?

WHITE: Something of the order of $25/hour*1k core slice. Most laboratories can run happily in one core slice, so that for a relatively uncomplicated problem you can plan on $25 per connect hour.

REFERENCE

1. Macy, Josiah, Jr. Hybrid computer techniques for physiology. Ann. N.Y. Acad. Sci. 115:568-590, (1964).

was Xerox's Extended FORTRAN IV, which was incompatible with our operating system for several reasons. These specific reasons may be summarized as conflicts with the hardware configuration and the operating system philosophy. The hardware configuration is designed to have the program's high input/output areas allocated to one memory area and the program's executable code assigned to another -- this scheme reduces the problem of input/output cycle stealing and subsequently aids in the precise allocation of time to program modules. The main conflict between standard Xerox FORTRAN and the operating system philosophy was in the area of program size and accurate estimation of program segment execution time.

The MLCC operating system was designed to support multiple real-time applications programs. To do this, all user programs must be core resident, with no automatic system rollout or checkpointing since for data arriving asynchronously and at high rates from multiple laboratories there is no time for automatic system rollout or checkpointing of user programs. This real-time criterion, coupled with limited memory (initial system configuration was 48k 32-bit words) capability, required that the application programs be small, efficient modules. Xerox's FORTRAN, like most other standard languages, makes certain assumptions about system resource availability and consequently produces program modules too large to fit within the MLCC's criteria. It should be noted that Xerox's FORTRAN is quite good for batch number-crunching and is probably adequate for single program real time systems. In fact, the number of successful installations using FORTRAN attest to its adequacy given the proper environment (i.e., IBM's 1800/MPX, Xerox's RBM, Data General's RDOS). With a limited programming staff whose primary mission was applications programming and only secondarily system programming, the MLCC did not quickly embrace the idea of taking on such a formidable task as implementing a higher-level programming language. Recognizing the long range advantages of having a higher level language for real-time use but with pressing applications to be programmed, the MLCC staff proceeded programming all real-time programs in assembly language and laying the ground work for designing and implementing a higher level language which would interface to the current system. The ground work took the form of analyzing the code produced by Xerox's FORTRAN compiler, building our own

chapter 2

DESIGN AND DEVELOPMENT OF A REAL-TIME PROGRAMMING LANGUAGE: RPL

M. P. WHITE, J. MACY, JR., AND J. D. GERARD

Division of Biophysical Science
University of Alabama
Birmingham, Alabama

WHY A NEW LANGUAGE?

With all of the available higher level languages what could motivate a research computer group to design and write a new one? Other than intellectual curiosity (which is hard to fund) the reasons are either a) existing languages are deficient for the application, or b) no acceptable existing language is compatible with the operating system or environment. The proliferation of new programming languages in and of itself serves no useful purpose to the neurobiological scientist. In the context of the Multiple-Laboratory Computer Center, described in "Design and Development of a Multiple Laboratory Computer Complex for Biomedical Research" in detail, a viable multiple laboratory computer system which had met its original design goals of guaranteeing system resources and data integrity still lacked one of its follow-up objectives, that of a compatible higher level programming language. The only acceptable higher-level language available for our computer system

---

The work described in this chapter was supported in part by the U. S. Public Health Service, National Institutes of Health Grant RR00145.

real-time math subroutine library, and establishing subroutine linkage conventions.

WHAT TYPE OF LANGUAGE AND HOW?

There are a variety of systems which are classified by their authors as languages. All these systems can be characterized by one of several general categories. The first category is parameter driven programs; the second, a set of macros; the third, packages with rigid statement style so that writing in the language is reduced to filling out a set of forms (i.e. Report Program Generators, RPG); and the fourth, languages which take on a procedural form have a syntax and grammar, and are generally flexible enough to solve a wide variety of problems. The parameter driven package approach and the rigid form type of languages are usually problem oriented. Since the macro type of language fails to provide so many of the features that are now considered basic to procedural higher-level languages and in general lacks cohesiveness, we were left with designing a procedural language, which would interface to the system (MLCC operating system) and to the environment (real-time programming).

Deciding to design and implement a procedural type of language meant that the on-line programming, for the most part, would be done by programmers -- not researchers. Considering the organization and technique awareness required to implement real-time programs, this is a case in which a good professional programmer can be a valuable asset to a researcher. Cooperation with the programmer allows the researcher to concentrate on his research and requires him only to specify the precise analysis he wishes to accomplish.

One event made the implementation of a procedural type higher-level language highly feasible -- the University of Washington wrote a compiler generator (XPL/S) for the Xerox Sigma computer series and subsequently submitted it to the Xerox User's Group Library. The University of Washington compiler writing system is a fairly straightforward implementation of the translator writing system described by McKeeman, Horning and Wortman (2). The model system was designed for and implemented on the IBM 360 computer series. The XPL/S compiler writing system allows a unique compiler to be assigned and implemented within a fairly reasonable amount of time (The University of Washington estimates four man months). There are naturally certain restrictions and inefficiencies as well as advantages in the compiler generator approach which will be considered in the next section.

DESIGN AND IMPLEMENTATION OF RPL

The design emphasis of the RPL language was the support of and interfacing to existing system features and environment, with style and language innovation considered secondary. The style selected would have to be conducive to the efficient implementation of algorithms. The following objectives defined the crucial design criteria:

> 1. Generation of compatible object code for the system loader and program library.
> 2. Support of the real-time executive system's input/output and overlay features.
> 3. Support of the system design criteria of separation of program executable code and input/output buffer areas.
> 4. Production of small executable programs (typical size 2k, minus input/output buffers and large arrays).
> 5. Subroutine capability.
> 6. Automatic generation of instruction execution time information for coding produced by RPL statements, an important capability in computing program and loop execution time.
> 7. Simultaneity between user input/output and user computation.
> 8. Production of information to assist in debugging of programs.
> 9. Production of a relatively efficient executable code.

In addition to meeting these requirements the XPL/S compiler generator system provided the following positive features without any corresponding implementation effort:

> 1. State of the art programming control structures and logical execution control, allowing Structured Programming. At present, RPL does not have a "GO TO" statement.

2. Conventions concerning small details which eliminated the need for the MLCC to create these conventions and allowed the programming time to be spent in more crucial areas, such as subroutines linkage conventions.
3. Requirement for the RPL language to be specified in an accepted (BNF) form for programming languages.
4. Generation of most of the brute force routines required for the compiler, eliminating a tremendous effort of design, code, and debugging of the standard compiler processor. This feature concentrates the compiler implemenation on the unique areas, such as code generation.
5. Facilitation of extensions to RPL and improvements in compiler efficiency.
6. Overall organization for both language design and implementation.

FORTRAN, ALGOL and PL/1 constitute the major procedural oriented languages whose notation is, unlike natural languages, succinct. FORTRAN, the first developed, was designed to handle numerical scientific problems at the statement level with little regard for file systems or character data. ALGOL followed approximately ten years later and raised its sights from the translation of the mathematical statement to the computational process. ALGOL's emphasis on the algorithmic approach to problem solving resulted in a very clean, powerful problem solving language, but it also lacked the ability to interface to a file system and to manipulate character data. PL/1 was introduced with all of the control structures of ALGOL, but containing so many special features and exceptions that, while very broad and powerful, it lost clarity. In fact, PL/1 programming can allow bit fiddling to the point where the programmer would be better off writing in assembly language. This background sketch of the primary procedural languages establishes the context for comparison with the form and features of RPL inherited from its predecessors.

RPL is a subset of PL/1, with non-standard extensions to interface with the MLCC operating system and real-time environment, and to provide some of the current control structures. The need to be close to the real-time environment necessitates some features which, like some in PL/1, are likely to produce programming errors. However, there has been an effort to limit these to functions recognized as necessary for real-time control. In designing RPL, each feature was included only after assessing its value to real-time programming. The MLCC real-time system provides to on-line laboratories computer support which directly affects the course of the researcher's experiment. The MLCC real-time system makes no attempt to provide conventional time-sharing or any batch type computation services. Because it is a dedicated system, RPL needs only to provide for the data types, commands and declarations required to support this defined environment.

RPL statements are free form, with keywords which require using spaces; where no special character separates two strings; and statements end with the semicolon as a terminal symbol. Within each procedure, data, buffers and declarations must be defined prior to the presentation of executable statements. The allowed data types are: integer (full word, half word and byte) and real. Data may be generated in the local procedure or mapped into a buffer (where no initialization is available) or referenced at run time through a pointer. When variables are dimensioned, only one dimension is allowed, with zero-origin indexing. Expression evaluation is like FORTRAN with the scan going from left to right except for exponentiation. Mixed mode arithmetic is allowed and the customary built-in arithmetic functions (i.e. SQRT, COS, SIN, etc.) are available. Structure is provided by external and internal procedure statements, grouping capability, iteration control and logical-flow control [IF (p) THEN (q) ELSE (r)]; system services such as input/output requests, overlays and date-time functions are built-in procedures. One of the unique real-time control statements is the WAIT UNTIL which suspends processing until an event occurs. Another specific real-time feature is the CONNECT function which links a laboratory interrupt to variables. In-line assembly code may be generated where circumstances require this type of extension.

The above description gives an overall view of the general make-up of the RPL language. To create a language with the flexibility to solve real-time problems efficiently, the elegance is somewhat degraded. Whether the language meets its objectives, will have to be reviewed after extensive use; however, it is a relatively easy language to extend, so the intial release should not be considered a crystallized product. The language design has been definitely slanted towards making it an effective tool for real-time programmers. The language is relatively easy for any programmer who understands real-time to learn. An understanding of real-time programming, not the language, is the major obstacle confronting the researcher. What the researcher desires, in order to participate more directly in the computer assisted investigation process, is a problem-oriented language. Blocking the implementation of the problem oriented language is the variety of approaches used by neurobiological scientists which have yet to be abstracted or defined.

Figure 1 outlines the process of generating a compiler with XPL/S compiler generation system. The system consists of a two-stage process. The first is the development of an unambiguous syntax aided by a program called the Analyzer. The Analyzer verifies that a

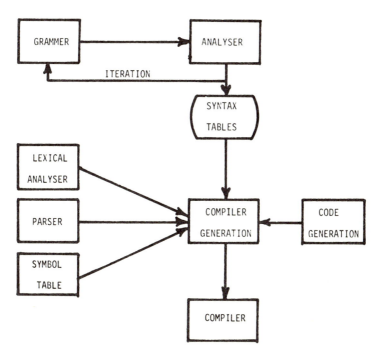

Figure 1. Steps in compiler generation.

grammar is consistent with proper syntax and that it adheres to the compiler generator's rules. Once a satisfactory syntax has been constructed, the second phase, generation of the desired compiler, can begin. The inputs to the XPL/S compiler are the elements of a compiler written in XPL. This includes the syntax tables developed by the Analyzer. Because the XPL/S compiler is written in XPL itself, the elements may be lifted, in whole or in part, from XPL/S. The degree depends on how close the target language is to XPL. The XPL language is basically a subset of PL/1, with extensions which are desirable for compiler writing. This characteristic of being a self-compiling compiler is the foundation of the compiler generating system and makes it especially useful. In the RPL compiler development the MLCC was able to make use of approximately the following sections of the XPL/S compiler, to the extent specified:

```
PARSER              100%
LEXICAL ANALYZER     80%
SYMBOL TABLE         60%
CODE GENERATOR       40%
```

XPL/S produces a compiler with many of XPL's advantages; unlike its IBM counterpart, it produces relocateable object modules. This type of object code facilitates building libraries and including external subroutines via the loader (linkage editor). The XPL/S compiler is a one pass compiler which makes it fairly rapid (2000 cards per minute) but reduces its capability to perform code optimization. We are currently analyzing ways of optimizing code generation that will not cause violent disturbances of the compiler generator system and will alow the compiler to take advantage of efficient hardware techniques.

The RPL language has met the previously stated goals. As indicated in the previous paragraph we are not totally satisfied with code generation, but it is temporarily satisfactory. Figure 2 illustrates a sample program for data acquisition from a satellite PDP-8 computer, using double buffering techniques to handle the input. The program uses various data types including pointers, and it contains some extra statements to illustrate computation and user-defined function capability. Certain unfamiliar substitutions (i.e. ¬ for the logical NOT and the % for the double quote") appear,

PROGRAMMING LANGUAGE: RPL

```
A00     RPL     11:43  APR 06, '75
   1                              /*
   2                                   THIS SAMPLE PROGRAM ILLUSTRATES VARIOUS FEATURES OF RPL
   3
   4                              EXAMPLE: PROGRAM;
   5                              BUFFERS(3);
   6                              DECLARE (A,B)(127) HALF BUFFER(1, 0), (SATBUF, FLAG) HALF POINTER;
   7                                  /* THESE ARE BUFFERS FOR DATA FROM A SATELLITE PDP-8, PLUS THE
   8                                     ASSOCIATED POINTERS */
   9
  10                              DECLARE BUFLIM LITERALLY '2047', FILEA LITERALLY '%0A%',
  11                                  TRUE LITERALLY '1', FALSE LITERALLY '0';
  12
  13                              DECLARE RADBUF(BUFLIM) HALF BUFFER(2, 0);
  14                                  /* THIS IS THE RAD OUTPUT BUFFER */
  15
  16                              DECLARE (I, J, STATUS, TYSTAT, FLIP) FULL,
  17                                  (P, Q, R) FULL INITIAL(1, 5, %FF%),
  18                                  (X, Y, Z) REAL INITIAL(1.5, 3.0);
  19                              DECLARE ERRMSG(15) BYTE INITIAL('RAD WRITE ERROR');
  20
  21                                  /* THE FOLLOWING ARE JUST MISCELLANEOUS EXPRESSIONS */
  22    34.0      C               Z = SIN(X) ** LOG(TAN(Y) / 5.2);
  23     8.9     21               P = SHL(R, 8) | %7F%;
  24     0.0     25
  25     0.0     25                  /* FUNCTION SUBROUTINE EXAMPLE */
  26     0.0     25               OPERATE: PROCEDURE(AX, BX) FULL;
  27    22.5     25                  DECLARE (AX, BX, P) FULL;
  28     9.1     2D                  P = AX + Q;
  29     0.0     31                  /* P IS DIFFERENT FROM MAIN-LINE P, BUT Q IS THE SAME AS MAIN-LINE Q */
  30    40.9     31                  RETURN BX / P;
  31     0.0     38               END OPERATE;
  32     0.0     38
  33     4.6     38               P = OPERATE(Q, R);
  34     4.6     3D               Q = OPERATE(P, Q);
  35     0.0     42
  36    14.1     42               SATBUF IS A; FLAG IS A(127);  /* INITIALIZE POINTERS */
  37    11.7     49               B(127), A(127) = -1  /* INITIALIZE FLAGS */;
  38     3.7     4E               FLIP = TRUE;
  39     0.0     50
  40     0.0     50                  /* DOUBLE BUFFER FROM A AND B INTO RADBUF */
  41     0.0     50               DO WHILE J < BUFLIM;
  42    42.6     53                  WAIT UNTIL FLAG #= -1 /* WAIT FOR FLAG TO CHANGE */;
  43     0.0     59                  DO I = 0 TO 127;  /* MOVE SATELLITE BUFFER TO RAD BUFFER */
  44    27.7     60                      RADBUF(J) = ASHR(SATBUF(I), 4); /* FIX DATA */
  45     5.5     66                      J = J + 1;
  46     1.9     69                  END;
  47     6.5     6A                  FLAG = -1;  /* RESET FLAG */;
  48     5.5     6D                  IF FLIP THEN DO;  /* SWITCH POINTERS */
  49    19.1     70                      FLAG IS B(127); SATBUF IS B; FLIP = FALSE;
  50     0.0     7A                  END;
  51     1.9     7A                  ELSE DO;
  52    17.8     7B                      FLAG IS A(127); SATBUF IS A; FLIP = TRUE;
  53     0.0     84                  END;
  54     1.9     84               END  /* OF DOUBLE BUFFERING OPERATION */;
  55     0.0     85
  56    19.5     85               CALL RADWRITE(FILEA, RADBUF, STATUS);
  57    22.3     8D               WAIT UNTIL STATUS #= 0; /* WAIT FOR WRITE TO FINISH */
  58    20.4     92               IF STATUS = -1 THEN CALL TYPE(ERRMSG, 15, 0, TYSTAT);
  59     3.2     96               CALL SIGNOFF(0, 1, 0); /* LOG OFF FILE A */
  60     0.0     97               EOF
60 CARDS, 37 STATEMENTS
FIN
```

Figure 2:   Sample program and statements in RPL.

due to the limitations of the MLCC line printer which does not have a full character set.

Figure 2 shows a sample output in RPL, demonstrating many of the features of this language.

A rigorous definition of RPL appears in Section 3.1 of the UAB MLCC Users Guide. (4)

CONCLUDING REMARKS

Development of specific application languages which must function on different computer systems is subject to a major obstacle -- implementation. Since the implementation of languages is generally left to computer manufacturers, either the languages are never developed or they are developed with non-standard extensions or as deficient subsets of the original design. While compiler generator systems should not be considered a panacea for language development, they could make an important contribution to the implementation of a language design for neurobiologists or other unique scientific applications. In fact, it would benefit us all if all computer manufacturers were required by law to supply a standard compiler writing system for government-funded systems.

DISCUSSION

CLARKE: I'd just like to make two comments with respect to our experience at Stanford with a similar system. I think there's no question that satellite processors work well. In our system, an attempt was made to use a central facility for the actual real-time data collection and the effects on system reliability were rather devastating in the sense that one user could easily crash everybody out. Secondly, I'd like to ask whether you intend to decentralize the software the way you've decentralized the hardware. I gather your staff spends a fair amount of time on applications programming.

WHITE: Right. The computer center acts as a hardware and software resource to the researchers in the community. They can operate independently if they want to, and we also will provide programming for people who have gone out and bought their own computer, if they need it. It takes probably a year for a person fresh out of undergraduate school to really learn how to be a good research programmer, participating with the researcher in implementing his ideas. I think that he's really helped by working with senior people whom he can ask questions as he learns. It's a question of economics: you can buy your own cow, but sometimes it's good to be a veterinarian if you have to service that beast, and that's the same problem a lot of researchers run into.

REFERENCES

1. Leach, Geoffrey C. "The University of Washington XPL Compiler." Proceedings of the XDS Users' Group 19th International Meeting, 2: XDS Users' Group (1972).

2. McKeeman, W. M.; Horning, J. J.; and Wortman, D. B. *A Compiler Generator*, Englewood Cliffs, N.J.: Prentice-Hall (1970).
3. Sammet, J. E. *Programming Languages: History and Fundamentals,* Englewood Cliffs, N.J.: Prentice-Hall (1970).
4. *Multiple-Laboratory Computer Center Users' Guide*, section 3.1.

chapter 3

DESIGN CHARACTERISTICS OF BIOMEDICAL RESEARCH INFORMATION SYSTEMS

B. J. RANSIL
Clinical Research Center
Beth Israel Hospital
Boston, Massachusetts

INTRODUCTION

During the past two decades an abundance of time, talent and funds has poured into the construction of computer systems designed solely for biomedical information storage, retrieval, processing and problem-solving. This singular development has occurred both as a response to urgent problem-solving needs on the part of research investigators on the one hand and as a unique opportunity for information systems research and development on the other, spurred on by the development of compilers and computing languages and by the remarkable advances and declining costs in hardware.

The general goal of these systems has been to bring state-of-the-art computer technology to the service of biomedical research in its many facets -- data retrieval, reduction, storage, processing and reporting; problem-solving; computer diagnosis; health care delivery -- under sets of constraints that vary markedly from one design situation to another. Sufficient experience has accrued with such systems already in operation or in their developmental stages. Of prototypic relevance to the development of a general purpose language for neurobiological research are three research information processing systems: OMNITAB, the PROPHET System and CLINFO.

OMNITAB

OMNITAB (1) was designed and implemented at the National Bureau of Standards in the late 50's and early 60's for a target population of physical scientists and engineers who needed both data reduction capability and computing power for data bases that by their very nature (multivariate functional dependencies) led logically to generating data in tables or arrays on "worksheets" from which processing could proceed. OMNITAB is one of the earliest examples of a user-oriented, user-specific program designed by computer-knowledgeable scientists for their less fortunate colleagues. It pioneered in the use of a high-level interpretive language constructed in simple English sentences designed to make the computer a tool easily learned and used by computer-naive investigators. Its unique design features -- mathematical and statistical analysis of data from tables or arrays by using a simple English command language -- arose from a realistic evaluation of the characteristics of the target population and its computing needs. At the same time the OMNITAB is limited in its usefulness because it simulates desk calculator computing. The instructions "control the flow of calculations in a manner highly analogous to the logic which prevails in carrying out computations on a desk calculator," thereby replacing "the desk calculator, mathematical tables and multicolumn worksheet" (2). Since then desk calculators have been replaced by remarkably versatile programmable calculators with large memory capacity, occasioning a quiet revolution in computing methodologies which renders superfluous or obsolescent those aspects of OMNITAB that simulate and perpetuate the drawbacks of mechanical calculators. Nevertheless for its time OMNITAB represented a marked step forward in user design.

THE PROPHET SYSTEM

A decade or so later the PROPHET (3) system was developed in response to the perceived need and a desire among pharmacological and pharmacologically-oriented investigators to apply state-of-art computer technology to pharmacological research. The target population embraced those individuals concerned with the behavior of any drug or metabolite in living systems, which essentially included the entirety of in vivo and a large portion of related in vitro biomedical research. All but a small fraction of this population were recognized to be computer-naive and incorrigibly so. Consequently its designers gave a top priority to PROPHET's being user-oriented and user-responsive to a hitherto unheard-of degree. It meant that the PROPHET language not only had to be simple -- that is, easy to learn and use for perpetual novices -- but at the same time able to accommodate the needs of the sophisticated user as well. Further, while it was expected that the bulk of applications would deal with numerical data, the need for a degree of textual entry and handling was anticipated and implicitly accommodated.

These system characteristics led to the concept of a multi-level language employing data types based on a table-file structure, which essentially sums up the major design features of the PROPHET system. The multilevel language feature enables the investigator to communicate with the computer at three levels:

- high-level English commands and command lists
- procedure written in PL/PROPHET, a language syntactically modelled on PL/1
- conventional programming languages.

In the author's experience, the design of PROPHET's English command language is so simple and yet so powerful that with about one hour's person-to-person orientation the average clinical investigator can proceed on his own, with the PROPHET manual (procuring assistance by telephone from the systems representatives when he runs into difficulty), to make effective use of PROPHET's full computing armamentarium which, with interactive graphics and the recent incorporation of MLAB, makes it one of the most powerful and versatile numerical analysis sytems available to biomedical research investigators today. A brief description of the PROPHET system command language together with some of its features is given in Chapter 24 (4).

While the PROPHET system language possesses a very efficient alphabetical and numerical sort and a text making capability, it is not currently as text-oriented as MUMPS, such that those clinical applications requiring textual sorts, searches, correlations and other manipulations are not presently accommodated at command language level. However the capability is implicit in the design and needs only explicit development.

Another user-oriented PROPHET design feature is its accessibility -- by virtue of its low-profile terminal hardware and operational simplicity -- such that the most naive investigator can quickly learn to use the system and come to view it as a personal tool for handling and analyzing his own research data without need for in situ systems middlemen. In this respect it is almost as "friendly" as a programmable hand calculator (but unfortunately not as portable or as inexpensive!), while possessing far more computing power and potential.

CLINFO

The CLINFO (5) project has arisen in response to the computing needs of the many Clinical Research Centers (6) at medical centers and teaching hospitals across the country. Its design (7) owes much to both a year-long survey of the needs of representative Clinical Centers and of the PROPHET experience at two Clinical Center sites. Both in its aplications and its target population attributes, CLINFO closely overlaps those encountered in the PROPHET system, so much that the two systems' design specifications seem identical in all but a few respects, which appear to represent more a difference in emphasis than difference in kind. One of these is a greater need in certain Clinical Centers for text entry and handling in a file form.

CLINFO's answer to the very similar design problems as those posed by the PROPHET user community is to create a "study data file" which guides the investigator in his data entry much in the same spirit of the MAKE TABLE dialogue in PROPHET. Once the data is entered in the "study data files", data for analysis is retrieved by abstracting it from the files into labeled worksheets via another dialogue process ("Create Worksheet") which searches, abstracts and tabulates the desired data. When this is completed, each worksheet may be analyzed by calling an individual program which operates upon the worksheet and producing output that is stored and displayed in a predefined way. CLINFO is currently in the process of prototype development.

THE DESIGN PROCESS

In all these examples, basic design flows from the characterization of target population and its computing requirements, much as form follows function in architecture. The basic design process is one of painstaking definition and description of goals, target population and applications, together with a delineation of priorities, resources and constraints, usually by a representative group of users in collaboration with computer systems specialists. System design begins with developing the best available answers to such questions as:

> What are the design goals?
> What is the intended purpose of the system?
> Who constitutes the target population; what are its problem-solving needs; what are the data characteristics?
> How and with what means is the project to be accomplished?

In addition to accumulating information of this nature, it is necessary to identify as well as possible the system constraints, that is, those factors which in some way or other impose real limitations upon the design, its realization and the ultimate functioning of the system itself. As detailed information of this nature is collected and correlated, design priorities can be identified and ordered. Where conflicts emerge between priorities and constraints, trade-offs can be formulated. Through such a process an overall system design gradually emerges which may then be compared feature by feature against existing resources for capabilities that conceivably might be borrowed or adapted, thereby saving developmental costs, and accelerating implementation. Assuming that the concept of a general purpose language for neuroscience meets a real need that cannot be realized by existing resources and that the needs and target population are sufficiently well-defined to make its development feasible, the design will be influenced by a number of factors such as the population characteristics, the type of data that is generated and its processing requirements, and the attributes desired for the language.

POPULATION CHARACTERISTICS

The target population for which the system is intended needs as precise a definition as possible because project justification and design are usually related to population size, specificity and expertise. The greater the number to benefit, the better justification for expenditure of time and effort. The broader the design specifications be and the less expert at computer manipulation the population be, then the greater the software and hardware interfacing requirements between the system and its user must be. Some of the population characteristics that enter into design considerations are:

- population size
- population geographic distribution
- population stability
- population composition and homogeneity
- previous computer/programming experience
- educability...time, opportunity and motivation to learn and use the facility
- variability of these characteristics from one institution to another, or from one application to another.

DATA CHARACTERISTICS

The design of information systems is markedly influenced by the quantitative, qualitative and temporal characteristics of the data. These include:

1) Qualitative -- kinds of quantities to be manipulated
    a) data types: number, signal, image, text, descriptors, table, array, graph, sample, cluster, molecule.
    b) analog, digital.
2) Quantitative -- amount of data to be manipulated
3) Temporal -- time dependent characteristics of the data
    a) on-line monitoring, sampling, reduction.
    b) off-line batch processing or ad lib calculating.

PROCESSING REQUIREMENTS

Among the capabilities frequently desired for information handling systems are:

- data screening
- data sampling
- data reduction
- data storage
- data retrieval
- data sorting
- display
- data compilation and data bank construction
- error checking
- data analysis and modeling

All of these entail a large number of unique operations (such as table-making, curve-fitting and graphing), many of sufficient stereotypy from application to application to permit command formulation. For a command language to be possible and practical in a given situation it must be possible to identify and define specific operations that meet the following criteria: stereotype of application, consistency of application, universal need, large user group.
For the most part the operations implemented by high level commands in OMNITAB, the PROPHET system and CLINFO are concerned with specific storage, retrieval, sorting, analysis, modeling and display operations such as FIT LINE, FIT POLYNOMIAL, MAKE GRAPH, etc. On the other hand the applications frequently encountered in neuroscience involving signal and image analysis, variable sampling rates and on-line error checking fall outside the current high level command language capability of PROPHET. But such application are not precluded by the PROPHET design. Once a set of operations of sufficient stereotypy are identified for neuroscience applications, the process of implementation via PROPHET command language (or any other high-level language), while not without its headaches and hazards should be relatively straightforward.

LANGUAGE ATTRIBUTES

In addition to types of applications and the characteristics of the target population and data, there are certain characteristics pertaining to the language itself which possess considerable design importance because they directly affect user acceptance. These include:

- ease of learning, use, and recollection
- flexibility or adaptability, i.e. responsivity to system stresses.
- versatility, i.e. ability to serve many functions.
- compatiblity with other languages and systems.
- exportability, i.e. may be readily implemented at different sites.
- hardware and configuration independence.
- multi-level capability, i.e. capable of serving both naive and knowledgeable users.
- logical sytax oriented toward the user's applications.
- brevity
- capability of growth with the branch of science it serves.

While the computer engineer and computer-oriented scientist are willing to sacrifice many of these conveniences in order to solve their problems, the large majority of investigators involved in life-science research won't. Accordingly, systems that are intended to make computing technology accessible to a computer naive target population should be "user-oriented" rather than engineering-oriented, that is, as many of the above attributes as is compatible with other design priorities should be incorporated into the system design.

CONCLUSION

Although the foregoing is by no means an exhaustive survey of the kinds of information desired to establish design specification and priorities for a general purpose neuroscience language, it does hit the major highlights. While the application of state-of-the-art computer technology to neurological science is not only a necessity but indeed a foregone conclusion, the question of how best to foster and finance this development precedes systems and language development. Does neurological science per se constitute a specific, characterizable identity with a unique set of needs such that a specific information system or specific general-purpose language is called for? Are the routinely encountered experimental operations identifiable and unique enough from application to application to justify a high level language? If so, can already existing systems be adopted to meet many of the needs, or is a de nouveau approach required? Are sufficient funds available to embark upon a global, de nouveau approach? Answers to these questions should help bring the project into clearer focus and help accelerate the decision-making process.

DISCUSSION

CAPOWSKI: I want to cry out in the wilderness. You would like to have transportability from one computer center to another. I've pretty much given up on that. We have a microscope and television system and a three-dimensional refreshed graphics system. Once a week someone comes up to me and says, "Joe, I have this program that's going at UCLA. Can I run it here?" Sometimes I can get it running if it doesn't require any special hardware. But whenever there's any sophistication required, it's hopeless. So I've pretty much given up on transportability.

RANSIL: Let me share the PROPHET experience on this particular point. The designers got together and decided on the features that the total target population would need, and these were implemented. Then we told the users, "You can develop any programs tha you want which are applicable to your own needs, and this can be done in any language you wish to use. Feed it in to us. We will put a PROPHET front end and a PROPHET rear end on it, and adapt it for the system." We now have something like 25 different user-developed programs which are not only tranportable, they are used extensively by everyone else in the system, simply because this is a design feature in the system: to make it portable in this way.

KEHL: It seems to me you left out a major important difference between OMNITAB and PROPHET, which is very important for consideration by this group. And that is, PROPHET represents a system that does file structuring. OMNITAB is a system that never did any file structuring at all. It was one of the most rudimentary systems imaginable.

RANSIL: I was going to mention that tomorrow, Ted. I think you're absolutely correct.

KEHL: It seems that that generation, that jump, from OMNITAB to PROPHET, is a significantly important step, often neglected.

RANSIL: Right, it's a quantum jump.

REFERENCES

1. OMNITAB: *A Computer Program for Statistical and Numerical Analyses*. National Bureau of Standards Handbook 101 (1966).
2. Op. Cit. p. 13
3. PROPHET is a chemical/biological information handling system sponsored by the Chemical/Biological Information Handling Program of the Division of Research Resources, NIH, designed and developed by the Medical Computer Systems Group of Bolt Beranek and Newman, Inc., Cambridge, Mass. and operated by First Data Corporation, Waltham, Mass. Cf. Castleman, P. A., C. H. Russell, F. N. Webb, C. A. Hollister, J. R. Siegel, S. R. Zdonik and D. M. Fram. Implementation of the PROPHET system. Proc. Natl. Comput. Conf. 43:457-468, (1974); Hollister, C. A. Users Manual for the PROPHET System. Bolt Beranek and Newman, Inc., periodically updated; Public Procedures Notebook. Technical Document No. 47, Bolt Beranek and Newman, Inc; Ransil, B. J. Applications of PROPHET in human clinical investigation. Proc. Natl. Comput. Conf. 43:477-483, (1974); Ransil, B. J. Use of table file structures in a clinical research center. Fed. Proc. 33:2384-2387 (1974).
4. B. J. Ransil. The table as a basis for data analysis. In: *Computer Technology in Neuroscience*. By: Paul B. Brown (ed.). Washington: Hemisphere Publishing and John Wiley & Sons, 1976.
5. The CLINFO Project for the development of a data management and analysis system for clinical investigators is sponsored by the Division of Research Resources, NIH, and is in the process of design and prototype development by the Rand Corporation, Santa Monica, California.

6. The Clinical Research Center Program is a nation-wide clinical research resource supported by the Clinical Research Centers Branch of the Division of Research Resources, NIH.
7. W. L. Sibley, M. D. Hopwood, G. F. Groner, and N. A. Palley. *A Prototype Data Management and Analysis System for Clinical Investigators: An Initial Functional Description* R-1621-NIH, Rand Corporation, (1974).

chapter 4

SNAX: A LANGUAGE FOR INTERACTIVE NEURONAL MODELING AND DATA PROCESSING

D. K. HARTLINE

Biology Department
University of California at San Diego
La Jolla, California

INTRODUCTION

This chapter will describe a general purpose I/O-oriented language written for the purpose of processing spike train and intracellular wave-form data from simple neuronal systems, for controlling aspects of an experiment and for simulating in quantitative detail

---

I wish to thank William Roberts for major contributions to the development of SNAX. Betsy Jaffe and David Schmitt also contributed to the programming. I am grateful to my colleague, Howard Warshaw, for his permission to include accounts of some of his work (especially figures 3b and 4) and to David Russell for making available information on the central connections of the stomatogastric ganglion in advance of its publication. I also appreciate the dedicated technical assistance provided by Donald Gassie, Denise Juppe and Claire Sirchia.
  This work was supported by grants from NSF (GJ-4317 and GB-28620), NIH (NS-09477 and NS-10064), the Alfred P. Sloan Foundation, and the University of California Academic Senate.

the activity in these systems. The first part of the chapter will be concerned with the programming language itself and the particular ways in which it is suitable for the tasks it was designed for. The second part will present two models which have been used: one for simulating properties of repetitively firing neurons, and one for simulating interactions among individual cells in simple neuronal networks. Finally, I will discuss the concept of the use of physiologically accurate models as a potentially powerful tool in neurophysiological research.

Writing the language was motivated by the need for a versatile, but rapid, I/O-oriented language for a PDP 11/45 computer (none existed at the time, although other fairly suitable options are now available on this machine). In addition, a simple language that would be readily picked up by people relatively inexperienced in computer programming was desired. In fulfilling this need, I was much influenced by my experience with SNAP, written by William Simon for the PDP-8 (Simon, 1966: "SNAX" stands for "SNAP extended"). SNAX was thus patterned after SNAP, but with several additions to allow it to better serve my applications.

INSTRUCTIONS

The basic instruction set is similar in pattern to that of other digital computer languages such as FORTRAN, BASIC, or FOCAL. Instruction mnemonics of up to six characters (see Table 1) are followed by a string of arguments (constants or variable-names) separated by spaces or commas. This format holds for virtually all instructions at present,* which greatly facilitates its use. Parentheses are lacking in instruction formats, which also simplifies learning the language, but means that complex algebraic expressions take several lines to represent. Also, it means there are no subscripted variables or (for the most part) complex numbers

---

* Currently the "ASCII" statement is the only exception: if separators occur in the character string, the string must be enclosed in angle brackets " <> ". Statements using an " = " as in "A = B+C" are contemplated for future implementation.

Table 1. SNAX Functions.

Instruction	Function
1. Arithmetic	
EQUL, ADD., SUB., MUL. DIV., SIN, COS, ATN, EXP LGN, SQRT, INTEGR, FRACT TRUNC, NEG, ABS, FIX, FLOAT	Standard arithmetic and trigonometric
2. Program Flow Instructions	
CMPARE	Compares 2 values, and exits according to <, =, or >.
IFPOS, IFNEG, IFZERO	Tests value and exits according to true or false.
IFNEAR	Tests for value within specified range of another and exits according to true or false outcome.
IFEXCD	Tests $|A|>|B|$ and exits according T/F.
IFSWCH	Tests specified switch and exits T/F.
IFFLG	Tests specified program flag (64 + available).
SETFLG, CLRFLG	Sets/clears specified program flag.
XCHFLG	Exchanges lower 32 flags with specified variable.
GOTO	Unconditional transfer.
3. Subroutine Instructions	
GOSUB	Subroutine call.
SUBRTN	First instruction in subroutine (holds dummy arguements).
RETURN	Return to calling program (location specifiable).
4. Loop Instructions	
BRSU	Set specified loop counter (5 available).
BRET	Decrement count and exits according to = or ≠0.
COUNTR	Read counter values.
5. Input Instructions	
TTY	Wait for teletype input of variable.
ADC	Do scaled A/D conversion on specified channel and set variable T = time of conversion.
SWR	Set variable to number keyed into specified switch register grouping.
TIME	Set variable to clock time.
IFDGI	Test specified digital input bit, exit T/F.

Table 1. (*continued*)   SNAX Functions.

Instruction	Function
6. Output Instructions	
PRINT	Teletype (or scope) output of up to 5 variables.
SPACE, CRLF, COLON, BELL	Special teletype instructions.
ASCII	Character string output.
SETPNT, SETPLT, SETPP	Sets ASCII output mode for teletype, scope, or both.
SETDSP	Initialize scope output position for ASCII.
PLOT	Scaled point plot of X, Y coordinate pair.
PLOTI	Scaled incremental point plot of ΔX, ΔY pair.
YPLOT	Scaled plot of value with autoincremented X value.
SETYPT	Set up for YPLOT setting autoincrement step size.
ORIGIN	Set origin.
AXIS	Plot coordinate axes with specified tic interval.
SCALE	Set default plot scale.
SETV	Set scaled voltage on specified DAC channel.
STEP	Change voltage on specified DAC channel
DAC	Read voltage currently imposed on DAC channel
SETDGO, CLRDGO, IFDGO	Set, clear, test specified digital output bit
7. Table Instructions	
TABLE	Fetch value from table at pointer and increment pointer
LOADTB	Store value in table at pointer and increment pointer
TBIND	Fetch value from specified location in table
PITB	Store value at specified location in table
RUBTB	Clear table
TBSIZE	Find size of table
PTR, BPTR, EPTR	Find position of pointer, begin-pointer, end-pointer
INCPTR, INCBPT, INCEPT	Increment pointer, begin-pointer, end-pointer
RESETB	Reset table pointers to specified positions
SETBSZ	Set sizes of all tables
TBLOOK	Find fractional table position corresponding to value in monotonic table (linear interpolation)
INTERP	Find value in table corresponding to specified fractional position (linear interpolation)
PRINTB	Print specified table positions
PLTINC	Plot table with given X-increment
PLOTB	Plot 2 tables one against the other as X,Y pairs

44

Table 1. (*continued*) SNAX Functions.

Instruction	Function

8. Stack Instructions

PUSH, PUSHL, POP, POPL	Push/pop top/bottom of stack
CYSTK	Rotate stack (popping one end & pushing other)
STKCNT	Find stack size
PSHPTR, POPTR, PSHFLG, POPFLG, PSHCNT, POPCNT, PSHAXS, POPAXS	Push/pop table pointer, flags, loop counter scope origin

9. Bulk Storage Instructions

READ, WRITE	Reads/writes specified table, file, device
DELETE	Deletes file
RDREC, WRTREC	Read/write one record
WAITIO	Wait for I/O device to finish
CLOSE	Close file
SETNAM	Set file name
GETNAM	Reads file name from teletype

10. Other Instructions

HISTO	Increment appropriate table location as histogram bin
BAR	Plot histogram bar at specified position and height
DELAY	Generate specified delay in execution
GOSNAX, MONRET	Enter/leave SNAX program
CONV	Takes the convolution of table-stored function with another
FFT	Takes the Fourier transform of a table-stored function

permitted. SNAX is not the language of choice for doing sophisticated "number crunching".

Many (and potentially all) instructions have useful and simplifying default options if an argument is not specified. For example PLOT X, Y will plot the variable Y vs the variable X on the scope face, with the limits specified in a "SCALE" instruction as full-scale values. On the other hand, PLOT X,Y,A,B will plot Y vs X, but with B and A respectively as full scale values.

Mnemonic labels of up to six characters for instruction lines can be specified (rather than simply numbers).

Variables and Constants

Numbers are internally represented for the most part in floating point format. This obviates the need for keeping separate integers and noninteger variables. Teletype input and output may, however, be in any format desired. The floating point processor of the 11/45 allows rapid manipulation of floating point numbers. Variables are stored preceded by a six character ASCII header containing their name, to allow header output in a "PRINT" statement, and to permit a monitor search when required.

Arithmetic and Trigonometric Instructions

These are quite standard functions. However, as mentioned above, only a fixed number of arguments can be used with each.

Program Flow and Loop Instructions

These instructions control the logical flow of the program. They are typified by the CMPARE instructions (similar to a FORTRAN "IF") and the loop instructions. The loop instructions, similar to those of SNAP, allow setting of a counter, then autodecrementing the counter and exiting to different locations depending on a non-zero or zero count. Program flow can also be altered using instructions which test switches (IFSWCH), bits in the digital input (IFDGI) or output registers (IFDGO), or a set of 64 program flags (IFFLG), the first 32 of which can be exchanged with any desired variable (XCHFLG) to give an unlimited number of flags. Flags and digital I/O bits are set or cleared by other instructions (Table 1).

Input-output Instructions

The "TTY" instruction hangs the computer up until a numerical value, assigned to its argument, is typed on the keyboard (alternative keyboard input can be effected using the SNAX monitor functions, see below). The "ADC" instruction samples and scales the voltage on the specified channel, and at the same time sets the variable "T" equal to the clock time of the sample. Specified groupings of switches on the console can be read as numbers using the "SWR" instruction.

Numerical values for any variable, as well as strings of alphanumeric characters, can be output to teleprinter or scope (e.g., PRINT, ASCII). Position and size of scope messages is controlled with the "SETDSP" instruction.

Scope (plot) modes include point-plotting of scaled X,Y pairs (PLOT), incremental plotting of scaled X, Y pairs, plotting of a sequence of Y values with autoincremented X spacing (YPLOT), plotting two tables one against the other (PLOTB) or of a single table with fixed X increment spacing (PLTINC). Default plot scale factors are established with the "SCALE" instruction, and the origin for any plot can be shifted with the "ORIGIN" instruction.

Scaled voltages on several DAC channels can be specified with the SETV instruction, or incremented with the STEP instruction.

Digital lines can be set or cleared with the SETDGO or CLRDGO instructions.

Table Instructions

At the moment, six tables of specifiable length (SETBSZ) are available to the SNAX user. Any desired location may be accessed or changed directly (TBIND, PITB) with a special "end-exit" from the instruction used if a location outside the table limits is specified. Alternatively, a set of pointers can be set delimiting a particular portion of the table and access made via a pointer with pointer autoincrement by a specified number of locations upon each access (TABLE and LOADTB instructions). Again, when a location beyond those set as limits is specified, special exit (or if defaulted, an error message) is made. This instruction permits a loop to operate on a data set of unspecified size but to terminate properly when the end is encountered.

A table interpolation instruction allows the "fractional" location corresponding to a desired value to be found in a table of monotonic numbers (TBLOOK); another allows a value corresponding to a fractional location to be read from a table using linear interpolation (INTERP).

Bulk Storage

Instructions storing or retrieving table data on bulk storage devices (DECtape, disk) utilize the software drivers made available by the disk operating system supplied by the manufacturer (Digital Equipment Corp.). Thus, files generated may be manipulated with the operating system software, and are compatible, for example, with FORTRAN on this system. Two classes of instructions exist, one for reading and writing single-record files (easier to use) and one for multi-record files.

Miscellaneous Instructions

Stack instructions (PUSH, POP, PUSHL, POPL,) push or pop numbers on the top or bottom of a stack. Loop counters, flags, table pointers, and scope plot origins can also be saved and restored in this fashion. Standard subroutine calls can be made, and subroutines may be nested to any level. Instructions exist for obtaining the clock time (TIME), generating a specified delay in the program (DELAY) and generating random numbers (RAND).

Special Purpose Instructions

A histogram instruction (HISTO) treats a specified number of table locations as bins of specified width for a histogram. The instruction increments by 1 the location corresponding to a value given as input. The BAR instruction plots a histogram bar of specified size on the oscilloscope face.

Convolution and Fourier transform instructions operating on SNAX tables are available as well; however, extensive "number crunching" is more appropriately done in FORTRAN.

SNAX MONITOR

Certain rudimentary "monitor" functions can currently be effected during execution of a SNAX program:

on keyboard command any variable may be printed out or changed, then computation resumed.

SNAX APPLICATIONS

In addition to the modeling applications to be discussed in the next sections we use SNAX to digitize spike-train and continuous-voltage (e.g. membrane potential) data both on-line during experiments and off-line for later analysis. Graphs, scatter plots and histograms are readily generated with the language, and knob-controlled ADC, especially, allows ready interaction with the experimenter. Standardized tables of spike times are easily manipulable with the table instructions and with a few subroutines geared to the spike storage format. Model output in the same format permits analysis programs to operate on either type of data, to give direct intercomparisons between the model and the physiology (e.g., figure 2).

Of particular value to this sort of program generation has been the variety of simple but useful I/O instructions. The stack, the program flags, the autoincrementing table pointers and the default table (to give error messages) exit capabilities are also features which have been particularly useful. Finally, the ease with which additions and changes can be made to SNAX itself has been an invaluable asset in tailoring it to our particular needs.

SIMULATIONS

Repetitive Firing Simulations

The core of these programs consists of two subroutines, one calculating the kinetics of membrane potential and threshold during the interspike interval of a (potentially) repetitively firing neuron, and one calculating the abrupt changes which are assumed when a spike is fired, the latter being used either when membrane potential exceeds threshold, or when an "antidromic" spike is generated. The approach of the model (which I call the "active pacemaker model") is to find a kinetic desciption of certain state variables such as membrane potential (V), threshold (Θ), adaptation (A_r, $A_{1,2}$), and conductance (g_m) at the trigger zone of a neuron without requiring voltage clamp experiments, which are as a rule technically not feasible in nerve nets of interest. Following experimental results of Hodgkin's (1948, see figure 1) the membrane potential trajectory (V) between spikes is viewed as consisting of two components, a passive component representing the recharging of the membrane capacitance, plus, above a threshold value (Θ_{pm}), a slowly developing active component (P) which eventually brings the membrane to firing threshold (Θ). The passive component, for convenience, can be broken into the steady-state passive value (X), which the membrane potential would exhibit in the absence of a spiking mechanism, plus a spike perturbation factor (W) representing deviation from X produced by the shorting or reset effects of a spike. This perturbation decays exponentially between spikes with the membrane time constant. The basic equation describing the membrane potential time course is thus given by:

$$V = X + W + P \tag{1}$$

$$X = E_0 + \frac{E + J - A_r}{g_m} \tag{2}$$

where E_0 is the resting potential, E is the effect at the trigger zone of impinging synaptic influences, J is the contribution to the passive potential of current injected through a microelectrode, and A_r is an adaptation quantity affecting the membrane potential, as for example, through an electrogenic sodium pump. The total membrane conductance, g_m, relative to rest is the sum of contributions from spiking (g_s) and synaptic inputs (g_e): $g_m = 1 + g_s + g_e$.

Most state variables are taken as exponentially decaying functions of time, governed by the general equation

$$\frac{dy}{dt} = -\frac{1}{\tau_y}(y - y_\infty) \tag{3}$$

where y_∞ is the asymptotic value for y (which itself, in general, is a function of other variables) and τ_y is the time constant for the approach (usually constant, but in the case of P, dependent on various conditional relations). Table 2 lists the state variables and the symbols or expressions used to denote their τ_y and y_∞. Note that many variables approach $y_\infty = k_y \times$ (some excitation), making the asymptotic value a function of sustained excitation levels, hence giving the model accommodative properties.

Table 2. State variables for repetitive firing simulations.

Variable (y)	Function	τ_y	y_∞	Δy_0	y_{max}
J	Current contribution to X	τ_m/g_m	I	–	–
g_s	Spike-induced conductance	τ_g	$K_g X'$	Δg_0	g_{max}
A_r'	Resting potential adaptation	τ_{AR}	$K_{AR}(X'+A_r)$	ΔA_{r0}	$A_{r\,max}$
A_1	Active response adaptation 1	τ_{A1}	$K_{A1} X'$	ΔA_{10}	$A_{1\,max}$
A_2	Active response adaptation 2	τ_{A2}	$K_{A2} X'$	ΔA_{20}	$A_{2\,max}$
$\theta-\theta_\infty$	Threshold deviation	τ_θ	$K_\theta X'$	$\Delta \theta_0$	θ_{max}
P	Active pacemaker response	τ_{pm} (a)	$K_{pm}(X-A_1-A_2)$ (c)	∗	∗
		τ_m/g_m (b)	0 (d)	∗	∗
W	Passive spike-perturbation	τ_m/g_m	0	†	†

Notes: g_m = membrane conductance; I = current imposed by microelectrode;

τ_m = resting membrane time constant; $X' \equiv E_0 + E + J - A_r$;

θ_∞ is the asymptotic value approached by θ when $X' = 0$.

∗ P is set to zero by each spike.

† W: After a spike is determined by resetting V as with standard state variables, then set $W = V - X$.

(a) $P_\infty > P$

(b) $P_\infty < P$

(c) $V < \theta_{pm}$

(d) $V > \theta_{pm}$ or $P_\infty < 0$.

When a spike is generated, certain "standard state variables", which obey Eq. (3) are augmented according to:

$$y_{new} = y_{old} + \Delta y_0 \frac{y_{max} - y}{y_{max}} \qquad (4)$$

that is, the change in the variable is proportional to its distance from a maximum (strictly, a maximum or minimum) value. Symbols for these quantities are also listed in Table 2.

Figure 1 diagrams the active pacemaker mode. This shows the basic kinetic components of the model. Membrane potential is considered to be composed of a passive component summing with an active component. Both components develop exponentially toward asymptotic values determined by excitatory influences. Deviations of model trajectories from crayfish stretch receptor trajectories are indicated by dashed lines at early and late phases in the interspike interval. When the exponentially decaying threshold crosses the membrane potential, a spike is

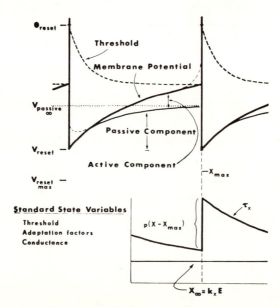

Figure 1. Diagram of the active pacemaker model.

generated. Thereupon the active component is set to zero, threshold to Θ_{reset} and membrane potential to V_{reset}. Various standard state variables (threshold, adaptation, conductance) are reset by each spike and between impulses, decay exponentially toward levels dependent on the excitation.

Application of the Active Pacemaker Model

The model has been tested on its predictions for changes in spike firing pattern changes produced by injected currents delivered at various phases in an ongoing spike train (see figure 2). These predictions have been compared with results of introducing the same perturbations through an intracellular microelectrode in the slowly adapting stretch receptor neuron (MRO) of the crayfish. Figure 2A shows a "dot" pattern for the first

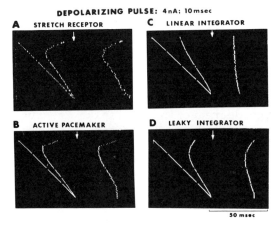

Figure 2. Reactions of stretch receptor and models to depolarizing pulses.

    A. Reaction of crayfish slowly adapting stretch receptor neuron ("MRO").

    B. Reaction of active pacemaker model.

    C. Reaction of linear integrator.

    D. Reactions of leaky integrator.

two impulses generated by the MRO after a 4 nA, 10 msec depolarizing pulse of current was inserted at a progressively increasing phase (top to bottom) into a regular ongoing pattern of 20 impulses per second (diagonal line of dots indicates beginning of stimulus). When the phase of the stimulus was very close to zero, relatively little decrease in interspike interval (leftward shift of pattern: unperturbed interval indicated by arrow) was observed. As phase increased, a rapid change in pattern was observed, reaching a minimum interval at a phase of about 0.3, then a gradual increase as the phase progressively increased to 1.0.

Figure 2B shows the same experiment applied to the model, with qualitatively, at least, the same results. The initial change in pattern at early ϕ is greater than it should be, and the overall shape lacks the smoothness of the MRO response. Parameters optimized within limits allowed by physiological measurements: τ_m = 10 msec (same value used for all models); Θ_{pm} = -5 (arbitrary units: one unit corresponds roughly to the effect of 1 nA of current membrane); K_{pm} = 100; τ_{pm} = .25 sec; τ_{Ar} = 10 sec; Θ_∞ = 5; τ_Θ = 4 msec; V was reset to -15 by each spike, Θ to +10, and A_r was incremented by .002 units, all other parameters not used in this simulation.

By contrast, figure 2C shows results from a linear "integrate-and-fire" model of the sort that has previously been used (and quite successfully) to simulate stretch receptor firing (Sokolove, 1972). Following an initial large decrease in interspike interval at ϕ = 0, the interval steadily increases as ϕ does. The "scoop" shape of the MRO pattern does not occur. Note failure of second spike following stimulus to parallel the pattern of the first, as expected for a non-leaky integrator. Figure 2D shows results from an optimized "leaky" integrator model, which are noticeably better than the linear integrator, but with some important discrepancies. Some of the "scoop"-shaped characteristics of the MRO pattern are mimicked by this model, but in addition to the too-large initial jump, the scoop does not develop fast enough, nor is it deep enough. The time-constant of the integrator was 15 msec.

The active pacemaker model is a definite qualitative improvement, but it still leaves quite a bit to be desired in simulating responses to certain other types of stimuli (see Hartline, 1976, for details).

Network Simulations

The basic unit for programs simulating a group of interacting neurons is a subroutine which operates on a stored data-tree to calculate one time-increment of cell kinetics for each neuron in the network. This uses the repetitive firing subroutine previously discussed to simulate events at the trigger zone, and to generate the output train of impulses. Excitatory drive to the trigger zone subroutines is supplied by two types of sources: impulse-mediated "synaptic" input, and low-pass "tonic" (electrical or chemical) input. In the present version, all inputs are calculated in terms of their effects on the trigger zone of the postsynaptic cell. Thus the need to directly take account of electrotonic spread through a dendritic tree is for the moment side-stepped.

The excitatory synaptic input for each cell is stored for a fixed number of time-increments into the future in a table which is shifted one place every time-increment. When an impulse occurs in a presynaptic cell, a synaptic potential, preceded by a fixed delay, is calculated and added into this storage area. The normalized shape of the synaptic potential is assumed constant for any given synapse, but that shape is arbitrary: it may be measured from intracellular records and stored in a data table. Synaptic potential amplitude is calculated based on facilitation kinetics which takes total facilitation to be the sum of contributions from some number of "compartments", each based on an hypothesized reaction:

$$A \underset{k_2}{\overset{k_1}{\rightleftharpoons}} X \qquad (5)$$

where X is an "active" complex and A an inactive form (for example representing "internal $[Ca^{++}]$" or "releasable transmitter"). Each impulse is considered to momentarily change the rate constants k_1 and k_2 causing an abrupt change (increase or decrease, corresponding to positive or negative (="anti") facilitation) in $[X]$. concentration of "X" will then approach its asymptotic value of 1 exponentially with rate constant k_2. Complex facilitation and antifacilitation kinetics can be simulated by using several compartments. The facilitation calculation is performed by a subroutine, and could readily be changed to accommodate different facilitation kinetics (e.g. Friesen, 1975). The reversal potential for a synaptic potential is taken account of by reducing the PSP amplitude in inverse proportion to the distance between the starting-point voltage and the reversal level. Although this simplified calculation is only approximate, provision is made for a more accurate calculation based, for example, on the formulation of Rall (1964) should it be required.

The excitation present in the "synaptic input" table area may be reduced abruptly by a fixed percentage by a spike in the post-synaptic cell, to simulate impulse invasion and reset of dendrites.

Tonic influences (electrotonic; slow chemical or "chemotonic") from one cell onto another are taken to follow simple kinetics characterized by a single time constant and an attenuation factor. The asymptotic level of the postsynaptic influence is given by the product of the attenuation factor with a presynaptic membrane potential composed of a selected fractional contribution from both the input area of the presynaptic cell and its trigger zone voltage. Clearly there are many simplifications in this formulation.

A final source of excitation for each cell is an imposed current from an external source, for example a microelectrode in the cell. This is read from a table, fed to the trigger-zone kinetics subroutine and handled as described above.

The calling program we use in our current network simulations (Warshaw and Hartline, 1974; 1976; and see below) reads in various data tables and permits alteration of desired parameters. Following this it calculates and plots network states at small temporal increments (e.g., 1 msec). An "event" table is checked at each time increment for firing of externally generated spikes representing either antidromic activation of one of the network's units, or activity in an input axon originating outside of the network being simulated (e.g. a command fiber or regulatory axon). A "stimulator" capability exists for producing trains of stimuli of a knob controlled value (in terms of the number of units activated) to a selected group of neurons at knob-controlled frequency, train duration, and train repetition rate. Under switch and knob control, direct current may be passed through teletype-assigned cells.

Applications of the Network Model

Howard Warshaw and I have been applying this model to the simulation of network interactions in the stomatogastric ganglion of the spiny lobster. This network of some 30 cells has had the activity and interactions of almost all of its cells described in some qualitative detail (Maynard, 1972; Mulloney and Selverston, 1974; Hartline and Maynard, 1975). It has been of some interest to us to find to what extent the observed activity patterns of this system can be simulated given as yet mostly qualitative and simplified information on the system, and then to observe the improvement of the simulation as more physiological details become available (figure 3).

Figure 3A shows a record of activity in one subgroup of 12 "pyloric cycle" units classed in three types and

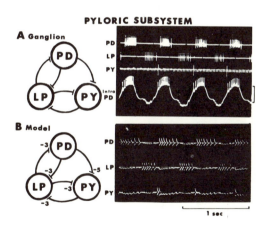

Figure 3. Simulation of pyloric subsystem of stomatogastric ganglion.

A. The diagram to the left shows the physiologically determined interactions among the three principal groups of units in this subsystem.

B. Simulation of system based on connectivity of A. Relative synaptic strengths given by each synapse.

interconnected as shown in the synaptic "wiring" diagram to the left. Activity consists of successive bursts in PD, LP, and PY units. There are 3 units in the PD group (technically 2 PDs and 1 AB), 1LP, and around 8 PYs. All synapses are inhibitory; the record shows activity from each group isolated in one recording channel, and an intracellular trace from a PD cell. Note the IPSPs which occur 1:1 with LP impulses. Note also the large oscillation with spikes riding on the peak, thought to be an endogenous rhythmic driver potential driving PD activity. In this particular example, only a single PY impulse occurs each cycle. Often there is a substantial burst of PY spikes.

This system was simulated with the model using three "cells" as representatives of the three cell types of the pyloric cycle; with appropriate interactions as indicated in the diagram to the left of figure 3B.

The PD-PY synapse had to be made stronger then the PD-LP one in order to obtain the correct sequencing. Other than in strength of synaptic interactions, all cells had identical properties. Each line shows the threshold (upper of two traces) and membrane potential. Spikes are generated when membrane potential exceeds threshold. They are seen as abrupt resets of threshold and membrane potential. Parameters: τ_m = 20 msec; θ_{pm} = -3.5; k_{pm} = 10; τ_{pm}=100 msec; τ_{A1}= 400 msec; θ_∞ = 0; τ_θ = 8 msec; V was reset to -10 by each spike, θ to 20, and A_1 was incremented by 0.6 units. Synaptic delays were 2 msec and PSP time course was a rounded one with rise-time of 15 msec and an exponential decay time-constant of 36 msec. Time increment 1 msec. Reversal potentials and other features of the model were not used in this simulation. We found that qualitatively at least, the appropriate activity pattern was exhibited by the model over a substantial range of parameters. We did find that the cells had to undergo adaptation, allowing their inhibitory influence to wane with time, and allowing postsynaptic cells (recovery from their own adapted state) to resume firing in their turn. In addition we found that the PD inhibition on the PY's, and thus delay PY activity until the appropriate part of the cycle. Other than that, the wiring itself was sufficient to produce a qualitatively correct activity pattern despite a number of simplifying assumptions, including the absence in the simulations of an endogenous bursting capability strongly suspected of occurring in PD cells.

An even more complex wiring has been worked out for the subgroup of neurons which participate in the gastric mill cycle. This includes two neurons, the "E" cells, found in the commissural ganglion (Russell, 1976). Figure 4 shows a simulation of neurons participating in this activity.

This shows a simulation run involving five groups of neurons (each containing only a single cell, except E and LG, which have 2), interacting in a complex manner. Inhibitory interactions are indicated by straight bars at synapse; excitatory by forked symbol, electrotonic by "register" symbol. Basic activity consists of alternating bursts in two sets of groups. Parameters: τ_m = 20 msec; θ_{pm} = -3; k_{pm} = 10; τ_{pm} = 100 msec; τ_{A1} = 400 msec; θ_∞ = 0; τ_θ = 8 msec; V and θ reset to -10 and +20 by a spike; A_1 incremented by 0.6 units; PSP rise time 10 msec; decay time constant 24 msec;

Synaptic Strengths:

		postsynaptic				
		E	I1	LC	GP	LG
	E	-	-	+3	0	+3
pre-	I1	-10	-	-10	-10	-
synaptic	LC	-	-10	-	-1	-10
	GP	-	-10	-1	-	-10
	LG	-	-	-	-10	-

Time calibration 2.5 sec.

Again there is qualitative agreement between the model's activity and activity observed in physiological preparations.

The qualitative agreement between observed activity and the model's behavior is gratifying but until more

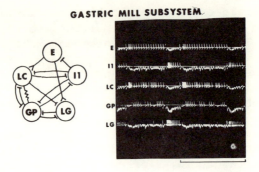

Figure 4. Simulation of gastric mill subsystem. Time calibration 2.5 sec.

exact quantitative data are available, it must be treated with caution. Although parameters for the models were chosen to approximate what quantitative estimates are available, these are far from complete. In addition, known properties of the cells (e.g. endogenous driver potentials, equilibrium potentials, accommodative adaptation) have been ignored in these simplified simulation runs. It remains to be seen how much input of more accurate quantitative data from physiological experiments (now under way) will modify these results.

USE OF PHYSIOLOGICAL MODELS

One expects a good model of a neuronal system to simulate successfully the actually observed output of the system under a variety of conditions. Given sufficient accuracy of the model, the success of the simulation is important confirmation that one's understanding of the system's kinetics is correct within limits. Some difficulty arises in deciding just how accurate such a simulation must be before one concludes he has the desired confirmation. The model itself can be of use in making this decision, since if estimates can be made of the possible error in determining a given parameter, or in omitting consideration of other parameters, the range of agreement between the model and the physiology for variation in the uncertain parameters gives an estimate of the certainty of a positive conclusion.

Of perhaps even more value to the study of a neuronal system is the use of the model when discrepancies occur between its predictions and physiological observations. In this case the nature of the discrepancy may suggest an area where understanding of the physiology is incomplete. For example, in undertaking a simulation of the time-course of the postsynaptic membrane potential in lobster cardiac ganglion, it was found that the simulation did a poor job of predicting the potential at the beginning of the periodic burst of ganglionic activity (Hartline and Cooke, 1969). It seemed that this discrepancy might be due to inaccuracies in the simulation of facilitation kinetics derived from limited observations. Since then Friesen (1974; 1975), working in my lab, has confirmed the existence of much more complex facilitation kinetics. Similarly, discrepancies found later in the burst were hypothesized to be due to turning off of an active response in the motor cells (Hartline and Cooke, 1969). Since then, substantially more evidence has been adduced for the presence of such active responses in cardiac ganglion motor cells (Tazaki, 1971; Cooke and Hartline, 1975; unpublished observations). However, it should be noted that for these discrepancies to have been picked up, reasonable accuracy in the original model had to be attempted.

Once a reasonably accurate model of a system can be established, the model may serve as a tool for the further understanding of the system. It is possible in a model, in ways not possible in physiological preparations, to alter physiological parameters (e.g., synaptic potential strength; adaptation properties) or even add new ones (e.g., inserting a synapse between cells where none exists). This should give insight into just why certain properties or connections exist in a system and not others, as well as answering just how stable a system is likely to be under variation of these parameters. Warshaw (personal communication), for example, has noticed that the addition of a synapse from the PY cells onto the PD cells destroys the correct sequencing in the pyloric system simulation of figure 3. Although the model is not yet accurate enough to allow us to predict a similar effect in the real ganglion, it does focus attention upon the absence of that synapse as being of potential interest.

An accurate model may provide a means for inferring something about presynaptic events from postsynaptic firing times. Thus, although a given output pattern might be produced by more than one input pattern, still it should at least restrict the number of different input patterns that could have produced it, and hence yield information about the system.

Certain properties not directly observable, such as a synaptic interaction between cells in which interactions cannot be directly demonstrated, might be inferred or at least suggested by simulation studies if a reasonable model for the rest of the system could be constructed. With less information about system connectivities or neuron kinetic properties, limitations on these variables might be deduced by studying the range of output patterns obtainable for a given set of assumed values for them. For example, in the model of figure 3 we have observed that hyperpolarizing the PD cell results in alternate bursting between the LP and PY's. This behavior is not seen in the real ganglion, nor is it found in the model if the strength of the interactions between the two groups is reduced substantially. We might suggest that in the real ganglion, the effective strength of one or both of the LP-PY interactions is less than we have used in figure 3. We are presently testing this hypothesis experimentally.

Finally, it seems possible that given an accurate model of a system based on intracellular studies in which the normal range of parameters for the model has been established, thereafter, based on much simpler, perhaps even just extracellular experiments, the correct values for model parameters could be uniquely determined, thereby reducing the need for intracellular recording in the further study of the system. This could be of particular value in larger systems of neurons, many of whose neurons have very similar properties, where intracellular recording from only a small fraction is possible in any given preparation.

DISCUSSION

BROWN: What is the minimum hardware configuration for this language?

HARTLINE: The language has been designed around the system that I have, and I haven't been able to devote much thought to exportability. It uses all-DEC hardware, the AD-01, DEC DACs, and so forth. We have a disk, required for the disk operating system. Depending on how many instructions you want to implement at one time and how large a program you wanted, it could take varying amounts of memory. We have the full 28k that is possible on this system, and some of our programs even use that much. But you could get away with half of that, anyhow. It also has not been optimized, since we had extra space to play with.

KEHL: Would you comment on how desirable it is to program in SNAX as opposed to BASIC and FORTRAN?

HARTLINE: For me, there are several considerations. First of all, I have all the input-output instructions which I don't have in FORTRAN, although I have in fact implemented some of the SNAX instructions as FORTRAN subroutines, so some of SNAX could be done in another language. I don't like FORTRAN, because of all the dirty syntax. I find SNAX an easier language to use. Some table instruction, the stack, and things like that, I find nice to work with. There's no doubt that given the I/O capability, you could use FORTRAN for the same sort of thing.

KEHL: Is SNAX really faster to program, or is it a matter of personal preference?

HARTLINE: I suspect it's largely a matter of personal preference, although I'm sure I could teach somebody SNAX a lot faster then you could teach them FORTRAN, because its simpler.

CLARKE: Could you comment on real-time capabilities of SNAX?

HARTLINE: At the moment that's certainly not optimized, but we hope to be able to take in spike trains on digital inputs. At the moment I sit in a loop and sample voltages until a threshold is crossed and store that information as a spike. The stretch receptor data I showed you was sampled in that way, and another program took that data and sorted it out and plotted it according to the phase of the stimulus, and calculated statistics from that. But at the moment I've not tried to develop a really powerful real-time capability. My problem is that I'm the one who has to do the programming, and I'd really rather spend my time on neurophysiology. That's one of the reasons that I'm particularly attracted to the basic concept of this conference; at least if I had had a language that I could really use - and I don't consider FORTRAN appropriate because it's not I/O oriented - I could have saved myself a lot of time.

BROWN: In fact, I gather SNAX, which relies on A/D conversion, is only suited for relatively slow systems, rather than, say a mammalian vestibular or auditory system. What's your timing resolution?

HARTLINE: I believe an A/D conversion takes something like 200 μsec., but I could shorten that. The conversion scales and sets the clock time. I think we could go faster.

BROWN: How fast could your program keep up with?

HARTLINE: If you're just storing the point in a table, you could do the whole thing in less than 50 μsec.

BROWN: That's reasonable for most systems, although only marginal for auditory work or for some electroreceptors.

REFERENCES

1. Cooke, I. M. and Hartline, D. K. Neurohormonal alteration of integrative properties of the cardiac ganglion of the lobster *Homarus americanus* J. Exp. Biol. 63:33-52 (1975).
2. Friesen, W. O. "Physiology of the spiny lobster cardiac ganglion" Thesis, Univ. Calif. San Diego (Neurosciences) (1974).
3. Friesen, W. O. Antifacilitation and facilitation in the cardiac ganglion of the spiny lobster *Panulirus interruptus*. J. Comp. Physiol. 101:207-224 (1975).
4. Hartline, D. K. "Simulation of phase-dependent pattern changes to perturbations of regular firing in crayfish stretch receptor", Brain Res. (in press) (1976).
5. Hartline, D. K. and Cooke, I. M. Postsynaptic membrane response predicted from presynaptic input pattern in lobster cardiac ganglion. Science 164:1080-1082 (1969).
6. Hartline, D. K. and Maynard, D. M. Motor patterns in the stomatogastric ganglion of the lobster, *Panulirus argus*. J. Exp. Biol. 62:405-420 (1975).
7. Hodgkin, A. L. The local electric changes associated with repetitive action in a non-medullated axon. J. Physiol. 107:165-181 (1948).
8. Maynard, D. M. Simpler networks. Ann. N.Y. Acad. Sci. 193:59-72 (1972).
9. Mulloney, B. and Selverston, A. Organization of the stomatogastric ganglion of the spiny lobster. III.

Coordination of the subsets of the gastric system J. Comp. Physiol. 91:53-78 (1974).
10. Rall, W. "Theoretical significance of dendritic trees for neuronal input-output relations". In *Neural Theory and Modeling*, ed. F. Reiss, pp. 73-79, Stanford University Press (1964).
11. Russell, D. F. "Rhythmic excitatory inputs to lobster stomatogastric ganglion", Brain Res. 101: 582 (1976).
12. Sokolove, P. G. Computer simulation of after-inhibition in crayfish slowly-adapting stretch receptor neuron Biophys. J. 12:1429-1451 (1972).
13. Tazaki, K. The effects of tetrodotoxin on the slow potential and spikes in the cardiac ganglion of a crab, *Eriocheir japonicus*. Jap. J. Physiol. 21:529-536 (1971).
14. Warshaw, H. S. and Hartline, D. K. A network model of the pyloric rhythm in the lobster stomatogastric ganglion. Abstr. Soc. Neurosci. 4:467 (1974).
15. Warshaw, H. S. and Hartline, D. K. Simulation of network activity in stomatogastric ganglion of the spiny lobster, *Panulirus*, Brain Res. (in press) (1976).

chapter 5

SIMULATION OF NERVE CELL KINETICS USING INTERACTIVE SIMULATION LANGUAGE

R. D. BENHAM
Interactive Mini-Systems, Inc.
Kennewick, Washington

D. K. HARTLINE
Biology Department
University of California at San Diego
La Jolla, California

INTRODUCTION

Interactive Simulation Language (ISL) can be used by scientists and engineers without prior programming experience for the solution of non-linear algebraic and differential equations. ISL permits hands-on interactive operation in which the user can monitor the course of a computation (via scope, printers, etc.) and change parameter values during execution (via knobs, switches, keyboard). ISL functions well either in an all digital system or in a system utilizing an analog interface. Since ISL is modular, computing functions can be added or deleted for tailoring the language to many applications in data collection, data analysis, curve fitting etc.

Support by Interactive Mini-Systems, Inc. and grants from NSF (GJ-43177), the Alfred P. Sloan Foundation, and the University of California Academic Senate to DKH.

ISL processes simulations quite rapidly by using a single-word mantissa and single-word exponent (mini-point). In addition, extremely large problems can be solved on a small computer. ISL is currently operational on PDP 8/12, PDP 7/9/15, PDP 11, NOVA, EAI 640, PACER100, DDP/PRIME, and Hewlett Packard computers.

This chapter shows application of ISL to neurobiology by:

>presenting an overview of ISL programming;
>illustrating the techniques for the simulation of a repetitively firing nerve cell;
>presenting comparative results for ISL and SNAX.

ISL PROGRAMMING OVERVIEW

It is easiest to consider the ISL system as providing a large number of operational elements such as those found on an analog computer. These elements include integrators, adders, multipliers, and many others (see Table 1). An ISL program is formed by "interconnecting" these elements to satisfy the equation being solved.

Drawing the Block Diagram

To illustrate the programming technique consider the equation for adaptation (A) of a nerve cell:

$$\frac{dA}{dt} + 0.6(A - A_{INF}) = 0 \quad A_{(0)} = 1.0$$

The first step in programming is to equate the highest derivative to the lower order terms. For the example:

$$\frac{dA}{dt} = -0.6(A - A_{INF})$$

The simulation diagram (using symbols from Table 1) is:

NERVE CELL KINETICS 69

Table 1. ISL functions. Each three-character function mnemonic is followed by a string of "input" arguments, NK, representing line numbers for elements whose outputs are to be operated upon. The result (R) of the operation is given in the right hand column. Symbols used are mostly standard as defined by the society for computer simulation (4).

function	instruction			remarks	
ARITHMETIC					
absolute value	ABS	N1	N1 —[ABS]—	R = |N1|	
add subtract	ADD	±N1, ±N2,... ±N9	N1 —[ADD]— N2	R = ±N1 ± N2 ± ... N9	
constant	CON	Nk		R = Nk	
divide	DIV	N1, N2	N1 —[DIV]— N2	R = N1/N2	
equate	EQT	N1, N2	N1 —[EQT]— N2	R = 0 : N2 = N1	
multiply	MUL	N1, N2	N1 —[MUL]— N2	R = N1 × N2	
pot	POT	N1, y.yyyE y		R = N1 × y.yyyE y	
TRANSCENDENTALS					
sine	SIN	N1	N1 —[SIN]—	R = sin N1	radians
	SND	N1		R = sin 2πN1	degrees
cosine	COS	N1	N1 —[COS]—	R = cos N1	radians
	CSD	N1		R = cos 2πN1	degrees
logarithm	LOG	N1	N1 —[LOG]—	R = log(10) N1	
	LNX	N1		R = log(e) N1	
exponential	TXP	N1	N1 —[EXP]—	R = (10) exp N1	
	EXP	N1		R = (e) exp N1	
square root	SQT	N1	N1 —[SQT]—	R = square root N1	
SUBROUTINE					
subroutine	SBR	N1, N2 ... N8		N1 = block transfer no. N2 ... N8 are inputs SBR can reference all blocks except 0 to 10 in the main program	
return	RTN	K		return to block N + 1 of main program from SBR K is meaningless	

Table 1. (*continued*) ISL functions.

function	instruction		remarks	

INPUT/OUTPUT

function	instruction		remarks
heading	HDR	T1, ... T5	label TTY/L.P. columns with up to 6 chrs. each
print/plot	PRI	N1, ... N5	print/plot to TTY/L.P.
display* (scope)	DIS	N1, N2, N3, N4	X - axis = N1/N2 Y - axis = N3/N4
digital* to analog conv	DAC	N1, N2, K	R = N1/N2 K = channel no.
analog* to digital conv	ADC	K	R = ADC output K = channel no.

CONTROL

function	instruction			remarks
change mode	CHM	K		reverses mode IC to C or C to IC
count	CNT	K	⟨CNT = K?⟩ → Nj+2, Nj+1	when iteration count = K skip next instr and reset count to K
end	END	K		reverse mode when in I and go to block 0 otherwise terminate comp
finish	FIN	N1, N2	N1 ─┐ N2 ─┤ FIN	If N1≥N2 take next instr otherwise up-date time and go to block 0
go to	GTO	Nk		go to block Nk
transfer	TFR	N1, N2, N3	N1 ─⟨TFR⟩→ Nj+1 N2 ─ ↓ Nj	if N1≥N2 go to N3

LOGICAL

function	instruction			remarks	
input relay	IRL	N1, N2, N3, N4	N3 ─┐N1 N2 N4 ─┤ IRL	R = N3 R = N4	N1≥N2 otherwise

limiter	LIM	N1, N2, N3	N1
			N3 ─ LIM
			N2
				R = N1
				  = N3
				  = N2
					N3>N1
					N1 ≤ N3 ≤ N2
					N3<N2

step size change	SCH	N1, N2, N3, K		if N1≥N2	S = N3 and T = K

Table 1. (*continued*) ISL functions.

function	instruction			remarks	
		SPECIAL FUNCTIONS			
comment*	CCC	EXAMPLE COMMENT		one line of comment	
function generation	FNG	N1	N1 —[FNG]—	$R = f(N1)$	
increment	INC	N1, N2	N2 —[N1]▷—	$R = \sum N2 + N1$ $= N1$ in I C mode	
integrate	INT	N1, N2, ... N9	N2 —[N1]▷—	$R = \int (N2 + \ldots N9) dt + N1$	
variable time delay	TDL	N1, N2, N3	N2 N3 N1 —[TDL]—	$R = f(N1)(T - 20N3/N1)$ $= f(N1)(T - 20S)$	$N3 \geq S \cdot N2$ $N3 < S \cdot N2$
time	TME		[TME]▷—	each iteration advances time by S	

Next by adding an integrator to the diagram the function (A) is available:

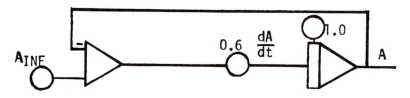

Notice the initial condition is shown as an auxiliary input to the integrator. To complete the simulation diagram it is necessary to assign numbers to each operational element, that is to form the derivatives and then integrate the highest derivative first. Thus:

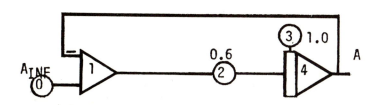

The final step in preparing the simulation diagram is to form operational blocks that:

> Label headings (HDR statement)
> Identify printing or plotting variables
>   (PRI statement)
> Allow the program to iterate and determine when computation should stop (FIN statement)

To aid in the interpretation of tabulated output data, a provision exists for specifying headers above each column. These headers will be the names given to the variables identified by the operational equivalent in the print (PRI) statement. Up to five names (having six characters) can be included in one heading statement (HDR). The HDR automatically labels the first value as TIME before using the defined labels. Accordingly the print statement gives the value of time. A header and print statement for the example could be:

```
6 HDR A
7 PRI 4
```

A finish (FIN) statement, as shown below, should be incorporated into each ISL program. This statement allows program iteration and also provides the user with a means of manually terminating a run with a keyboard interrupt. Its diagram would be:

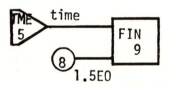

When the value of the first input (time) is less than the second input, then for a new iteration, the FIN statement transfers control to statement 0. When the first input is greater than or equal to the second input, the next numerical statement is executed. In this example we will use an END statement to terminate the run.

Preparing the Listing

The program is entered into the computer by using the instructions shown in Table 2. The user area of memory is first erased with the "K" command. Then by

Table 2. Definitions of constants used in the simulation.

	program preparation and modification
A	Append program
E	Enter program from external device
Fn, m	edit line m of FNG n
J	enter Job (program) title
K	Kill (erase) current program
L	List current program
Ln	List statement n
Ln, m	List statements n through m
Mn	Modify statement n
Mn, m	Modify statements n through m
R	print Remaining memory locations
W	Write to external device
space	define a CON with current block no.
	exit append
CR	terminates all program statements

	program operation
B	Back to numerical listing from plot
C	enter Compute mode
I	set Initial conditions
H	go to Hold (sense switch A for EAI computers)
P	Parameter read and set
Qn	integrator selection (Quadrature)
	$n = 0$ EULER
	$n = 1$ RUNGE-KUTTA 2
	$n = 2$ RUNGE-KUTTA 4
S	select integration Step size
T	select Typing frequency
Un	select output Unit
	$n = 0$ teletype
	$n = 1$ line printer
V	read numerical Value
X	select plotting routine and set range

* *Numerical entries are terminated with a space, comma or carriage return (CR).*

typing "A" (for APPEND) the teletype responds by typing (0) for the first line number. The user may then type in the program as shown below:

```
 0 CON = 1.0E-1
 1 ADD - 4, +0
 2 POT 1, 6.0E -1
 3 CON - 1.0E0
 4 INT 3, +2
 5 TME 5
 6 HDR A
 7 PRI 4
 8 CON = 1.5E0
 9 FIN 5, 8
10 END
```

For each line of code the teletype prints a line number, following which the user types a three-character function code, followed by its arguments (separated by commas). Arguments are either "inputs" to the elements (integers) or numerical values assigned to them (floating point notation for POTs).

If a spelling error is detected while entering the program the printer responds by typing a question mark (?) and retyping the statement number. It then waits for the correct information. The computer will not catch "patching errors." For example if a 7 were typed for a 4 in statement 1, the computer would not recognize the error. However the user can correct any numerical error at any time prior to terminating the entry with the "slash" (/). This signals the computer to cancel the entry and accept a new one.

Any statement can be modified or re-entered with the MODIFY (M) command shown in Table 2.

Running the Program

The program is run by selecting an integration step size and print interval with the "I" command. For the current example a step size of $S = 1.0E-2$ and a print interval of $T = 10$ will allow tabular output as shown on the following page.

NERVE CELL KINETICS 75

ADOT = -0.6(A - AINF)   ADAPTATION EXAMPLE

S = 1.000E-2   T = 10

TIME	A
0.000E 0	1.000E 0
9.999E-2	9.474E-1
2.000E-1	8.979E-1
3.000E-1	8.513E-1
4.000E-1	8.074E-1
5.000E-1	7.661E-1
6.000E-1	7.272E-1
7.000E-1	6.906E-1
8.000E-1	6.561E-1
9.000E-1	6.236E-1
1.000E 0	5.931E-1
1.100E 0	5.642E-1
1.200E 0	5.371E-1
1.300E 0	5.116E1
1.400E 0	4.876E-1
1.500E 0	4.649E-1

By using the X command of Table 2 information is plotted by the teletype as shown below. Notice the range for 50 spaces is set to 1.0 as defined by the RANGE = 1.000E 0 comment.

ADOT = -0.6(A - AINF)   ADAPTATION EXAMPLE

S = 1.000E-2   T = 10
RANGE = 1.000E 0
A = A   ∅....1....2....3....4....5....6....7....8....9....∅....1....2
```
                                                              A
0.000E 0                                                   A
9.999E-2                                                A
2.000E-1                                             A
3.000E-1                                          A
4.000E-1                                       A
5.000E-1                                    A
6.000E-1                                 A
7.000E-1                              A
8.000E-1                           A
9.000E-1                        A
1.000E 0                     A
1.200E 0                  A
1.300E 0                A
1.400E 0             A
1.500E 0          A
```

Of much more interest is display of the results on a scope and repetitively running the simulation while varying AINF with an analog pot. This may be done by modifying the following lines with the "M" command:

```
0 ADC 0              (use pot 0 to change AINF)
6 DIS 5,8,4,7        (plot A versus TIME on the scope)
7 CON = 2.0E0        (normalized value of A)
8 CON = 1.5E0        (normalized value of TIME)
10 CHM 10            (allows repetitive operation)
11 END
```

With scope output and parameter variation through an analog pot and an ADC channel, better interactive communication is established than with just a TTY printer or a line printer. With this form of interaction one can begin to appreciate the speed and power of a simulation using ISL. Solutions will appear about 250 times faster than possible with interactive languages like BASIC. In addition an 8k computer equipped with ISL can solve up to 100 non-linear equations. Over 1000 equations can be solved with 32k of memory available.

SIMULATING A REPETITIVELY FIRING NERVE CELL

This section illustrates the use of ISL on the more complex problem of a repetitively firing nerve cell. A complete description of the model is given in Chapter 4. The model is summarized with the simulation diagram of figure 1 and the listing of figure 2. Notice that the ISL simulation diagram combines both the mathematical and logical flow diagrams for the nerve cell. As such it is an aid (mathematical analog) in visualizing the operation of the nerve cell. The model is composed of 8 non-linear differential equations that account for:

Conductance associated with each spike (g_s)
Current injected with a microelectrode (I)
Resting potential adaptation (A_r)
Short- and long-term active response adaptation $(A_1$ and $A_2)$
Active pacemaker response (P)
Threshold (Θ)
Perturbation variable (W)

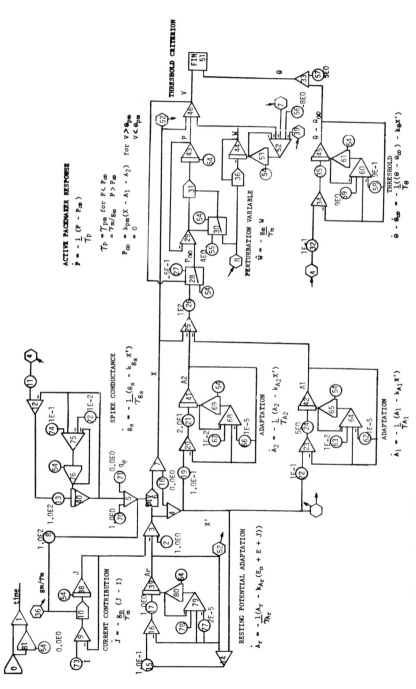

Figure 1. ISL simulation diagram for active pacemaker model.

```
 0  TME    0                  =  0.000E  0     43 INT   54, 31                 =  0.000E  0
 1  ADD    0, 81              =  0.000E  0     44 INT   53, -36                =  0.000E  0
 2  CON    2                  =  1.000E  0     45 INT   61, 35                 =  0.000E  0
 3  ADD   38, -39, 2          =  0.000E  0     46 ADD   7, 43, 44              =  0.000E  0
 4  ADD    3, 18              =  0.000E  0     47 DAC   1, 37, 1               =  0.000E  0
 5  ADD   70, 40, 71          =  0.000E  0     48 DAC  33, 82, 2               =  0.000E  0
 6  DIV    3, 5               =  0.000E  0     49 DAC  46, 82, 2               =  0.000E  0
 7  ADD    6, 18              =  0.000E  0     50 DAC  56, 84, 1               =  0.000E  0
 8  POT    5, 1.000E 2        =  0.000E  0     51 FIN  46, 33                  =  0.000E  0
 9  ADD   73, -38             =  0.000E  0     52 ADD  -44, -53, -7, 56, -39   =  0.000E  0
10  MUL    9, 8               =  0.000E  0     53 INC  54, 52                  =  0.000E  0
11  POT    4, 1.000E-1        =  0.000E  0     54 CON                          =  0.000E  0
12  ADD   11, -40             =  0.000E  0     55 CON                          =  4.000E  0
13  POT   12, 1.000E 2        =  0.000E  0     56 CON                          = -8.000E  0
14  ADD    4, 39              =  0.000E  0     57 CON                          =  5.000E  0
15  POT   14, 1.000E-1        =  0.000E  0     58 POT  45, 9.000E-1            =  9.000E  0
16  ADD  -39, 15              =  0.000E  0     59 CON                          =  9.000E  0
17  POT   16, 1.000E 0        =  0.000E  0     60 ADD  -61, 59, 45, -58        =  0.000E  0      60
18  CON   18                  =  0.000E  0     61 INC  54, 60                  =  0.000E  0      61
19  POT    4, 1.000E-1        =  0.000E  0     62 POT  42, 1.000E-5            =  0.000E  0      62
20  ADD   19, -41             =  0.000E  0     63 CON                          =  1.000E-2        63
21  POT   20, 2.000E 1        =  0.000E  0     64 ADD -65, 63, 42, -62         =  0.000E  0      64
22  POT    4, 1.000E-1        =  0.000E  0     65 INC  54, 64                  =  0.000E  0      65
23  POT   22, -42             =  0.000E  0     66 POT  41, 1.000E-5            =  1.000E-2
24  ADD   23, 5.000E 0        =  0.000E  0     67 CON                          =  1.000E-2
25  ADD    7, -41, -42        =  0.000E  0     68 ADD -69, 67, 41, -66         =  0.000E  0      68
26  POT   25, 1.000E 2        =  0.000E  0     69 INC  54, 68                  =  0.000E  0
27  CON                       = -5.000E-1       70 CON                         =  1.000E  0
28  IPL   46, 27, 26, 54      =  0.000E  0     71 CON                         =  0.000E  0
29  ADD   28, -43             =  0.000E  0     72 POT  40, 1.000E-2            =  0.000E  0
30  IPL   54, 29, 8, 55       =  0.000E  0     73 CON                          =  0.000E  0
31  MUL   29, 30               =  0.000E  0     74 CON                          =  1.000E-1
32  POT    4, 1.000E-1        =  0.000E  0     75 ADD -76, -72, 40, 74         =  0.000E  0      75
33  ADD   45, 57                =  0.000E  0    76 INC  54, 75                 =  0.000E  0
34  ADD   32, -45               =  0.000E  0    77 POT  39, 2.000E-5           =  0.000E  0
35  POT   34, 2.500E 2        =  0.000E  0     78 CON                          =  1.000E-1
36  MUL   44, 8                =  0.000E  0    79 ADD -80, 78, 39, -77         =  0.000E  0      79
37  CON                        =  1.000E  0    80 INC  54, 79                  =  0.000E  0
38  INT   54, 10                =  0.000E  0    81 INC  54, 0                  =  0.000E  0
39  INT   80, 17                =  0.000E  0    82 CON                          =  5.000E-1
40  INT   76, 13                =  0.000E  0    83 CHM  1                      = -1.000E  0
41  INT   69, 21                =  0.000E  0    84 CON  84                     =  0.000E  0
42  INT   65, 24                =  0.000E  0    85 END  0
```

Figure 2. ISL listing for nerve cell simulation. This shows the teletype listing corresponding to the simulation diagram of Figure 1. First and last columns: Element number; second column: Element type (see table 2); third column: Inputs/values for elements; fourth column (following "="): Output value of element.

NERVE CELL KINETICS

The basic equation for each term is similar to the equation for adaptation used for describing ISL programming in the last section. However there are additional complexities, as can be seen by close inspection of figure 1. A cell will fire (spike) when the membrane potential (V) becomes equal to or greater than the threshold (Θ). When this happens, new initial conditions are automatically calculated and imposed for the terms g_s, A_r, A_1, A_2, P, W, and Θ. The form of the IC up-date equation is given for the adaptations:

$$A_{new} = A_{old} + \frac{\Delta A_0 (A_{max} - A_{old})}{A_{max}} = A_{old}\left(1 - \frac{\Delta A_0}{A_{max}}\right) + \Delta A_0$$

The new value is a fixed fraction of the distance from the old value to a maximum value. The ISL implementation for this equation is:

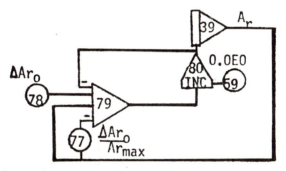

The initial condition calculations appear after the FIN statement (no. 51) so that they all are calculated when $V = \Theta$ and t = time of the spike. Since ISL includes a second IC iteration the calculation must not be updated during the second pass. The solution to this is the INC block which allows correction during the compute pass but not the second IC pass through the program. Since the INC block adds the new input to the old output, the INC output must be subtracted from the input to satisfy the above equation.

The Change Mode (CHM) instruction (block 82) allows simulation control for repetitive operation. ISL has two modes: "compute," in which normal program execution occurs, and "initial conditions" (IC), in which new initial conditions are set as shown in the preceding example. When the CHM is executed and ISL is in IC, the mode is changed to compute. Control is then transferred to block 0 (zero). Conversely when CHM is executed and ISL is in compute mode, the mode is changed to IC for an initial condition pass. The FIN statement then allows for the next instruction to be executed when in IC mode. This allows the CHM block to again change the mode back to compute from IC mode.

Notice that blocks 0, 81, and 1 are used for generating time. Each time there is a spike ($V = 0$), time as generated by the TME (block 0) is automatically reset to 0. Therefore by using the INC (block 81) the value of time (when $V = \Theta$) is added to the last INC value and remembered. Thus by adding the output of TME and INC the true accumulative time is made available.

Notice also that the input relays (IRL) as shown in blocks 28 and 30 are used to select values for τ_p and P_∞ for the active response equation (P) which are dependent on condition requirements on certain variables.

Statements 45, 46, and 47 show the ISL commands for outputting results to digital to analog converters which are in turn used to drive a scope (see figure 3).

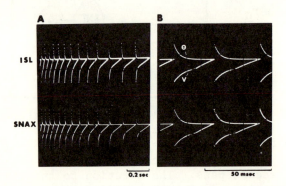

Figure 3. Comparison between ISL and SNAX simulations of a repetitively firing neuron. A and B show two different time scales for a simulation of an abrupt depolarization to the cell which is responded to by an initially high but exponentially declining firing rate. For each simulation the upper line represents threshold (Θ) and the lower line membrane potential (V). Impulse time-course is not simulated, but impulses occur at the discontinuities of Θ and V, representing resets of these quantities by the spike.

COMPARISON OF ISL AND SNAX

The SNAX language was developed for the PDP 11/45 computer and is currently used for math model development and testing of the nerve cell simulation. The purpose of the ISL implementation is to provide comparative information for the two languages.

Approximately four hours were required to study the math model and draw a preliminary flow diagram. This experience generated several questions that required consultation for correction. An additional seven hours were needed to enter, debug, test, and compare the model with the SNAX version. The same PDP 11/45 computer was used. ISL required 15 seconds of computer time to simulate the 14 firings shown in figure 3. The ISL program required about 520 memory locations to hold the 83 lines of code and computer output. Although provision exists for using hardware multiply/divide and floating-point processor hardware, these features were not used for the comparison run. ISL could have processed the information substantially faster if this additional hardware had been used.

The SNAX simulation requires 230 lines of code. SNAX processed the simulation in 14 seconds by using the floating-point processor available on the PDP 11/45. The results for ISL and SNAX were found to be identical, as can be seen by comparing figure 3A and B. To verify the comparison of ISL and SNAX a run was made where SNAX results were written over the ISL results (saved on a memory scope) and no appreciable difference could be detected.

CONCLUSION

ISL is an efficient general purpose language suitable for simulation of analog computer functions on mini computers. Its block diagram representation scheme aids in visualizing the processes being simulated. For an on-line interactive language it is rapid and conservative of memory. Its availability on a number of different minicomputers makes ISL programs "exportable" for many of the computer systems currently employed in neurobiology research.

DISCUSSION

D.O. WALTER: I'd like to ask Dr. Hartline to comment on ISL and this comparison with SNAX.

HARTLINE: I think ISL is a very pretty language. I was delighted to learn about it. I think that for some of my applications, it would not be appropriate. We have only simulated a single nerve cell here, and in order to simulate a pair of cells, we would have to double the amount of ISL code, whereas in SNAX it's done with subroutines. To simulate a network in ISL, I don't know what sort of complexity would be involved, but I think it might get out of hand whereas in SNAX you would only need a few more lines of code.

BENHAM: I don't know how much memory SNAX uses, but ISL used about 600 words of memory to solve that. With regard to what Dr. Hartline says about repeating the code: If there's an integrator involved, the code has to be rewritten in ISL. We do have subroutine capabilities, however, for things like derivatives.

CLARKE: Do you have the capability of adding instructions to your language?

BENHAM: Yes, there are 13 slots available for user definitions. If there were sufficient need, we could put in a provision for subroutining the integrator. It would just mean putting in an additional table for each simultaneous integration.

DUFFY: Have you met your goal of getting ISL to run on most minis?

BENHAM: That's a good question. It's available for all of DEC's 12-bit machines, all the PDP-11/45s, all Data General NOVAS, the PDP-7,9,15, Electronics Associates' EAI 640, PACER 100, analog computers like the 680 and the 7800, we have it on the Honeywell 116 through 716 series machines, and the PRIME. We just about have it for the Hewlett Packard 2100 series.

CLARKE: Can the user write subroutines, or must they come from you?

BENHAM: We're not anxious to let the listings out. We'll sell them, but we'd rather not. We do recognize the need for people to put their own individuality in. It's just a matter of knowing about three locations to define a mnemonic, where to put it, how to pass arguments, and how to exit. We provide that. We have a system in Germany that has put in about 8 or 9 instructions for their own purposes.

MCINTYRE: Do you have any way of making the system compatible with existing operating systems? It seems tedious to have to re-enter a program from scratch every time you want to run it.

BENHAM: You don't have to. You can enter it from tape. The user programs can be stored on tape, and will run on any machine with ISL.

CLARKE: Can you describe the hooks in ISL for adding subroutines? Are similar hooks available for insertion of text?

BENHAM: Yes, there's a commenting command. Also, we're putting the system in at the University of Washington and they had a lot of good ideas. They convinced us we could make some improvements. For example, we're making it DOS-compatible so that a person could use the PDP-11 editor and use DOS file manipulation or anything else he could do with DOS.

DUFFY: Do you have capability for multi-user time-sharing applications?

BENHAM: Not yet. We have systems at Walla Walla College and Evergreen College in the state of Washington that are going in that direction, but we haven't done it yet.

REFERENCES

1. *ISL Interactive Simulation Language Programming Manual*, Interactive Mini Systems, Inc., Kennewick, Wisconsin (1974).
2. Hartline, D. K. Simulation of phase-dependent pattern changes of perturbations of regular firing in crayfish stretch receptor, in preparation (1975).
3. Hartline, D. K. "SNAX": a language for interactive neuronal modeling and data processing. In *Computer Technology in Neuroscience*, ed. P. B. Brown, Hemisphere Publishing and John Wiley and Sons, New York (1975).
4. Proposed IRE standards for analog computers, Simulations 2: no. 1 pp. S-2 to S-12 (1964).

chapter 6

A CELL KINETICS SIMULATION SYSTEM

C. E. DONAGHEY

Department of Industrial Electrical Engineering
University of Houston
Houston, Texas

In the field of cell cycle kinetics there have been a number of computer simulation models that have been developed and reported in the literature (1),(2),(3). However, this valuable tool is often not applied in this area because of two basic problems:

(1) The cell biologists who are engaged in cell kinetics research usually do not have the computer background required to develop these programs. Thus, in most cases, if they desire to use computer simulation they must communicate their models to a computer programmer, who usually does not have a biological background.
(2) Because most cell models assume rapid cell proliferation, a large amount of computer memory is required to store the information on the ever-increasing number of cells in the models. The alternative is to only simulate a short time frame and this reduces the usefulness of the model.

This chapter presents a cell kinetics simulation system which is intended to solve these problems. The user interfaces to the system through a simple problem-oriented language which uses the conventions and the vocabulary common to the cell biologist. Therefore, a cell biologist should be able to develop his own computer simulation models after very little instruction. An added dividend to the use of this language is that the assumptions and the logic of the model are apparent to other researchers who examine the program. This is not true of the simulation models that have been developed to date using a variety of computer languages. This simulation system uses a number of built-in collection algorithms that automatically re-group and re-assign cells when the cell population approaches the memory limit of the computer system on which it is being run. These algorithms have been designed so that they cause a minimum of disturbance to the models being simulated. Thus, a researcher can simulate a cell model for any length of simulated time without concern for exceeding memory limits.

The name given to this simulation system is CELLSIM. Figure 1 gives a sketch of the classic four state cell model and a program written in CELLSIM to simulate it.

The first instruction in the program is a CELL TYPES

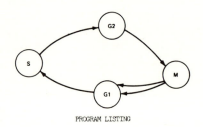

PROGRAM LISTING

```
CELL TYPES  TYPA;

STATES(1) G1, S, G2, M;

FLOW(1) G1-S(ALL),  S-G2(ALL),  G2-M(ALL),  M-G1(ALL);

TIME IN STATES(1)  G1: NORMAL(5, 1),   S: NORMAL(10, 2),
                   G2: NORMAL(3, .6),  M: NORMAL(2, .4);

PROLIFERATION(1) M: 2;

INOCULUM(1) 1000;

REPORTS 1.0 HOURS;

GRAPH TOTAL,  S(1)/TOTAL*100.0;

SIMULATE 50 HOURS;

                INFORMATION ABOUT THE PROGRAM

CELL NAME    INDEX    STATE  NAMES AND INDECIES

   TYPA        1      G1  1    S   2    G2   3    M    4

NUMBER OF GROUPS IS NOT SPECIFIED FOR CELL TYPE  1  DEFAULT OPTION IS 100

GRAPHS SPECIFIED:
              1  TOTAL
              2  S(1)/TOTAL*100.0
```

Figure 1. Four state cell model, and CELLSIM simulation program.

instruction and the user names the cell population in his model TYPA. He can have as many as nine populations in a model and each one would be given up to a four character arbitrary name.

In the next instruction the user names the states he wants for each population. In this model he assigns to the first (and only) population the state names of G1, S, G2, and M. The names of the states are again arbitrary.

The next instruction specifies the desired cell flow for the model. The user wants all of the cells leaving G1 to go to S, all cells leaving S are to go to G2, etc. He has the capability of specifying different proportions of existing cells going to separate states. For example, if there were an absorbing state G0 in the model, the flow from M might be M-G1(.90), M-G0(.10).

The next instruction specifies the time the cells are to spend in each state. This can be a constant time or it can be a random value from a probability distribution. In this model the time a cell spends in G1 is to be a random value from a normal distribution with a mean of 5.0 hours and a standard deviation of 1.0. The time in S is also to be a random value from a normal distribution with a mean of 10.0 and a standard deviation of 2.0. The next instruction specifies that state M is to be the only proliferation state in the model, and the proliferation rate is 2.0. The proliferation rate does not have to be an integer and it does not have to be a value greater than 1.0. If he had specified the proliferation rate to be .80, it would cause 80 cells to exit M for every 100 that entered. The proliferation rate can also be a random variable.

The next instruction specifies the initial cell inoculum in the model. There are a number of options available for specifying how cells are to be initially distributed throughout the cell cycle. They may be distributed evenly throughout the cell cycle, all may be clustered at a single point, or the times remaining in the various states may be random values from probability distributions. In the example the user merely says he wants 1000 cells initially in the model. The system will distribute these cells according to theoretical distributions for proliferating cell populations. (4)

The next command specifies that reports on the status of the model are to be made every hour of simulated time. The system will print the number of cells in each state for each population at each report. The next command specifies that the user would like a graph of the total number of cells in the model vs. time and he would also like a graph of the percent of cells in the S state vs. time.

The final command of the program specifies that the user desires the model to be run for 50 simulated hours.

The remainder of figure 1 shows the program information furnished by the CELLSIM interpreter when the program is executed. The system has divided the 1000 cells into 100 groups of 10 cells each. As the model moves through time the number of groups may increase as the cell population increases. When an upper limit on available memory is reached the system will automatically re-assign all the cells that are currently in the model back into 100 groups. All of memory in CELLSIM is dynamically allocated so the point at which this re-assignment will take place is dependent on the number of populations and the number of states in the model. The user has an option of specifying how many initial groups he wants in his model.

Figure 2 shows some of the reports that would be furnished for this model. For example, at 1.00 hour there are now 1020 cells in the system, 330 are in state G1, 510 are in S, 150 are in G2, and 30 are in M.

Figure 3 shows the graph of the total number of cells in the system for the entire simulation. Figure 4 shows the graph of the percentage of cells in S. This entire simulation took less than 2 minutes on an IBM 360/44.

To illustrate some of the other capabilities of CELLSIM again refer to figure 4. It can be seen that there are three peaks for the percentage of cells in S. These peaks occur at t = 3.0, 22.0, and 41.0. Suppose the researcher has a drug which will instantaneously kill 90% of the cells that are in the S state and will not affect cells that are in the other states. Hydroxyurea would be an example of such a drug. He would like to test the effect of administering this drug at these three times. Figure 5 is a listing of the program to simulate this administration. The program is exactly the same as the previous program except that now there are three KILL commands included. These commands specify when the kill is to take place, the state that is affected, and the proportion of cells that are to be destroyed in that

REPORT NUMBER	REPORT TIME	TOTAL CELLS IN THE SYSTEM	CELL TYPE	TOTAL CELLS FOR THIS TYPE	NUMBER OF CELLS IN EACH STATE TYPA		
1	0.0	0.1000E 04	1	0.1000E 04	0.3200E 03	0.5100E 03	0.1300E 03	0.4000E 02
2	1.000	0.1020E 04	1	0.1020E 04	0.3300E 03	0.5100E 03	0.1500E 03	0.3000E 02
3	2.000	0.1040E 04	1	0.1040E 04	0.2800E 03	0.6000E 03	0.1000E 03	0.6000E 02
4	3.000	0.1050E 04	1	0.1050E 04	0.2200E 03	0.6300E 03	0.7000E 02	0.1300E 03
5	4.000	0.1100E 04	1	0.1100E 04	0.2900E 03	0.6100E 03	0.1000E 03	0.1000E 03
6	5.000	0.1160E 04	1	0.1160E 04	0.3200E 03	0.6500E 03	0.1500E 03	0.4000E 02
7	6.000	0.1200E 04	1	0.1200E 04	0.3800E 03	0.6100E 03	0.1800E 03	0.3000E 02
8	7.000	0.1210E 04	1	0.1210E 04	0.3600E 03	0.6000E 03	0.1800E 03	0.7000E 02
9	8.000	0.1240E 04	1	0.1240E 04	0.4000E 03	0.5900E 03	0.1500E 03	0.1100E 03
10	9.000	0.1300E 04	1	0.1300E 04	0.4200E 03	0.6300E 03	0.1400E 03	0.1000E 03
11	10.000	0.1350E 04	1	0.1350E 04	0.4400E 03	0.5400E 03	0.2700E 03	0.1000E 03
12	11.000	0.1380E 04	1	0.1380E 04	0.3800E 03	0.6100E 03	0.2900E 03	0.1300E 03
13	12.000	0.1460E 04	1	0.1460E 04	0.5000E 03	0.6200E 03	0.2100E 03	0.1800E 03
14	13.000	0.1490E 04	1	0.1490E 04	0.4800E 03	0.6300E 03	0.2000E 03	0.1500E 03
15	14.000	0.1630E 04	1	0.1630E 04	0.6600E 03	0.6900E 03	0.1300E 03	0.1300E 03
16	15.000	0.1680E 04	1	0.1680E 04	0.6400E 03	0.7700E 03	0.1400E 03	0.6000E 02
17	16.000	0.1780E 04	1	0.1780E 04	0.7800E 03	0.7700E 03	0.1700E 03	0.1200E 03
18	17.000	0.1790E 04	1	0.1790E 04	0.7000E 03	0.8000E 03	0.1700E 03	0.1700E 03
19	18.000	0.1830E 04	1	0.1830E 04	0.6600E 03	0.8600E 03	0.1900E 03	0.7000E 02
20	19.000	0.1920E 04	1	0.1920E 04	0.6200E 03	0.1020E 04	0.2100E 03	0.1200E 03
21	20.000	0.1960E 04	1	0.1960E 04	0.6000E 03	0.1000E 04	0.2400E 03	0.1500E 03
22	21.000	0.2010E 04	1	0.2010E 04	0.5200E 03	0.1140E 04	0.2200E 03	0.1500E 03
23	22.000	0.2070E 04	1	0.2070E 04	0.5600E 03	0.1180E 04	0.2200E 03	0.1200E 03
24	23.000	0.2160E 04	1	0.2160E 04	0.6000E 03	0.1220E 04	0.2600E 03	0.1400E 03
25	24.000	0.2220E 04	1	0.2220E 04	0.6400E 03	0.1220E 04	0.2600E 03	0.1800E 03
26	25.000	0.2280E 04	1	0.2280E 04	0.6000E 03	0.1260E 04	0.2000E 03	0.1800E 03
27	26.000	0.2420E 04	1	0.2420E 04	0.8000E 03	0.1300E 04	0.2000E 03	0.1600E 03
28	27.000	0.2460E 04	1	0.2460E 04	0.7600E 03	0.1240E 04	0.3000E 03	0.1600E 03
29	28.000	0.2520E 04	1	0.2520E 04	0.6800E 03	0.1300E 04	0.4200E 03	0.1200E 03
30	29.000	0.2620E 04	1	0.2620E 04	0.7600E 03	0.1200E 04	0.5400E 03	0.1200E 03

Figure 2. CELLSIM report output for program in figure 1.

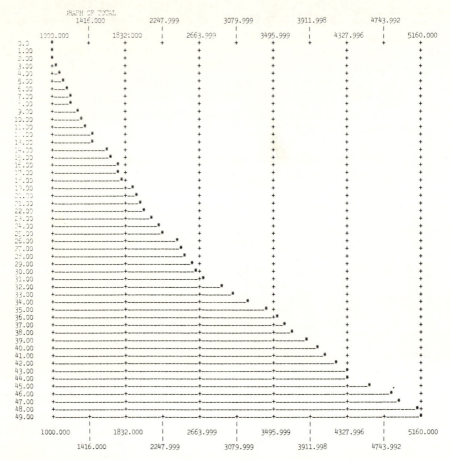

Figure 3. Graphical output for CELLSIM program of figures 1 and 2: total cells.

state. Figure 6 shows the graph of the total number of cells in the system for this model. Note that the total number of cells at $t = 49$ is now 1719 as opposed to 5160 in the previous model. Figure 7 shows the graph of the percentage of cells in S. The administration of the drug at $t = 3.0$ and $t = 22.0$ has caused an extremely high peak of cells in S to occur at $t = 32.0$. At $t = 32.0$ there are over 94% of the cells in S. This would suggest a better drug schedule of 3.0, 22.0, and 32.0. Figure 8 shows the graph of the total number of cells in the system when the drug is administered at these times. Notice that there are only approximately 300 cells in the system at $t = 49.0$ with this administration.

Some examples of other commands available to a CELLSIM user will be shown below:

SUSPEND (1) BETWEEN 25.0 AND 28.0 S ;

This command would cause all cells in S at $t = 25.0$ to be halted in place. They will remain in place until $t = 28.0$ when they will continue normal development.

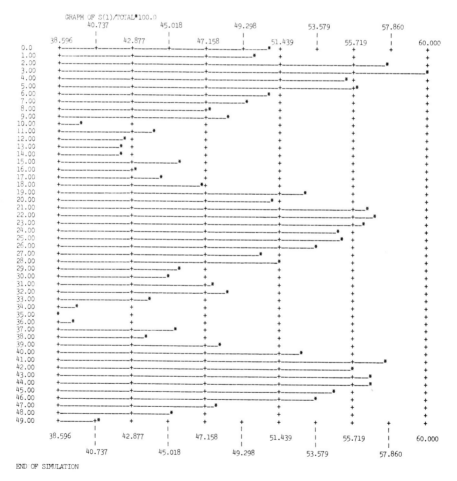

Figure 4. Percentage of cells in S.

BLOCK EXIT (1) BETWEEN 5 AND 15 M;

All cells that attempt to leave the M state between 5 and 15 will be blocked. At $t = 15$ all the cells that have accumulated at the exit of M will be released.

TRANSFER CELLS BETWEEN 30 AND END FROM (1) S : .90 TO (3) G2;

CELL TYPES TYPA;

STATES(1) G1, S, G2, M;

FLOW (1) G1-S(ALL), S-G2(ALL), G2-M(ALL), M-G1(ALL);

TIME IN STATES(1)   G1: NORMAL(5, 1), S: NORMAL(10, 2),
                    G2: NORMAL(3, .6), M: NORMAL(2, .4);

PROLIFERATION(1) M:2;

INOCULUM(1) 1000;

KILL(1) AT TIME = 3.0   S: .90;

KILL(1) AT TIME = 22.00 S: .90;

KILL(1) AT TIME = 41.00 S: .90;

REPORTS 1.0 HOURS;

GRAPH TOTAL, S(1)/TOTAL*100.0;

SIMULATE 50 HOURS;

INFORMATION ABOUT THE PROGRAM

CELL NAME   INDEX   STATE NAMES AND INDECIES

   TYPA   1   G1   1   S   2   G2   3   M   4

NUMBER OF GROUPS IS NOT SPECIFIED FOR CELL TYPE 1 DEFAULT OPTION IS 100

CELL TYPE = 1 ONE TIME KILL SPECIFICATION AT TIME = 3.000 FOR STATES 2

CELL TYPE = 1 ONE TIME KILL SPECIFICATION AT TIME = 22.000 FOR STATES 2

CELL TYPE = 1 ONE TIME KILL SPECIFICATION AT TIME = 41.000 FOR STATES 2

GRAPHS SPECIFIED:
            1 TOTAL
            2 S(1)/TOTAL*100.0

Figure 5. Listing of program to simulate Hydroxyurea effects: drug administered at t= 3.0, 22.0, 41.0.

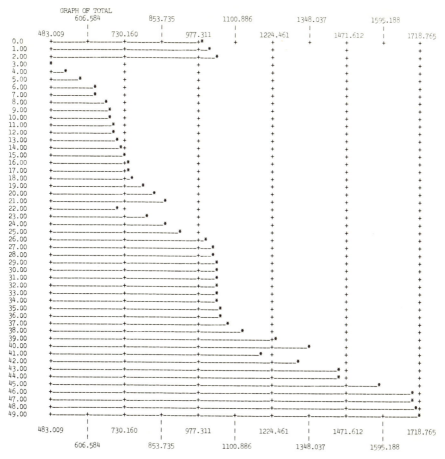

Figure 6. Graph of total number of cells in model system of figure 5.

At $t = 30$, 90% of the cells exiting from the S state in population 1 will enter into the G2 state of population 3. This split of the population will continue until the end of the simulation.

CHANGE TIME IN STATES (1) AT TIME = 35 G1 : NORMAL (8,2) ;

This command would cause the time cells spent in state G1 to change to a random value from a normal distribution with a mean of 8.0 and a standard deviation of 2.0. This change would take place at $t = 35$. The other times in states would remain as they were previously specified.

The user can, by using CHANGE commands, alter the cell flow, the proliferation rate, or the report period.

The CELLSIM interpreter has been entirely written in FORTRAN IV. It contains over 7000 source statements, but the program has been executed with less than 25k of words using an overlay structure. The present commands allow a wide range of cell kinetics models to be simulated. It does assume independence between cells which is the more general case in most models. Future work will hopefully add other commands to increase the capability of the language.

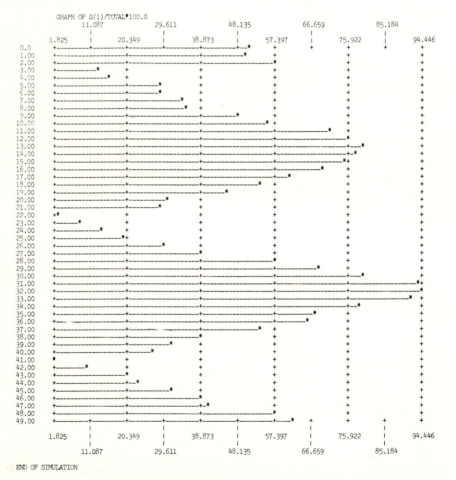

Figure 7. Graph of percentage of cells in S in model of figures 5 and 6.

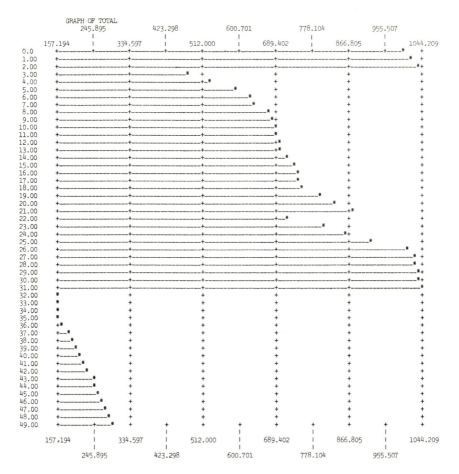

Figure 8. Graph of total number of cells in the system, with drug schedule t = 3.0, 22.0, and 32.0.

DISCUSSION

KEHL: Your system is a discrete simulation system whereas the previous chapters were continuous simulation systems. This has always been a classic problem. Discrete systems are very good when populations are small, because they track individual transactions.

DONAGHEY: Of course here, the population size really has no bearing on it because of the grouping that we do.

KEHL: Well, yes and no. The transactions always seem to grow out of bounds. You use up more memory than you can afford if you try to stay in discrete mode. And what you do is just start clumping them together. The classic problem I was talking about is to convert from a discrete to a continuous system. Has there been any progress on that problem?

DONAGHEY: We expect this in CELLSIM III, where the user can furnish differential equations and so on, and build that into the model.

KEHL: So far that hasn't been done, has it?

DONAGHEY: Well, of course the GASP language does allow discrete and continuous modelling.

KEHL: In the same program?

DONAGHEY: The new version, GASP V, does. It has now been available for a few months.

MCINTYRE: Do you provide interactive capability? Can you watch the models grow, and twiddle variables?

DONAGHEY: You could do this. We have not provided capability for the user to halt a run and change parameters, but he can sit there and watch it grow.

HOWLAND: Your probabilistic distributions require slowing of time. Don't you have to take the time units into account?

DONAGHEY: No, the "clock" is continuous.

REFERENCES

1. Barret, J. C. "A mathematical model of the mitotic cycle and its application to the interpretation of percentage labeled mitoses data," J. Nat. Can. Inst. 37: No. 4, (1966).
2. December, N. F. "Growing bones on the computer - some pitfalls of a computer simulation of the effects of radiation on growth. Cell and Tiss. Kinet., 2: No. 1 (1969).
3. Wilson, R. and Gehan, E. "A Digital Simulation of Cell Kinetics with Applications to Lizio Cells," Computer Programs in Biomedicine, (1970).
4. Cook, J. R. and James, W. T. "Age Distribution of Cells in Logarithmically Growing Cell Populations," Synchrony in Cell Division and Growth, ed. Zeuther, E., Interscience Publishers, New York (1964).

chapter 7

RECORDING AND ANALYSIS OF 3-D INFORMATION FROM SERIAL SECTION MICROGRAPHS: THE CARTOS SYSTEM

E. R. MACAGNO, C. LEVINTHAL, C. TOUNTAS,
R. BORNHOLDT, AND R. ABBA

Department of Biological Sciences
Columbia University
New York, New York

INTRODUCTION

The study of the morphology of individual neurons and neurons in a nervous system requires abstraction and analysis of information from large numbers of electron micrographs of serial sections of nerve tissue. This information, which includes such parameters as location of branch points of fibers, and locations of ends of branches, is intrinsically three-dimensional. The recording and visualization of this data about neurons requires a 3-D notebook, and a computer graphics system can be used ideally for such a purpose.

In order to store, visualize and analyze such information about neuronal morphology, we have developed CARTOS, a system for Computer-Aided Reconstruction by Tracing of Serial Sections (1,2). Various premises have directed the design of the hardware and the software for this system:

1) The recognition and recording process must be done by the researcher. There are good reasons for this.

This work was supported by a gift from the RGK Foundation and Public Health Service Facility Grant RR00442.

First, there is a large amount of anatomical data on an electron micrograph, some of it irrelevant, much of it irregular and much of it ambiguous. Pattern recognition by machine is difficult for most specimens because of the complexity of the patterns. Second, the noise level, e.g., heavy atom "dirt", or differential distortion caused by the electron beam, is non-trivial and difficult to reduce. Other variations also occur, such as changes from micrograph to micrograph of the contrast level and loss of structures by tangential sectioning. It is much easier if the investigator does the recognizing and tells the computer what is relevant and should be stored, and what label should be attached to it.

2) The recognition of a feature of interest in sequential micrographs is greatly facilitated by the ability to observe such micrographs in alignment with each other. In effect, this just states that it is easier to follow a structure from picture to picture if our attention is fixed to a small region in the plane of the image, and the over-all pattern doesn't vary a great deal. We have developed a procedure for reshooting each electron micrograph onto a frame of a 35mm film strip. (This procedure is described in detail elsewhere (1,2) and will not be further discussed in this paper). This film strip can then be viewed at various speeds on a projection screen, allowing the viewer to follow easily a particular feature throughout a set of sections. It is such a film strip which is used by the CARTOS system as the source of primary data (3). The stored data must be available to the operator during the input stage and must be displayed in ways that aid the identification of related patterns and the recognition of mistakes. In many cases, the data being recorded may belong to a group of similar patterns; for example, one might be recording the pathway of one fiber in a fascicle which contains dozens of identical cross-sections of other fibers. The possibility of viewing already recorded information in superposition with the next micrograph allows one to determine without uncertainty which is the relevant structure. It is often also useful to view the data already recorded in various directions, e.g., from the side, in order to judge whether any errors have been made in the recording procedure or in identification of the relevant features.

4) The analysis and manipulation of the stored information must be done interactively with the operator

capable of viewing the data in various formats and in various absolute or relative positions in real or close-to-real time. Three-dimensional structures can be visualized in various ways: by rotating the structure in space, by means of stereo pair, by showing various projections simultaneously, or by somewhat complicated optical techniques (holograms). At least one of these techniques is necessary in order for the investigator to study the morphology of a cell, the relative position of branches, the relative distribution of synapses, etc. Moreover, the comparison of various neurons requires the ability to move the recorded data of one neuron with respect to another. Basically, we are at the stage where much of the study and analysis is visual, and hence the investigator must be afforded a means for straightforward visualization of the data. The CARTOS system was first designed and implemented around a powerful graphics computer, the ADAGE AGT/50, linked to a large processor, the University's IBM 360/91-75. This system has been described in some detail elsewhere (1,2). More recently, we have adapted CARTOS to the smaller DEC GT40 graphics terminal. The GT40 system can be run as a stand-alone device, or as a peripheral device on a DEC PDP11/45 computer, through which it can talk to the ADAGE or the IBM 360. The CARTOS software includes three types of programs: (1) recording programs that run on the ADAGE, the GT40 or the GT40-PDP11/45 coupled system; (2) viewing programs that run on the ADAGE; and (3) editing and analysis programs that run on both the ADAGE and IBM 360 simultaneously. Various aspects of the hardware and software of CARTOS will be discussed below.

HARDWARE

The computer of the Department of Biological Sciences includes: (a) an ADAGE AGT/50 Graphics Display Computer, with a high speed parallel data link connection to the multiplexer channel on the University's IBM 360/91, (b) a DEC (Digital Equipment Corporation) PDP 11/45 computer with 56k of memory, and (c) a DEC GT40 graphics terminal which includes a PDP 11/05 computer. Each machine has various peripherals which will be described below in more detail. In addition to the ADAGE link to the IBM 360/91, the GT40 is connected to the 11/45 by a 9600 baud line and the 11/45 talks to the IBM

360/75 (which is coupled to the 360/91) through a 1200 baud line. A 9600 baud line between the 11/45 and the ADAGE has recently been completed. The overall connection scheme of the facility can be seen in the diagram in figure 1.

Besides the connections to other machines, each computer has various input-output devices. Both the ADAGE and the PDP 11/45 have double magnetic disks, a

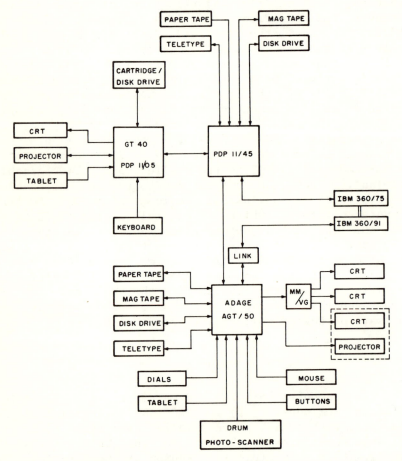

Figure 1. Block diagram of the Columbia Biology Department computer facility, including the University's IBM 360. The various peripherals are labeled in a standard manner except for the matrix multiplier (MM) and vector generator (VG) and the "mouse" (an x-y digitizer) on the ADAGE.

seven track magnetic tape coupled to a Tridata Cartrifile-20, and a relatively low speed storage device which has two cartidge magnetic tapes for program and data storage. Additional peripherals related to tracing of serial-section movies are described below for the ADAGE and the GT40.

The ADAGE Tracing Unit

Various devices (see figure 1) are tied to the ADAGE for the specific purpose of recording and analysing data from serial section movies. A Vanguard 35 mm projector (on an S-1 stand) is driven by the computer in a single-frame mode in the forward or reverse direction. The computer also stores the frame number and converts it to a z-coordinate in accordance with information about section thickness furnished by the operator. The film is back-projected onto a ground glass screen which is positioned at a right angle with respect to the large display screen of a CRT driven by the computer. A half-silvered mirror at 45° to both the CRT and ground glass allows the viewer to see both the film and the computer-generated display in superposition (see figure 2). The operator can communicate with the computer by means of teletype commands, by pushing buttons and turning dials which are periodically sampled by the computer, or by moving the "mouse", a small box with two wheels at right angles connected to two potentiometers which are read by the ADAGE as the x and y coordinates. The mouse, which fits into the hand of the operator and is moved on any flat surface, has three buttons whose function is assigned by the program and will be discussed in the software section. In total, 51 buttons and 18 dials are available, besides the keyboard. A second display unit is available for viewing and for photography. A hard-wired matrix multiplier and vector generator allow the display of data under rotation in real time.

The GT40 Tracing Unit

The GT40 with its PDP11/05 computer is less complicated and less expensive than the ADAGE, but it also has a much less sophisticated display system. The

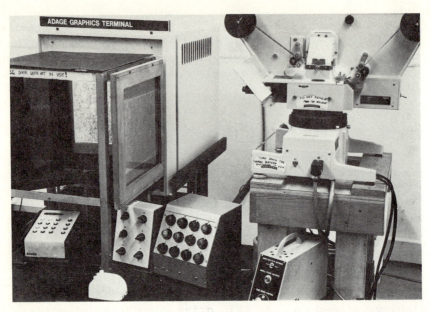

Figure 2. Photograph of the ADAGE Tracing Unit. The 35 mm projector Vanguard on its S-1 stand is at the right of the picture; it projects an image on the ground glass screen at the left. This screen and the 45° half-silvered mirror are held in the box which sits in front of the CRT, so that a viewer looking at the CRT sees an image of the film superimposed on the CRT. A box of button controls, two boxes of dials and the "mouse" are seen on the table.

tracing unit is in principle very similar to the one on the ADAGE described above. The Vanguard 35 mm projector in this case is mounted on a Vanguard S-13 viewing stand (see figure 3), so that the operator sees the projected film on a screen directly in front of him. The display on the screen of a small slave oscilloscope (Hewlett-Packard Model 1330A) is back-projected onto the ground glass of the stand by means of a lens and some mirrors. Hence the operator can view superimposed computer-generated displays and related films. The advantage of this over the ADAGE System is that the viewer can get much closer to the ground glass screen, thereby effectively gaining in ability to resolve small features. The projector can be run by the 11/05 computer either backward or forward by means of teletype commands, and the frame number converted to a z coordinate and stored. The x and y coordinates are provided by means of an electronic digitizer (Numonics model 224A) mounted directly onto the Vanguard stand. Besides position information, all other data and commands are entered by the operator through the keyboard on the PDP 11/05.

CARTOS SOFTWARE

The CARTOS software consists of three types of programs: (a) *recording* programs that are used in the acquisition of data from the serial-section film strips and run either on the ADAGE alone, on the GT40 - PDP

Figure 3. Photograph of the GT40 Tracing Unit. The 35 mm projector is at the left. A frame of the film is shown back-projected on the ground glass screen. The Numonics digitizer is mounted directly onto the viewing stand. The digital display on top of the GT40 shows the x-y position of the Numonics arm. The light microscope on the far left is used to observe original sections while recording data from a movie of light micrographs.

11/45 coupled system; (b) *display* programs that are used for viewing and manipulating the recorded data and run on the ADAGE alone or on the ADAGE-IBM 360 coupled system; and (c) *editing* programs, which are used for correcting, modifying or adding data files and run on the ADAGE-IBM 360 coupled system at present but which will be adapted to the PDP 11/45 in the future.

There are three groups of system support programs for CARTOS on the GT40: one to handle communication between the host (PDP 11/45) computer and the GT40, an I/O package for the cartridge tape, and a graphics package. Communication between the GT40 and the host is done over a high speed asynchronous serial data link. In addition there is a hard-wired program in ROM which performs the functions necessary to allow the GT40 to operate as a CRT test terminal on the host's monitor system (UNIX) and also loads and starts programs sent from the host. The communications package has three functions. It enables one to transmit programs from the host computer suitably coded for loading into the GT40 by the ROM loader. One may also read files from the host's disk and, of course, write or rewrite such files.

The GT40 has a fairly efficient 2-D display processor with a flexible instruction set. The display package builds and maintains a display list in user-supplied buffers suitably structured for input to the display processor. The third vector component is stored in a specially coded NOOP instruction. By means of program calls one can add and delete vectors and scan or chain display buffers.

The addition of the Tri-Data Model 20 Cartrifile system enabled the GT40 to be used stand-alone. It consists of two cartridge drives one of which is used for program loading and the other for data file manipulation. The cartridge file handling package was written specifically for CARTOS, and its file organization is tailored to tracing conventions (e.g., one brain per cartridge, 3 character file names, etc.). The I/O program allows storage, retrieval, and replacement of data files. The utility programs for the Cartrifile system include those for listing, dumping, condensing, and initializing data cartridges. In addition there are programs for copying files stored on cartridges onto the host computer's disc and creating program tapes from programs assembled on the host.

Recording Programs

The recording programs on the GT40 and on the ADAGE are basically similar, although more functions are available on the latter. The output image on the oscilloscope is optically mixed with the projected image of a film frame; a cursor on the oscilloscope is moved by the observer using the "mouse" on the ADAGE or the pointer of the digitizer on the GT40. The position of the cursor corresponds to the x and y coordinates of the overlapping feature on the film. The frame number on the film gives the coordinate. The operator, by depressing a combination of buttons on the mouse or of keys on the GT40 keyboard, can record the x, y and z coordinates of the cursor, add a label to that position, and draw a vector from the last recorded position to the current one. The operator can also advance the film in either direction, record more positional information and thus effectively "draw" in three dimensions.

From each serial-section movie various data files are generated:

(a) a *brain file*, which contains general information about the movie, such as magnification, section thickness, etc.; (b) a set of *nerve files*, each containing all recorded information about a specific neuron, such as contour drawings, branching patterns or synapse position and polarity; and a *nucleus file*, which contains the location and diameter of each cell nucleus found within the movie. Once a data file is generated, it can then be directly accessed by both display and editing programs.

On the ADAGE, data files are generated with the interactive recording program REC (which exists in various versions with different optional functions). Three different modes for recording data are available: (a) *contours*, where a series of line segments is used to draw the outline of the cross section of a cell on each relevant frame, (b) *tree*, where a series of connected vectors is drawn through the center of each branch of a neuronal dendritic tree, and (c) *synapses*, where the position and nature of each synapse of a neuron are recorded and displayed as a point and a visible character. These modes are selected by means of function switches. In addition, the three buttons on the "mouse" are assigned different specific functions in each mode, as for example in tree mode, where the buttons signify branch point, continuation point or end of branch. During a recording session any or all modes can be employed in any order, and the stored information recalled and displayed for observation and to check that no errors have been made (e.g., that while recording a neuron one doesn't mistakenly record another neuron's branch). Other features of the ADAGE recording program have been described in detail elsewhere (2).

On the GT40 there are two versions of the recording program: "GTREC", which operates on the GT40-PDP 11/45 coupled system, using the 11/45 disks for programs and data storage; and "TREC", which resides in and talks to the Cartrifile 20 cartridge tapes in the stand-alone GT40 system. All commands are entered by depressing single keyboard keys. The following commands are executed immediately:

    Projector Advance (one frame)
    Projector Back (one frame)
    Enter location of cursor
    Enter location of cursor and draw vector to it
      from last point
    Clear buffer
    Update on disk (or tape)

For some commands, the key acts as a switch; for example:

    Enter/exit windowing mode
    Rotate/return 90° about y-axis
    Display/suppress numerical information (e.g., current frame)

Other commands ask for additional information to be typed into the machine:

    Start new brain file (enter brain name)
    Start new nerve file (enter nerve number)
    Enter label for last recorded point (enter text)
    Enter window thickness (enter number)

The windowing commands allow the operator to view only a portion of the recorded data, hence uncluttering the picture but retaining information useful in the identification of features being recorded.

Display Programs

Data files generated either on the ADAGE, or on the GT40 but tranferred to the ADAGE, can be displayed and manipulated using various versions of the SHOW program. In general these programs permit the operator to view the files individually or in groups, rotating with respect to the laboratory coordinates in real time or under dial control by the operator, oscillating back and forth about an axis, or translated with respect to laboratory. In one version, GSHOW, the operator can also rotate and translate a group of data files relative to the other files in core. A special version of this option permits reflection about a y-z vertical plane in order to study bilateral symmetry by superposition of a mirror image of the left with the right half of a brain. Various other options allow the operator to window in three dimensions; to display stereo pairs; to display some files as dashed or dotted lines instead of continuous lines, so that the data that pertains to one file can be differentiated from another; and to display only contours, trees, synapses or any combination for a specific nerve file. For GSHOW, for example, the following functions are handled by means of buttons:

1. Overall Rotation, ON/OFF
2. Overall Translation, ON/OFF
3. Dot-Dash Mode, ON/OFF
4. If 3 On, DOT/DASH
5. Stereo Pairs, ON/OFF
6. Numerical Information, ON/OFF
7. Overall Rotation, Real Time/Manual Control
8. Reset Rotation to Initial Position
9. Windowing, ON/OFF
10. Contours, ON/OFF
11. Trees, ON/OFF
12. Synapses, ON/OFF
13. Oscillation, ON/OFF

The magnitude of some of these functions, as for example rotation speed, amount of translation, stereo pair separation, oscillation angle and window position are handled by means of dials (potentiometers). In addition, various teletype commands are used to select and retrieve files, and to locate them in pre-determined core locations under control of specific buttons. The versatility of these display programs has provided us with a powerful tool for the study of the morphology of individual neurons and the interaction of neurons in their complex three-dimensional configurations.

Editing Programs

Nerve file editing is done with coupled programs which run simultaneously on the ADAGE and the IBM 360. These programs allow nerve files to be edited , contours to be interpolated between others which are traced, contours to be generated by extrapolation from those which are already traced, and files to be altered in such a way that errors introduced by the Image Combiner can be corrected on the computer. In addition, the scale used for a file can be changed in any dimension and images can be translated with respect to each other. This entire set of programs can be used on nerve files which are generated on the ADAGE or on the GT40 and transferred to the ADAGE.

Our original purpose in generating these editing programs was to save the time of the tracer by making use of the fact that many more surface points are needed for an adequate display of a nerve than are logically necessary for defining a smooth nerve surface. Thus, a small number of contours can be traced and many more can be interpolated to give an adequate image. With the present system, an observer can see the interpolated files which are generated by the program; if they are satisfactory they can be added to the permanent file, and if not, then additional contours can be traced and used for generating a new interpolation.

Another major use of these editing programs has to do with the fact that different kinds of information can be obtained from high magnification and low magnification micrographs. It is frequently necessary to take a series of micrographs of a whole ganglion or portion of a brain at low magnification in order that the relative orientation of various structures can be established. The same structure can be traced into the computer from high and low magnification movies of the same sections. These two traced images can be aligned on the ADAGE and this information fed to the parallel 360 program so that the two files are organized appropriately in relation to each other, allowing a three-dimensional structure to be reconstructed by combining both high magnification and low magnification photographs.

Another application of the editing progam arises from the fact that if the Image Combiner is not accurately adjusted, a positional drift is introduced into the motion picture made from the successive serial photographs. Thus, even a small displacement of photograph N from N-1, which can take place if a mirror is very slightly out of alignment, can produce a substantial drift in the overall picture when many hundreds of photographs are combined. This drift can be corrected at the level of the computer files as long as there is some feature in the material traced which allows the investigator to identify a straight line through the structure. This can frequently be done in the vertebrate material using the spinal cord or the outer boundary of the brain segment or of the ganglion. It can also be done using the midplane in any of the bilaterally symmetric organisms we are now working with.

CONCLUSION

The CARTOS system has been used extensively in the past few years in our laboratory in the study of the anatomy and development of various organisms that form isogenic clones (3-5). Some of the three-dimensional reconstructions of neurons in *Daphnia* are shown in figure 4, which illustrates the three modes of recording data: contours, tree and synapse; and in figure 5, showing symmetric neurons on each side of the midplane of the brain of an animal, displaced with respect to each other. Much of the information we have obtained using the computer reconstructions would have been very difficult or impossible to obtain otherwise. For example, we have found (3) that during development of a part of the *Daphnia* nervous system, a cell that is touched by a growing fiber proceeds to wrap around the fiber in a characteristic manner. This wrapping or enfolding may well be a signal or a response to a signal which tells the target cell what its future identity will be. The geometric arrangement in this unique phenomenon was discovered in the three dimensional reconstructions from movies made from serial section micrographs of embryological tissue.

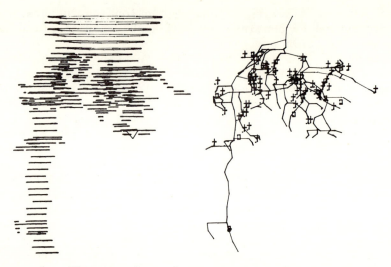

Figure 4. Photographs of a computer reconstruction of part of a neuron, showing contours, tree and synapse data (90° rotation about the axis). The cell is a lamina neuron in the *Daphnia* optic ganglion.

The implementation of CARTOS on the GT40 tracing unit, either as a peripheral device on a larger computer or as a stand-alone system, should make this technique available to a much larger group of researchers. The Columbia facility has been visited by a number of investigators who are now designing and building similar devices. We are prepared to make programs and a telephone data link to our PDP 11/45 available to other laboratories if they wish to use a system similar to the one we have developed.

DISCUSSION

SHANTZ: How large are the cells you're looking at, and how far out on the branches can you follow a cell?

MACAGNO: This particular animal is rather small, with neurons of 5-10 μ diameter at the soma. We're also working on the Mauthner cell, which is rather large. The distance we can trace on the branches depends essentially on the resolution of the micrographs. We usually have a

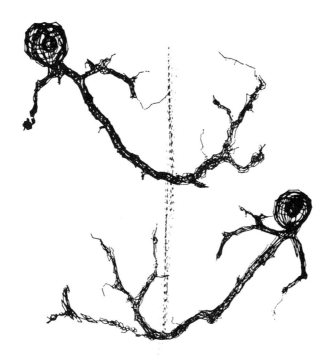

Figure 5. Photograph of a computer reconstruction of two symmetric neurons, shown crossing the midplane of the brain from opposite sides. The nucleus has been traced within each cell body. The cells were traced from different movies of sections of one brain and then positioned with respect to each other in their correct relative lateral positions, but subsequently displaced vertically to make the picture clearer. The neurons are medulla cells in the *Daphnia* optic ganglion.

resolution of about 150 Å. The slices are cut at 600–1000 Å. However, we've never found a synapse in a branch which is less than 2000–3000 Å, so I suspect we usually don't lose synapses.

DIERKER: How much do these two systems cost?

MACAGNO: The present cost of the ADAGE system would be about $300,000. The GT40 system is closer to $35,000. The vector generator in the GT40 is actually quite nice. You cannot do hardware rotations like you can in the ADAGE, and therefore it takes a lot of time to rotate things using software. But in general, the GT40 is quite adequate.

RANSIL: What was the rationale for using the GT40 with the 11/45?

MACAGNO: Originally we planned to have several GT40s attached to the one 11/45. Then we decided we wanted to be able to keep the GT40 portable, and to keep the 11/45 dedicated to it for that reason. The additional cost is only some 7% higher this way, so it's not too bad.

REFERENCES

1. Levinthal, C. and Ware, R. Three-dimensional reconstructions from serial sections, Nature 236:207-210 (1972).
2. Levinthal, C., Macagno, E. R., and Tountas, C. Computer-aided reconstruction from serial sections, Federation Proceedings 33:2336-2340 (1974).
3. Macagno, E. R., LoPresti, V. and Levinthal, C. Structure and development of neuronal connections in isogenic organisms: variations and similarities in the optic system of *Daphnia magna*. Proc. Natl. Acad. Sci., U.S. 70:56-61 (1973).
4. LoPresti, V., Macagno, E. R., and Levinthal, C. Structure and development of neuronal connections in isogenic organisms: cellular interactions in the development of the optic lamina of *Daphnia*. Proc. Natl. Acad. Sci., U.S. 70:433-437 (1973).
5. LoPresti, V., Macagno, E. R., and Levinthal, C. Structure and development of neuronal connections in isogenic organisms: transient gap junctions between growing optic axons and lamina neuroblasts. Proc. Natl. Acad. Sci., U.S. 71:1098-1102 (1974).

chapter 8

A MINICOMPUTER-BASED IMAGE ANALYSIS SYSTEM

MICHAEL J. SHANTZ
Information Science Department
California Institute of Technology
Pasadena, California

INTRODUCTION

Picture processing is a computer activity which requires the ability to handle huge amounts of data. A typical digital picture element (pixel) is one byte. This amounts to over a quarter of a million bytes for a two-dimensional picture. If we wish to represent a third dimension by using a stack of images, the data space required may exceed ten million bytes. This situation would be quite discouraging for the minicomputer user if it were not for the rising power and decreasing costs of the mini plus the availability of large, high-speed secondary storage. This paper describes a picture processing hardware and software system which runs on a mini-computer. The cost of such a basic hardware system could be kept under $80k.

HARDWARE

Figure 1B shows our present system for picture processing. The system hardware consists of a 28k

---

Dr. Ken Naka has provided many of the ideas in this chapter. Professor G. D. McCann has encouraged us during the course of this work. We thank Howard Frieden, W. D. Knutsen and Jerrie Pabst for help in system programming.
This research was supported by Public Health Service Grant NS10629.

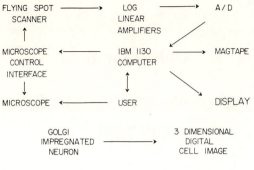

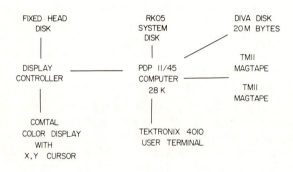

Figure 1. Diagram of automated light microscope system (ALMS). The computer controls the microscope stage and focus position, and drives the image plane scanner which scans the image. By sampling the image at sequential levels of focus, a stack of images representing the 3-dimensional structure of the neuron can be obtained. B: Diagram of our hardware for image analysis (see text).

PDP-11/45 computer, two TMI tape decks, a 20M byte DIVA disk system, a DEC RKO 1.2M word disk cartridge, a COMTAL color image display with a .5M byte refresh storage capacity, interactive cursor and pseudocolor hardware, and a Tektronix 4010 terminal. This system costs in the neighborhood of 150k dollars but a minimal configuration consisting of a PDP-11/45 with a floating point processor, an RKO5 disk cartridge for system storage and high-speed secondary data storage, a TM11 tape deck for long term, large scale storage, and a scan converter-display system with light pen could be purchased for under $80k. This is, I think, a minimal system. Long-term storage on magtape is a must, and high-speed secondary storage on disk is necessary due to the fact that an entire picture will not normally fit in core. A light pen or cursor is invaluable for many of the interactive procedures that enrich a picture processing system. This configuration does not include a method for digitizing images via TV camera, etc., or for hard copy output. These two items would add another 10-20k dollars to the cost of a system, depending on the quality desired.

For our work with Golgi-impregnated neuronal images, we use an automated light miscroscope system (ALMS) (figure 1A), which was developed at Caltech's Jet Propulsion Laboratory (9), to digitize our cell images. This digitization system uses computer driven galvanometers to move mirrors which scan the image seen through the microscope across a small hole behind which is positioned a photo-multiplier tube. A TV camera with computer interface and digitization hardware could also be used for data acquisition, with some sacrifice of resolution.

SOFTWARE (IAS)

Our Image Analysis System (IAS) runs under the disk operating system (DOS) provided by Digital Equipment Corporation. DOS provides an editor for creating source files, an assembler for assembling macro-coded routines, a FORTRAN compiler, and library facilities in addition to the DOS monitor which handles system programs and utilities, input/output transfer, device handlers, and file management. It also provides various commands for control of program execution and modification. We can therefore write FORTRAN callable subroutines in Macro programs by using a CALL macro. This enables one to encode the inner loops, or those portions of the program which need to run fast, in Macro and to use FORTRAN elsewhere for ease of coding and debugging. We find that little can be gained by macro encoding anything other than the innermost loops of a program. This system would benefit greatly by inclusion of a compiler for a more powerful high-level language such as PL1, PASCAL or LISP, but is very useful as it stands.

The core of IAS is composed of the basic input-output routines which provide buffered transfer of image lines to and from core. These routines were developed by Howard Frieden (2) and Kenneth Castleman at JPL to run on a PDP 11/40 computer. They provide double or single buffering so that a maximum of speed can be realized during operations on images too large to fit into core. These routines are used to get a line of an image from disk into core, to put a line from core to disk, to get an image label record, to put a label, and to transfer image lines from tape to core and core to tape. I have used this basic input-output package, modified to run on our configuration, as the underlying input-output for IAS.

Our basic application routines consist of the following:

1) FLT - This digital filtering routine inputs an image and a user specified two-dimensional filter array and computes the two-dimensional image which is the cross-correlation of the input image and the specified filter. It allows for user specification both of scaling parameters and of the area of the input image to be filtered.
2) FFT - This routine computes the image (whose elements are complex numbers) which is the complex Fourier transform of the input image. This routine can input either byte or complex images, and computes either the direct or inverse transform. The algorithm uses a one-dimensional, FORTRAN-coded, Cooley-Tukey algorithm which we obtained from IBM to transform each line of the image in place. The resulting image array is transposed and the lines are again transformed in place producing the final transformed image.
3) DIF - This routine subtracts one image from another to yield a difference image.
4) STR - This routine performs a user specified linear transformation on each element of the input image.
5) MAX - This routine computes the element maximum of two input images.
6) TOP - This routine computes the top view through a

stack of images representing a three- dimensional image of an object. The output image is obtained by selecting the maximum element in each vertical column down through the stack.

7) SVU - This routine computes an image which is the side view of an object which is represented by a stack of images.

8) FGEN - This program is a set of function generating routines. It provides facilities for the creation of grey level ramps, gaussian intensity humps, gaussian distributed random images, line drawings interactively created, sinusoidal images and variable size text on images.

9) NPT - This routine provides for the creation of a descriptive file (DF) through interactive feature extraction. This file contains a set of items, each item in the file has a name, a number, up to 90 byte attributes, and up to 100 real parameters. It can be updated by interactively taking measurements of various kinds from the display using the x,y cursor and entering these numbers in the file. The parameters represent an n-dimensional space in which each item is a point. Thus this descriptive file provides a means of collecting information of various kinds on a set of items such as neurons for the purpose of describing and comparing them, through cluster analysis.

10) NT3D - This routine provides for the creation of a three-dimensional point file consisting of those items from an n-dimensional DF which have a specific attribute. The user also specifies which of the parameters of the n-dimensional file are to be the x, y, and z values for the three-dimensional item. This allows an investigator to examine the clustering, in various descriptive three-dimensional spaces, Of items having certain attributes in common. For example, if the items are neurons the investigator may wish to create a three-dimensional file of all cells which are monostratified (assuming that this is one of the attributes in the n-dimensional file), and to have the x dimension represent the size of the dendritic field, y represent the density of processes at a given level in the retina, and z

represent the soma depth from the outer plexiform layer.
11) DSPPTS - This routine displays a 3-dimensional descriptive file as a scattergram, each point of which is the number of an item or cell. A two dimensional projection of the 3-dimensional point cluster appears on the screen. Facilities are provided for translating, scaling, and rotating the points in the file to aid in the visualization of the clusters in the three dimensions. These point transformations are performed using the method of homogeneous coordinates described in Newman and Sproull (7). These routines allow an investigator to collect varied information about a collection of items and to examine the way the items cluster in various descriptive spaces. This approach is useful in getting a feel for the data and in pattern recognition applications.
12) SCRUB - This routine provides a "paint brush" for creating or modifying images using the interactive cursor. The user specifies a grey level and a brush size and then moves the cursor around on the display screen as if it were a paint brush. In this way selected portions of the image can be erased to remove noise, or corrections can be performed on the image.

The above routines provide some basic analytical tools for analyzing images. Images can be analyzed in the frequency domain by Fourier transforming and performing operations such as multiplication or squared absolute value in the spatial frequency domain. The filtering facility allows great flexibility in linear filtering of pictures to detect edges, simulate optical defocus or to model retinal function.

APPLICATIONS

Three Dimensional Reconstruction

One of our image analysis problems is to reconstruct 3-dimensional representations of neurons from light microscope images. We prepare Golgi-impregnated neurons from the frog retina imbedded in Spurr plastic, then using the automated light microscope system, we digitize the neuron in the following way (see figure 1A). First the microscope is focused at a starting focal depth. The user specifies the area of the microscope image to be scanned and digitized, then the computer scans a digital image, automatically moves the focus to a new level, scans another image, and so on until a stack of images representing the 3-dimensional structure of the neuron is obtained. These digital images are stored on magnetic tape for later processing under IAS. This method produces a 3-dimensional image which is finely sampled in the x, y dimensions (typically 512 lines by 512 samples), but coarsely sampled in the z direction (typically 15 to 30 samples or focus levels). This representation gives the optical density of the Golgi stained material as a function $f(x,y,z)$, the magnification of which depends on the microscope objective used, ($10X$, $20X$, $40X$, or $100X$). The higher power objectives have a shorter depth of focus and thus give better axis resolution. We refer to the above method of obtaining 3-dimensional representations of microscope images as an Optical Sectioning procedure.

The function $f(x,y,z)$ has been degraded by the optics of the microscope to some extent, and this degradation can be represented mathematically as follows, using the approximation that the microscope optical system is a stationary, linear system (i.e., assuming that isoplanarity holds, or that the 3-dimensional defocus point spread function is spatially invariant) (3,4,8):

$$f(x,y,z) = \iiint_V w(p,q,r)g(x-p, y-q, z-r)\,dp\,dq\,dr$$

where $g(x,y,z)$ is the actual undegraded 3-dimensional distribution of optical density in the specimen, and $w(p,q,r)$ is the 3-dimensional defocus point spread function of the microscope with a given objective and given light source. Our problem is to obtain a close approximation to $g(x,y,z)$ given $f(x,y,z)$. We obtain $f(x,y,z)$, coarsely sampled in the direction, through use of our optical sectioning procedure. The next step is that of deblurring or correcting for the degradation produced by the method. This problem was partially solved heuristically by Castleman and Weinstein (9), and was improved and formalized by Castleman et al. (1). Figure 2A shows a top view of an example of $f(x,y,z)$

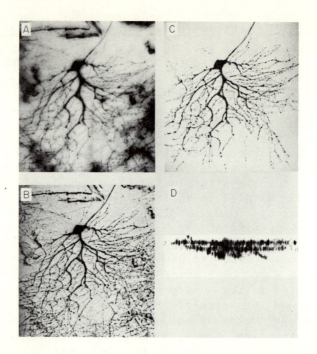

Figure 2. The process of removing defocused information and interactively removing objectionable Golgi material. A: Top view of a stack of images representing the 3-dimensional structure of a neuron before processing. B: The same cell after removal of defocused information. C: The result after interactive noise removal. D: A computed side view.

obtained with 20X objective. This is our raw data. We applied the method of Castleman et al. and obtained the image shown in figure 2B. This process has the major advantage of improving z axis resolution, for improved side view information and improved stereo pair generation. The following algorithm for this deblurring step produces $f'(x,y,z)$, the cleaned up image, using an approximation to the process shown in figure 3.

$$f(x,y;z_i) = f(x,y;z_i) - \max[(f(x,y;z_{i-1})*w(x,y;\Delta z)), f(x,y;z_{i+1})*w(x,y;\Delta z))]$$

Where z_i represents the ith optical section, $w(x,y,\Delta z)$ is the two dimensional point spread function of the microscope at z amount of defocus, * represents convolution and MAX obtains the pixel by pixel maximum of two images. We obtain the defocus point spread functions from experimental and also theoretical methods. Experimentally we scan a very small spot of Golgi-stained material at various levels of defocus. Then assuming that our spot approximates a delta function we take these defocused spot images to be an approximation to our desired $w(x,y;\Delta z)$. We have also computed these functions by using a theoretical procedure described by Hopkins (4) and improved by Stokseth (8).

After the preceding step the 3-dimensional image still contains many types of objectionable noise. Often

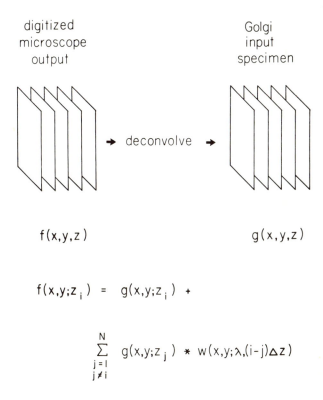

$$f(x,y;z_i) = g(x,y;z_i) + \sum_{\substack{j=1 \\ j \neq i}}^{N} g(x,y;z_j) * w(x,y;\lambda,(i-j)\Delta z)$$

Figure 3. The procedure for correcting degradation caused by the 3-dimensional defocus point spread function of the microscope. We use a gross approximation to deconvolution in order to recover an estimate of $g(x,y,z)$.

the Golgi method will stain objects lying adjacent to or on the object of interest. This type of noise is impossible to remove in a fully automatic manner without solving many tough problems in the field of pattern recognition. The human is quite good at distinguishing cell from garbage through use of subjective high level descriptors. It is very difficult to give the computer these capabilities, hence an interactive noise removal stage is called for. We have implemented this type of noise removal capability through use of a user controlled "paint brush" routine (SCRUB) described above. Using this routine a cell image can quickly be erased, leaving only undesirable Golgi artifacts in the image. This resulting "noise image" can be subtracted from each level of the stack of images, thus carving a vertical silhouette through the 3-dimensional representation. This procedure is very quick and removes all unwanted material except for that noise lying directly above or below the actual parts of the desired object. Figure 2C shows the resulting top view after this interactive noise removal step. Figure 2D shows the side view computed after the above steps were performed. This side view is considerably improved if a higher power microscope objective is used. The side view in figure 2D was computed from only 15 levels, i.e., 15 samples in the z direction.

After the above procedures are carried out we have a stack of images representing the clean, relatively noise-free neuronal object of interest, and we can proceed to compute top views, side views or stereo pairs and thus to form convenient representations from which to extract quantitative features for the description and classification of neuronal morphology.

Feature Extraction and Selection -- Creation of a Descriptive File

We are using our image analysis system to create descriptive files for the purpose of storing concise structural descriptors to be used in an attempt to objectively classify the various morphological types in the frog retina. Each neuron is an entry in the file and has a name, a number, up to 90 byte attributes, and up to 100 real parameters. We can view the clustering of these points in this space by selectively viewing 2- and 3-dimensional projections of these points. These

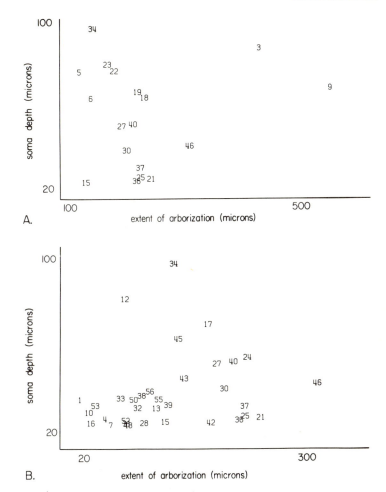

Figure 4. A scattergram for all monostratified neurons displayed in the space defined by their maximum arborization extent and the position of their soma. B: A similar scattergram of all neurons whose cell bodies lie in the inner nuclear layer.

facilities are provided by routines NPT, NT3D, and DSPPTS as described above. Figure 4A shows a scattergram of all neurons which have monostratified characteristics (attribute 1 represents monostratification).

These facilities greatly aid the process of examining a large data base for the purpose of discovering groupings or categories of data. They allow one to investigate the correlation between particular structural characteristics and particular functional characteristics in a statistical manner, and thus they provide for the objective examination of morphology.

Functional Modeling

In our studies of the structure and function of the frog retina a third application of image processing suggests itself. Since the retina is essentially a 2-dimensional transucer and image analyzer, we may use our computerized image analysis system to study this biological image analyzer and to create 2-dimensional models of retinal function.

We are using IAS to generate stimulus patterns for use in neurophysiological experiments. Figure 5 shows

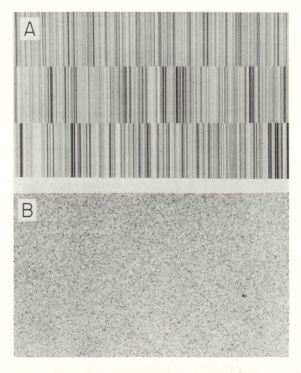

Figure 5. Two examples of white noise patterns to be used as stimuli in experiments to determine the spatial Wiener kernels which characterize a given layer of neurons in the retina.

spatial white noise patterns to be used in this manner. From spatial white noise experiments we plan to obtain the spatial Wiener kernels which characterize the system. IAS can be used in the computation of the Wiener kernels, using the Fourier transform facilities to perform the crosscorrelations and to model the system. The following expressions represent the nonlinear 2-dimensional response as determined by the first two spatial Wiener kernels (5,10):

$$g(x, y) = \sum_{n=0}^{\infty} G_n(h_n, f(x,y))$$

$$G_0(h_0, f(x,y)) = h_0$$

$$G_1(h_1, f(x,y)) = \iint_{-\infty}^{\infty} h_1(\sigma, \xi) f(x - \sigma, y - \xi) d\sigma d\xi$$

$$G_2(h_2, f(x,y)) = \iiiint_{-\infty}^{\infty} h_2(\sigma_1, \xi_1, \sigma_2, \xi_2) f(x - \sigma_1, y - \xi_1)$$

$$f(x - \sigma_2, y - \xi_2) d\sigma_1 d\xi_1 d\sigma_2 d\xi_2$$

$$- P \iint_{-\infty}^{\infty} h_2(\sigma, \xi, \sigma, \xi) d\sigma d\xi$$

$$h_n(\sigma_1, \xi_1, \ldots, \sigma_n, \xi_n) = \frac{1}{n!P^n} E\{[g(x,y) - \sum_{m=0}^{n-1} G_m(h_m, f(x, y))]$$

$$f(x - \sigma_1, y - \xi_1) \ldots f(x - \sigma_n, y - \xi_n)\}$$

where $f(x,y)$ is the spatial input to the system, $g(x,y)$ is the spatial output, $h_1(\sigma, \xi)$ is the first order spatial Wiener kernel, $h_2(\sigma_1, \xi_1, \sigma_2, \xi_2)$ is the second order kernel, P is the stimulus (spatial white noise) power level, and E denotes the expected value.

Spatial Wiener analysis can be used to describe or model the output of a 2-dimensional array of neurons subjected to a spatial input pattern. The first order spatial kernel can be viewed as the point spread function for a layer of neurons. Figure 7 shows qualitative

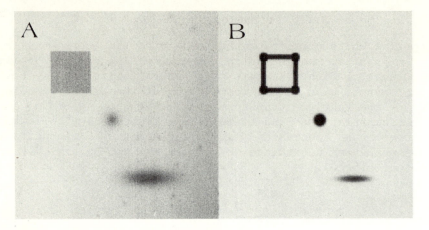

Figure 6. Example of the output (B) from a model of the layer of bipolar neurons when stimulated with the pattern in (A). Image (B) was produced by convolving image (A) with a 2-dimensional filter or point spread function, a cross-section of which is shown in figure 7C. This linear model of the bipolar later ignores gradual variations in background intensity, is excited by edges of objects, is more excited by corners of objects, and is most excited by small spots of a particular diameter.

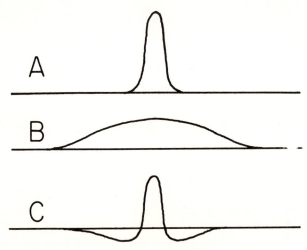

Figure 7. Speculative linear spatial kernels (or point spread functions) for (A) the layer of receptors, (B) the horizontal cell layer and (C) the bipolar cell layer. This qualitative model is based on evidence found by Naka (6) et al., which indicated that the horizontals low pass filter the input and that the bipolars subtract the horizontals from the receptors.

suggested point spread funtions for the layers of receptor, horizontal, and bipolar cells. The bipolar point spread funtion is obtained by subtracting horizontals from receptors. Figures 6A and 6B show respectively an input image, and the output image from the bipolar layer as computed by a model made up of the linear spatial kernel only. This speculative model was computed using the program FLT. It can be seen that if the bipolars subtract the horizontal cell output from the receptor cell output, and if the horizontal cells perform a spatial low pass filtering as suggested by Naka (6), The bipolars can perform as edge and spot detectors. They also remove slowly varying or low frequency components of the input image, such as variations in background illumination.

This IAS system shows promise for this kind of functional modeling and system characterization for spatial properties of neuronal arrays. Previously most neurophysiologists have looked only at the temporal transfer characteristics of retinal neurons. I think this temporal approach can only give limited results since retinal function and structure seem to be primarily designed to extract spatial information from the environment. Since IAS is designed to process images, it is an ideal tool for use in modeling the function of the vertebrate retina.

CONCLUSION

We are using our minicomputer-based image analysis system to perform 3-dimensional reconstruction of neuronal structure, to analyze this structure through use of structural descriptors and cluster analysis, and to model the nonlinear spatial characteristics of retinal cell layers. Due to the decreasing costs, and despite the huge data sets involved in image processing, useful image analysis can be done efficiently and relatively cheaply with a minicomputer. Such a system can be quickly developed by using a computer for which an operating system is available to provide high level language features (even if it is FORTRAN). This system has aided our attempts to answer questions concerning the function and structure of the vertebrate retina.

DISCUSSION

D.O. WALTER: Will that white noise image be time-varying?

SHANTZ: No. At the moment we're thinking of taking a

series of strips varying in white noise fashion, plaster it on the retina, maintain the background intensity at constant level whether the white noise pattern is there or not, shift it, record, shift, record, and obtain the tranfer characteristics in that manner.

CLARKE: In your digitization of the images, I noticed you had both the microscope stage and the flying-spot scanner under computer control. Would you comment further on their interactions?

SHANTZ: The computer controls the focus level, a discrimination level is set, and flying-spot scanner scans for digital representation. It's like a TV camera, generating separate pictures of different focal planes.

MACAGNO: How long does it take you to digitize one cell?

SHANTZ: About 15 minutes, to digitize it at 15 layers. It depends on resolution.

MACAGNO: So you digitize, then, at 5-10 microns?

SHANTZ: Typically at 2-5µ, depending on cell size and the objective used.

L. PUBOLS: How do you handle z-axis?

SHANTZ: I'm not sure what you mean. Z-level could refer to intensity or the geometrical axis. I assume you're talking about the intensity, or the grey-level. Each point in the image is digitized to a number from 0 to 225, proportional to its intensity.

MCINTYRE: How does the flying-spot scanner permit that much resolution?

SHANTZ: Oh, it doesn't. We just digitize to that level of precision. We probably get some 60 levels. That's more than you need anyway - a Golgi stained neuron's just a black object, much darker than the background.

REFERENCES

1. Castleman, K. R., Shantz M. J., and Fuchs, H. "Digital processing of three dimensional light microscope images", in preparation.
2. Frieden, H. (Personal communication).
3. Goodman, J. W. *Introduction to Fourier Optics*, McGraw-Hill Publishers, New York (1968).
4. Hopkins, H. H. The frequency response of a defocused optical system, Proc. of Royal Soc. (London) A, 231:91 (1955).

5. Lee, Y. W. and Schetzen, M. Measurement of the kernels of a nonlinear system by cross-correlation. Quarterly Progress Report No. 61: Massachusetts Institute of Technology Research Laboratory of Electronics.
6. Naka, K.-I., Mararelis, P. Z., and Chan, R. Y. Morphological and functional identifications of catfish retinal neurons. III. Functional identification. J. Neurophysiol. 38:1 (1975).
7. Newman, W. M. and Sproull, R. F. *Principles of Interactive Graphics*, McGraw-Hill Publishers, New York (1973).
8. Stokseth, P. A. Properties of a defocused optical system. J. Optical Soc. Amer. 59:10 (1969).
9. Weinstein, M. and Castleman, K. R. Reconstructing three-D specimens from two-D section images, Proc. of the SPIE, 26:131 (1971).
10. Wiener, N. *Nonlinear Problems in Random Theory*, Wiley, New York (1958).

chapter 9

AN ALGORITHM FOR THE ALIGNMENT OF SERIAL SECTIONS

M. L. DIERKER
Department of Anatomy
Washington University
St. Louis, Missouri

Over the past few years, I have worked with members of the Department of Anatomy at Washington University, Dr. Thomas Woolsey and Dr. Donald Wann, developing a computer system for reconstruction of Golgi-stained or dye-injected neurons. Our system is designed to allow neuroanatomists to enter the x, y, and z locations of selected topological features of the cell, and provides displays of reconstructed neurons.

Until recently, our system has had a major limitation: namely, the inability to use data from more than one histological section. We have recently added the capability to use serial sections. This had to be an interactive system, it had to be easy to use, and it had to be fast. As an index of our success, one operator recently reconstructed an entire Purkinje neuron in 45 minutes.

The system consists of a PDP-12, an RK08 disk drive with 108k words, a Zeiss Universal microscope, and associated hardware and software. The microscope has a motorized x, y stage, with motorized focus (2 axis) all controllable to within 0.5μ. A joystick is used to

---

Editor's note: This chapter is an edited transcript of the paper delivered by Mr. Dierker at the symposium; the author did not submit a manuscript. Any errors should be attributed to imperfect summarization by the editor.

indicate desired stage movements to the PDP-12, which then controls the motorized stage. Other controls are used to tell the computer to change focus, and to enter coordinate data for topological features which are in focus and centered on a cross-hair. These features include branch points and end points of dendrites, cell body, and dendrite origins. A keyboard for control of analytical procedures and display parameters and a display screen complete the system.

The joystick, controlled by the right hand, is used to control x,y positioning. A toggle switch (left hand) is used to change focus (z axis). Branch points are sampled with a pushbutton (left hand), sample points with a button on the joystick (right hand). A foot pedal is used to enter endpoints of dendrites. There are other controls for entering cell center and dendritic origin.

The machine keeps track of branch points as the operator proceeds onto higher order branches, and automatically returns the operator to branch points until he has entered all the branch data. This saves a considerable amount of operator time, especially when high magnifications are used. When a process is terminated by a section boundary, it will of course be necessary to connect it with the corresponding truncation on the adjacent section.

The three-dimensional perspective on the display has proved very useful for visualizing cell geometry. Rotation with perspective enhances the sense of depth. Individual dendrites can be blinked on and off, helping to separate them in complex displays.

Recently we have added the capability of aligning adjacent serial sections, for purposes of reconstruction. The problem has been to minimize errors in these alignments. The first step, before going to a new section, is to enter the upper and lower boundaries of the just- completed section by focussing on them. The program then locates truncation points in the current section by finding z-values of endpoints which are close to those of the upper and lower boundaries. The machine then tells the operator which surface(s) has (have) truncations. Before going on to the next section, the operator selects gross landmarks, such as vessels, and enters their locations. These are gross reference points which will exist in the next section.

The same reference points are then entered from the next section. The operator selects a translation and rotation which approximately superimposes the simultaneously displayed landmarks from the two sections.

The machine then moves the new section to the predicted locations of the truncations, based on the alignment and the locations of truncations on the old slide. The operator can then search in the vicinity of each predicted location and enter exact locations of the truncation endpoints on the new section. The machine then performs a hill-climbing algorithm to optimize the alignment translation and rotation, minimizing the net discrepancy of truncation locations in the two sections.

This procedure is continued from one section to the next, until, in the last section, no more truncations are found. Each alignment procedure takes perhaps five minutes.

DISCUSSION

DIERKER: Before starting the discussion, I'd like to make a quick comment. Most people like hardware displays, but we really find our PDP-12 software-refreshed displays have quite high quality. It turns out that depth effects really eliminate the disadvantages of dotted lines, for example.

BROWN: How long would it take to rotate a 100-point display?

DIERKER: I couldn't say, but we've displayed neurons with over 100 vectors, with no flicker. An entire set of transformations, rotations, translations, perspective, stereo pair generation, and display, is done in less than 1/30 sec. The most core required is about 2k for the program. The only other core required is that needed to keep all coordinates in core at once.

FRENCH: You know you can buy relatively cheap hardware to store and display vectors? About a dollar a vector, I believe.

DIERKER: The nice thing about this procedure is it's simpler to implement using core you already have. I know Optical Electronics produces analog modules for perspective calculation, rotation, vector generation, the whole thing for maybe $2000-$3000, if I recall.

SHANTZ: Why don't you use the homogeneous coordinates method in Newman and Sproull's text on computer graphics?

DIERKER: It turns out to be restrictive in what it allows, and isn't any faster than our own methods.

chapter 10

COUNTING HIGH CONTRAST CLOSED OBJECTS IN BIOLOGICAL IMAGES USING A 525-LINE RASTER SCAN TELEVISION CAMERA AND A MINICOMPUTER

D. F. WANN
Department of Electrical Engineering
Washington University
St. Louis, Missouri

We are using autoradiographic material in neuroanatomical research, and need to reconstruct the distribution of silver grains in the three-dimensional tissue. The grains are about 0.5 to 1μ in diameter, with an irregular distribution. When the high power objective required to resolve them is used, refocusing is necessary as we move from one part of a section to another. We wish to do all this automatically, as fast as possible.

There are two complex procedures involved: recognition of grains, and focusing the microscope. We will describe the system we use to accomplish this.

The selection of sensor and algorithm go hand in hand. We can narrow the problem down by considering the

---

Editor's note: This chapter is an edited transcript of the paper delivered by Dr. Wann at the symposium; the author did not submit a manuscript. Any errors should be attributed to imperfect summarization by the editor.

material we are working with. The grains are 1/2-1µ in diameter, and we need perhaps five scan lines through each grain to distinguish it from artifact. This means we need ten lines per micron, or 250 lines and 250 columns per 25µ square field, or about 62,000 picture elements. Each element is one bit, with 1 = black, 0 = white.

We could use an image dissector, which examines the field on a point-by-point basis, measuring light intensities. At a minimum, about 200 µsec./point is needed to accumulate enough light, or 12.5 seconds per 25 µ frame. A TV raster scan, on the other hand, has several advantages, including high signal/noise ratio, as well as the ability to integrate photons over the whole field simultaneously. Also, the operator can see the entire field on a monitor. The device is somewhat harder to interface, but much more economical. The image dissector costs $10,000 or more.

Since 250 lines are adequate, we use only the odd field of the video raster. We sample one column per raster scan, for 250 scans. This means the 250 x 250 data input is accumulated in about 8 seconds, somewhat faster than the image dissector, with much better signal/noise ratios. The brightness signal is compared with a threshold to obtain a 1-bit (intensity) axis.

The algorithm which detects grains runs fast enough to keep up with the video input, so this is done with software, providing data reduction on the fly. A grain is detected by detecting emergence of the scan from the grain. By examining adjacent scan rows or columns, concavities can be detected, thus avoiding artifactual multiple grain counts for single grains with concave margins.

Focusing is accomplished with a hill-climbing algorithm, which maximizes grain count over a given field.

DISCUSSION

CAPOWSKI: Why do you do your thresholding in hardware instead of software?

WANN: We could do it either way. We've found that it's not very sensitive in any event.

CAPOWSKI: We've found we get better results on silver grains using dark-field illumination.

WANN: This was done in bright field; we're now using dark field. Using a 40X objective you get a larger field of view, you can count large areas in shorter periods of time. But we didn't re-shoot the movie.

CLARKE: Could you re-describe the focusing algorithm?

WANN: It's a hill-climb. The simplest one we've used looks at the grey-level picture, scans a frame, and counts the number of points exceeding threshold. It then changes focus, and does that again. It keeps doing this, seeking a maximum number of dark points. Once that's found, it's in focus. In this particular application, the computer never sees the grey levels.

chapter 11

AN ALGORITHM FOR THE DISPLAY AND MANIPULATION OF LINES IN THREE DIMENSIONS

M. L. DIERKER
Department of Anatomy
Washington University
St. Louis, Missouri

INTRODUCTION

In the past few years there has been a large increase in interest in the gathering of various structural data about neurons with the aid of a computer, as illustrated by the number of papers which have appeared on the subject (3,4,5,6). Along with the attendant data gathering procedure, among the first programs usually developed is one allowing the researcher to view, usually on a CRT screen, a representation of a previously entered cell. Most of the systems described have the capability for a real-time three dimensional display with attendant controls for rotation about any of the three coordinate axes. Thus the introduction of the computer immediately gives the operator a useful analytical tool, the ability to enter the cellular structure and then view it from any desired vantage point. This alleviates the need of obtaining several sets of sections in different planes to examine the spatial distribution of cellular processes.

Although it has been observed that the capability to produce a three dimensional display is highly desirable, the researcher with a small computer may despair its implementation as beyond the capability of his machine.

This follows from the belief that while a small computer may be more than adequate for data gathering and analysis of neuronal structure, the time requirements of maintaining a real-time three dimensional display with such features as stereo pairs and dynamic rotation may seem to be beyond its capability. It has been illustrated (1), however, that if the small computer possesses at least a hardware multiply capability and 4k to 8k of core, then quite respectable stick figure representations of the cellular structure may be displayed.

In the following, the equations and techniques describing the data manipulations used successfully in a program providing the above mentioned display capabilities, shall be described. This particular system has been implemented on a DEC PDP-12 computer with 8k of core and the Extended Arithmetic Element option which provides, among other things, a PDP-8 hardware multiply capability. A description of this system may be found in the paper by Wann et al. (6) and a good description of various display techniques may be found in the paper by Barry et al. The only special hardware required is an interface to allow random point display upon a CRT.

ROTATION AND TRANSLATION IN THREE DIMENSIONS

The ability to dynamically rotate a structure in three dimensions tremendously enhances the usefulness of any display system. This feature not only allows other aspects of the structure to be brought into view, but also serves to dramatically enhance the three dimensional effect. Rotations, however, as general coordinate transformations, usually entail a series of multiplications, rivaling vector generation as the costliest (in terms of time) operation encountered in the generation of three dimensional displays. The implementation discussed here takes cognizance of this fact by reducing the number of multiplications needed for a true three dimensional rotation to a minimum at the expense of losing a fixed frame of reference about which the rotations are performed.

General rotation matrices can easily be defined for rotations about any one of the three coordinate axes (2). If a positive angle of rotation is defined in a manner consistent with a right hand coordinate system, then the following three rotation matrices can be obtained.

$$R_{x,\theta} = \begin{bmatrix} 1 & 0 & 0 \\ 0 & \cos\theta & \sin\theta \\ 0 & -\sin\theta & \cos\theta \end{bmatrix}$$: counter clockwise rotation through an angle θ about the positive x-axis

$$R_{y,\phi} = \begin{bmatrix} \cos\phi & 0 & -\sin\phi \\ 0 & 1 & 0 \\ \sin\phi & 0 & \cos\phi \end{bmatrix}$$: counter clockwise rotation through an angle ϕ about the positive y-axis

$$R_{z,\psi} = \begin{bmatrix} \cos\psi & \sin\psi & 0 \\ -\sin\psi & \cos\psi & 0 \\ 0 & 0 & 1 \end{bmatrix}$$: counter clockwise rotation through an angle ψ about the positive z-axis

A transformation producing the rotation of a point $p = \begin{bmatrix} x \\ y \\ z \end{bmatrix}$ through an angle ψ about the positive z axis would thus be represented as:

$$p' = \begin{bmatrix} x' \\ y' \\ z' \end{bmatrix} = \begin{bmatrix} \cos\psi & \sin\psi & 0 \\ -\sin\psi & \cos\psi & 0 \\ 0 & 0 & 1 \end{bmatrix} \begin{bmatrix} x \\ y \\ z \end{bmatrix} = \begin{bmatrix} x\cos\psi + y\sin\psi \\ -x\sin\psi + y\cos\psi \\ z \end{bmatrix}$$

One obvious way of obtaining a general rotation matrix for rotations in three dimensions is to perform successive rotations about each of the three coordinate axes. This is obtainable by simply multiplying the three two-dimensional rotation matrices obtained above together, yielding:

$$G_{\theta,\phi,\psi} = R_{x,\theta} R_{y,\phi} R_{z,\psi}$$

$$= \begin{bmatrix} \cos\phi\cos\psi & \cos\phi\sin\psi & -\sin\phi \\ \sin\theta\sin\phi\cos\psi - \cos\theta\sin\psi & \sin\theta\sin\phi\sin\psi + \cos\theta\cos\psi & \sin\theta\cos\phi \\ \cos\theta\sin\phi\cos\psi + \sin\theta\sin\psi & \cos\theta\sin\phi\sin\psi - \sin\theta\cos\psi & \cos\theta\cos\phi \end{bmatrix}$$

Performing a three-dimensional rotation in this manner has one major disadvantage: the net rotation is a function of the order in which the component two-dimensional rotations are performed. Examination of the manner in which the three-dimensional rotation is performed shows why this is true. The z axis about which rotation occurs will appear to be rotated through the succeeding y and x rotations, while similar reasoning holds for the axis of y rotation. It has been found that display systems using this method of rotation are still very useful, though obvious care must be exercised if one is attempting to rotate a structure to bring a specific aspect into view. Although a general rotation matrix can be constructed which produces a rotation of specified angle about a general axis of rotation, specifiable in three dimensions, it has not been used in this approach for two reasons. First of all this matrix would require two additional matrix multiplications, which would significantly decrease the number of points displayable without intolerable flicker, during a dynamic rotation. The second reason is an operational one, it being found that the lack of independence between the various rotations does not hinder the operator in bringing a desired aspect of the displayed structure into view. Thus there is no human engineering reason to go to a more complex implementation.

Occasionally the operator desires to have the structure rotated about a specified point other than the origin set at the time the neuron was entered. For this purpose it is useful to provide a feature whereby a coordinate translation to a given point in the structure may be specified point in the structure, p_0, and then translating the structure origin to this point and then performing the rotations. The equations describing this process are seen simply to be:

$$p^\star = \begin{bmatrix} x^\star \\ y^\star \\ z^\star \end{bmatrix} = G_{\theta,\phi,\psi}(p - p_0) = G_{\theta,\phi,\psi} \begin{bmatrix} x - x_0 \\ y - y_0 \\ z - z_0 \end{bmatrix}$$

In the system described here, operator control of the rotations is implemented via three potentiometers, and for operator feedback the angles of rotation are displayed on the CRT screen. Alternate ways of entering this data could be through the teletype console or through the switch registers available on most machines. Upon receipt of the numbers representing the three angles the corresponding values of the sine and cosine are obtained through a table lookup. This table is precomputed and core resident. It has been found that quantization of the sine into 256 values over the range 0 to 180 degrees is sufficient to provide continuous smooth rotations.

PERSPECTIVE TRANSFORMATION AND GENERATION OF STEREO PAIRS

In order to convince the operator that he is observing a three-dimensional object projected on a two-dimensional screen, the techniques of perspective transformation and stereo pair generation can be employed in much the same way that they are employed by the artist. The laws and principles of perspective were first clearly described by Leonardo da Vinci in his *Notebooks*, an outline of a course of study for the artist. Leonardo's description of geometrical perspective provides an exacting guideline for the mathematical description of this phenomenon.

> "Perspective is nothing else than the seeing of a plane behind a sheet of glass, smooth and quite transparent, on the surface of which all the things approach the point of the eye in pyramids, and these pyramids are intersected on the glass plane."

The procedure of stereo pair generation can be derived from similar considerations by taking into account the existence of the two viewpoints offered by the human binocular visual system.

From Leonardo's description a mathematical formulation of the effect of perspective shall now be derived. Consider the point (x_i, y_i, z_i) being projected onto a planar screen (e.g., a CRT screen) and being viewed monocularly at some point (a,b,c) from the opposite side of the screen. The observed position of the point upon the screen will be designated $(X_i, Y_i, 0)$. This configuration is illustrated in figure 1. From a purely geometrical argument using similar triangles we can establish the coordinates of the projected point as:

$$X_i = \frac{az_i - cx_i}{z_i - c}$$

$$Y_i = \frac{bz_i - cy_i}{z_i - c}$$

$$Z_i = 0$$

If the point of view is assigned to lie along the z axis (i.e., $a=b=0$), then the following simplified form describing the apparent position of the projected point may be obtained:

$$X_i = \frac{x_i}{1 - (z_i/c)}$$

$$Y_i = \frac{y_i}{1 - (z_i/c)}$$

$$Z_i = 0$$

Although multiplications are bad in terms of the cost in machine time, divisions are even worse and therefore the following approximation is made by using only the first term of the series expansion for $1/(1-x)$

$$X_i \approx x_i + x_i\left(\frac{z_i}{c}\right)$$

$$Y_i \approx y_i + y_i\left(\frac{z_i}{c}\right)$$

$$Z_i = 0$$

Although a mathematical formulation describing the generation of stereo pairs may be derived from purely geometrical considerations in the same manner as the formulation for perspective, a more intuitive derivation will be given here. The fact that the two eyes are separate and focus individually upon a point implies that the two images seen can be thought of as being related via small symmetric rotations about the y axis. The rotations are done with respect to the monocular line of sight established in figure 1. In order to simplify the result, and maintain consistency, the two viewpoints representing the eyes, separated by distance d, will be assumed to be oriented parallel to the x axis and in line with the object being viewed. This is illustrated in figure 2. Formulating the positions of the two projections as seen by the left and right eyes we have:

$$\begin{bmatrix} X_{iR/L} \\ Y_i \\ Z_i \end{bmatrix} = \begin{bmatrix} \cos \alpha & 0 & \mp\sin \alpha \\ 0 & 1 & 0 \\ 0 & 0 & 0 \end{bmatrix} \begin{bmatrix} x_i \\ y_i \\ z_i \end{bmatrix}$$

In the above the minus sign is associated with the right eye image, and the plus sign with the left image. If $d \ll c$ is assumed, then the small angle approximations hold and the formulation for the left and right stereo pairs reduces to the following:

$$\begin{bmatrix} X_{iR/L} \\ Y_i \\ Z_i \end{bmatrix} = \begin{bmatrix} 1 & 0 & \mp\alpha \\ 0 & 1 & 0 \\ 0 & 0 & 0 \end{bmatrix} \begin{bmatrix} x_i \\ y_i \\ z_i \end{bmatrix} = \begin{bmatrix} x_i \mp \alpha z_i \\ y_i \\ 0 \end{bmatrix}$$

The techniques useful in implementing perspective and stereo pair calculations may now be discussed. It has been found in practice that the multiplicative factor $1/c$ in the perspective approximation may be equated to an integral power of two, 2^{-p}. This allows the division operation to be implemented as an arithmetic shifting operation at a considerable saving of computation time. The same technique is seen to be useful in the generation of stereo pairs. The small rotation α can also be expressed as an integral power of two, 2^{-s}, again resulting in a saving of computation time. It should be noted that since the generation of stereo pairs results in a transformation which produces two points from one original point all other transformations (e.g., rotation, translations and perspective calculations) should be done

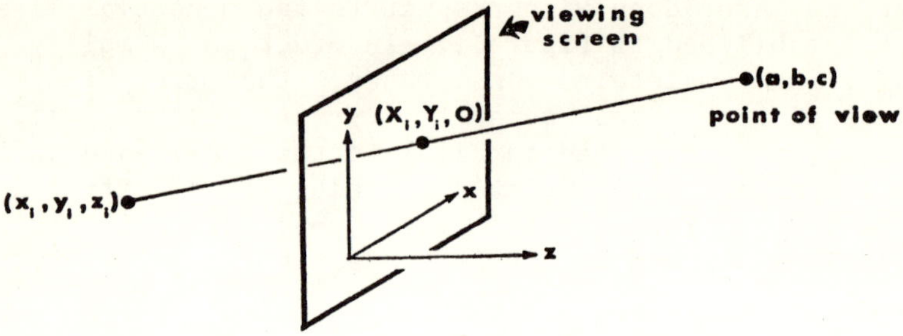

$$X_i = \left.\frac{az_i - cx_i}{z_i - c}\right|_{a=b=0} = \frac{x_i}{1-(z_i/c)}$$

$$Y_i = \left.\frac{bz_i - cy_i}{z_i - c}\right|_{a=b=0} = \frac{y_i}{1-(z_i/c)}$$

$$Z_i = 0$$

recalling that: $\dfrac{1}{1-x} = \sum_{j=0}^{\infty} (x)^j$

$$X_i = x_i + \frac{x_i}{c} z_i$$

$$Y_i = y_i + \frac{y_i}{c} z_i$$

$$\dot{Z}_i = 0$$

Figure 1. Derivation of perspective computation.

first. Finally, the addition of the capability to provide variable separation between the generated stereo images is necessary if the operator is to fuse the generated images together in his mind.

VECTOR GENERATION

As was mentioned earlier, programmed digital vector generation can easily rival the matrix multiplications required for rotations in terms of the time consumed. The primary cause for this is that some form of extra precision is required for a 12 or 16-bit machine to generate an arbitrary vector on a 10-bit by 10-bit addressable CRT screen if the vector is to appear as a continuous line. This can take the form of binary rate multipliers or simply the use of double precision. As an illustration of this point let us assume that one wishes to generate a vector between the two points (0,0) and (8,256) with a discrete point density such that the vector appears continuous. If we assume that $256=2^8$ points will be sufficient to satisfy this requirement, then it is easily seen that the incremental change needed along the x axis is only 1 step per display point. However, along the axis an increment of 2^{-5} steps per displayed point is required. To put it another way, five extra bits of precision are needed in addition to the 10 bits required for the display address. Using the same argument we can see that for a 10-bit by 10-bit addressable display a maximum of 10 bits additional accuracy could be needed over the 10 bits allocated for the coordinate address, thus forcing the use of some extra precision method if displays of high quality are to

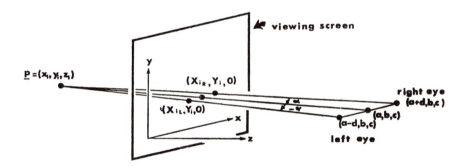

Expressing the two viewpoints as symmetric rotations through an angle α about the y-axis.

$$\underline{P}_{R,L} = \underline{R}_{y,\pm\alpha}\underline{P}$$

Taking the projection onto the viewing screen and using the small angle approximations we have

$$\begin{bmatrix} X_{iR,L} \\ Y_i \\ 0 \end{bmatrix} = \begin{bmatrix} 1 & 0 & \mp\alpha \\ 0 & 1 & 0 \\ 0 & 0 & 0 \end{bmatrix} = \begin{bmatrix} x_i \mp \alpha z_i \\ y_i \\ z_i \end{bmatrix}$$

Figure 2. Derivation of stereo pair computation.

be obtained (e.g., for photographic purposes). If however some restriction can be placed upon the maximum distance between terminal points then the necessity for extra precision may be removed, significantly increasing the vector generating speed.

Even though some form of double precision may be required, it is still possible to speed up the calculation of the needed increments in x and y, and to combine in this calculation those needed for the generation of stereo pairs. The first obvious simplification is indicated from the approximations used above for perspective and stereo pair generation, i.e., to let the number of points displayed between terminal points be of the form 2^n, replacing the division operation with an arithmetic shifting operation. The form of the increment thus becomes:

Increment = (terminal point - initial point) /2

The second result is obtained through a re-examination of the equations for stereo pair generation and those for the increment calculation in vector generation. Noting that the introduction of the stereo pair calculation only introduces a change in the value of the coordinate, i.e.,

$$X_{iR/L} = x_i \mp z_i 2^{-s}$$

it is seen that the calculation for the increment along the x axis is the only incremental calculation affected. The effect produced is seen as

$$\text{incremental } X_{iR/L} = (x_{it} \mp z_{it} 2^{-s} - x_{ii} \mp z_{ii} 2^{-s}) 2^{-n}$$

$$= (x_{it} - x_{ii}) 2^{-n} \mp (z_{it} - z_{ii}) 2^{-(n+s)}$$

Thus it is seen that the generation of stereo pairs is obtainable at a cost little more than that incurred in generating a single image, only the addition of a second incremental register containing the accumulating value of x for the second image.

CONCLUSION

The techniques discussed, while relatively simplistic in nature when implemented provide a valuable tool in the preliminary analysis of neuronal structure. The ability to display and rotate neural cells, in real time, helps provide the operator with valuable insight into the material under study, insight not provided by computer printout of various cell parameters. Although a system implemented in software along the lines in this chapter could not compete with the many hardware display systems currently in use, it does provide a relatively cheap and effective way to experience the benefits which a good display system may offer.

DISCUSSION

BROWN: What histological procedures have you found to minimize the types of distortion that are hard to compensate for, like stretching and wrinkling?

DIERKER: I've been working on the engineering end, although I've started getting into the histology; we haven't investigated that to any extent because these were Golgi-stained preparations mounted in celloidin, very stiff things. Unfortunately we have found some small problems. But the procedure has actually worked quite well. You can sit there and do multiple sections with no great difficulty. Its limited in that you're talking about stick-figure representations of the cell--you're not recording synapses. However, our primary interest is in structural analysis of the cells, for which this system is more than sufficient.

MACAGNO: How do you go from one slide to another?

DIERKER: Just a variation of the same process. The machine just needs the coordinates of the crude points of re-registration, whether you're going from one section to another on the same slide or on different slides.

MACAGNO: I see, you don't go back and forth between two sections.

DIERKER: Not generally. You might have to follow a sinuous process as it exits and re-enters a section, but we've never had to. We've thought about letting the operator track more than one cell at a time, incidentally.

CAPOWSKI: How do you keep the mapping of one section with respect to the next? Do you use a matrix?

DIERKER: Yes. The data that's really stored is the set of angles of rotation and translation that need to be used for alignment.

CAPOWSKI: How many sections per neuron do you usually have?

DIERKER: Two to four.

CAPOWSKI: Have you noticed cumulative errors developing over a series of sections?

DIERKER: We worried about that originally, but we don't see any signs of it on displaying the final reconstruction. We need to analyze the process further to see if that's really true or not.

BROWN: I should think that if you're just looking at the geometry of one neuron, it's not too important whether it's a little distorted, and if you're looking at connections between neurons, those relations will be preserved anyway.

DIERKER: Yes, but it is a reasonable question, particularly for cells that undulate between sections. We're working on it.

CLARKE: What's your software environment?

DIERKER: Really a rather crude system. We operate under a machine language operating system for the PDP-12 called DIAL, which we have modified to support 32k of core and an interactive language, FOCAL. We're currently evaluating a change-over to a FORTRAN IV system supported on the machine for greater exportability of software.

CAPOWSKI: What is the worst case you've seen of shrinkage of one section with respect to another?

DIERKER: We haven't seen any discrepancy between sections. They're all prepared together, and seem to be affected uniformly by processing.

CAPOWSKI: So you're happy with just six parameters.

DIERKER: Yes.

DUFFY: Were the rotations of the display which you showed in your movie done in real time or did you use time-lapse photography?

DIERKER: That wasn't done in software. It was done at the Computer Systems Lab, which has a hardware matrix multiplier and hardware vector generator. It is possible, though, to generate displays with dotted vectors for cells of this degree of complexity and refresh and compute a rotating display in real time.

SHANTZ: What types of structural descriptors are you interested in?

DIERKER: We're trying to look at angle between processes, average dendritic length, average dendritic segment length, spatial distributions of cells in polar coordinates; we're trying to develop programs to produce cluster analyses of morphological descriptors. We may be able to find correlations with things like cell area or soma geometry which will allow characterization of cells incompletely filled by intracellularly injected dyes. Maybe we can even use these soyts of correlations to give confidence levels in our classifications of cell types. The original characterizations could be done with Golgi-stains and then these data could be applied to dye-injected cells which had been characterized electrophysiologically.

REFERENCES

1. Barry, C. D., Ellis, R. A., Graesser, S. M., and Marshall, C. R. Display and Manipulation in Three Dimensions. In *Pertinent Concepts in Computer Graphics*. ed. M. Faiman and J. Nievergelt. Chicago, Ill.: University of Illinois Press (1969).
2. Korn, G. A. and Korn, T. M. *Mathematical Handbook for Scientists and Engineers*. McGraw-Hill (1968).
3. Glaser, E. M. and Van der Loos, H. A semi-automatic computer-microscope for the analysis of neuronal morphology. I.E.E.E. Trans. on Biomed. Engng. BME-12:22-31 (1965).
4. Levinthal, C. and Ware, R. Three-dimensional reconstruction from serial sections. Nature 236:207-210, 1972).
5. Reddy, D. R., Davis, W. J., Ohlander, R.B. and Bihay, D.J. In: *Intracellular Staining in Neurobiology*, ed. S. D. Kater and C. Nicholson. Springer Verlag, New York (1973).
6. Wann, D. F., Woolsey, T. A., Dierker, M. L. and Cowan, W. M. An on-line computer system for the semiautomatic analysis of Golgi-impregnated neurons, I.E.E.E. Trans. on Biomed. Engng. BME-20:233-247 (1973).

chapter 12

APPLICATIONS OF PLATO IN COMPUTER ASSISTED RESEARCH

D. WALTER AND R. McKOWN

Department of Physiology and Biophysics
University of Illinois at Urbana-Champaign
Urbana, Illinois

INTRODUCTION

In the past two years, we have been developing a graphics terminal system to provide interactive computer services at the University of Illinois, Urbana-Champaign (UIUC). This laboratory is investigating electrical behavior of a wide variety of excitable tissues using voltage clamp techniques. The host system providing computer support for this laboratory graphics terminal is the PLATO IV system which is currently under development at the UIUC Computer-based Education Research Laboratory (CERL). Our laboratory terminal system provides facilities for data acquisition, numerical analysis, and interactive modeling. Our data acquisition involves the recording of many series of simultaneous voltage and current waveforms from excitable membranes of biological

---

We would like to recognize Susan Dragich for her help in the documentation and construction of the interface equipment. We would also like to thank Dr. R. L. Johnson and Fred Ebeling for engineering consultations. Finally, we wish to note the encouragement we received from Drs. J. A. Connor, E. Jakobsson, and L. Barr at the Department of Physiology and Biophysics, UIUC.
    This work was supported by NSF GB-39 946 and PHS 5T1 GM 00 720 15.

preparations with limited experimental lifetimes. The analysis consists of the extraction of digital information from these analog waveforms with the usual need for further calculations to obtain descriptive parameters. Typically, our modeling activity is the integration of a differential equation model to simulate membrane electrical activity. We have found four basic impediments in our work which can be reduced through computer assisted research. They are:

1. Acquisition and digitization of large amounts of data for an accurate description of the experimental system.
2. The restricted lifetime of many biological preparations under the stress of experimental measurements.
3. The need for stimulus reproducibility in comparison studies (e.g., testing drug and temperature effects).
4. The need for interactive modeling facilities including visual displays and hard copies for permanent records.

We feel these problems are representative of those found in many areas of research, and we hope our solutions will be useful to others.

In this chapter we will briefly describe the PLATO IV system and its advantages as an interactive timesharing system, trace the development of our laboratory terminal system, and describe the hardware and software involved. Finally, we will discuss some of the problems still remaining and a solution based on an intelligent terminal system which is also being developed at UIUC (1).

OVERVIEW OF THE PLATO IV SYSTEM

The PLATO system has been under development since 1959 and is now in its fourth generation. A basic schematic of the system is shown in figure 1, which is used with permission of the author (2). Individual terminals are grouped in "sites" of 32 voice grade telephone lines which communicate with the system through a common "site controller". Each terminal receives 1260 bps from the system. The keypresses at all 32 terminals

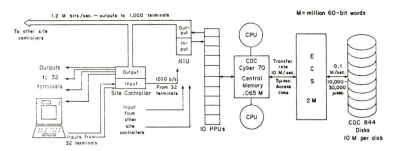

Figure 1. A schematic diagram of the PLATO IV system. Over 800 terminals are now in operation.

of a site are multiplexed into one 1260 bps line back to the network interface unit. There are approximately 860 terminals in the PLATO system; the largest concentration is at UIUC with more than 450 at 69 other educational campuses. Currently, there are approximately 3500 programs (termed lessons) available on the PLATO system. Approximately 105,000 terminal hours were logged on the system during the month of March, 1975.

The programming language used on this system is TUTOR. TUTOR was developed for the requirements of computer assisted instruction and has extensive answer judging and display features. Furthermore, TUTOR was intended to be used by teachers who were not expected to be extensively trained as computer programmers. System operation is designed to give each student the impression of individual attention by the computer. This is accomplished in part by having an average response time of 125 milliseconds for a keypress. These features also aid in the writing and use of interactive programs for computer assisted research.

EARLY DEVELOPMENTS

The initial work with PLATO for electrophysiology was a lesson providing simulations of voltage or current clamp experiments on a squid axon. The response of the simulated squid axon was calculated from the Hodgkin-Huxley equations (3). This lesson provides a simulated laboratory for a graduate course in cell physiology. A more complete description of this nerve simulation lesson may be found in (4). Since its conception, the lesson has undergone continuing modifications and is now used for simulations based on experimental models of nerve behavior developed in the laboratory (5).

Soon after this nerve modeling lesson was operational, other lessons were begun, this time with the express intention of providing research tools which would use the advantages of the PLATO system. These lessons initially consisted of a group of lessons for data acquisition, curve fitting, and a general differential equation simulator. Data could be entered to permanent storage either via the keyboard or by using a computer generated axis and cursor system superimposed on projected 35 mm oscilloscope photographs. The filmstrip projector was mounted in place of the standard microfiche projector in the terminal top. Limitations at this time were the amount of permanent storage available and the considerable time required to digitize and manipulate data from manually specified experiments.

The curve fitting lesson used an algorithm based on that described by Marquardt (6). An expression with up to five parameters can be fit to data which has either been entered into a temporary memory or recalled from a permanent memory. Temporary data may be loaded into the permanent storage area at the user's option. Other user options include: performing algebraic manipulations with blocks of data; plotting single or multiple data blocks on appropriate graphs; using linear, semi-log, or log-log scales; labeling displays for hard copies. This curve fitting lesson has been very useful and emphasizes the utility of interactive computer services with good graphics displays.

The differential equation modeling lesson accepts up to six simultaneous differential equations from the user and numerically integrates them by a modified Euler predictor-corrector method. It has been used for both research and teaching applications. Additionally, a routine for calculating convolution integrals is available but is seldom used because of the limitations of a time sharing system.

THE PRESENT LABORATORY

Our next step toward an automated laboratory was the design and construction of interface hardware to allow

APPLICATIONS OF PLATO IN COMPUTER ASSISTED RESEARCH   157

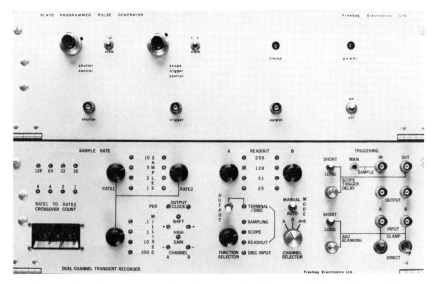

Figure 2. Front panel of two devices which interface a PLATO student terminal to an electrophysiology laboratory.

control of experiment execution and data acquisition through the PLATO terminal. This equipment, the Programmed Stimulator and the Transient Digitizer, is now in operation (see figure 2). A second laboratory is being equipped with duplicate equipment. Interactive computer control of experiment execution and of data digitization is accomplished by execution of our data acquisition program, a standard PLATO lesson. Currently, due to data transmission limitations, analog data is stored on magnetic tape during an experiment and replayed through the Transient Digitizer at a later time. The next two sections will describe the design of the Programmed Stimulator and the Transient Digitizer, leaving further discussion of the operation of our present laboratory interface for a later section.

DESIGN AND OPERATION OF THE PROGRAMMED STIMULATOR

The Programmed Stimulator is a digital-to-analog interface device (see figure 3A) which receives 16 digital control lines from an auxiliary output of a Plato

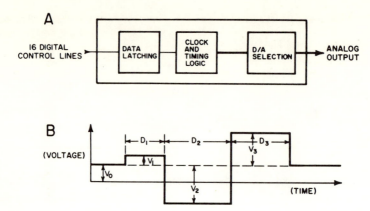

Figure 3. A. A block diagram of the Programmed Stimulator. The analog output is used as the command potential for voltage clamp circuitry.

B. The output waveform of the Programmed Stimulator. Voltage amplitudes $V_0 \ldots V_3$ have 8-bit resolution and pulse durations $D_0 \ldots D_3$ have 7-bit resolution for each time base. The time base has a selected range of 4 decades (10 Hz to 10 kHz).

IV terminal and provides a train of analog signals to control laboratory equipment used in our experiments. The Programmed Stimulator's principal output is an analog signal which becomes the command potential for conventional voltage clamp circuitry. This programmed multistep waveform is of the type illustrated in figure 3B and is constructed from information sent as 5 15-bit words to the Programmed Stimulator via the 16 digital control lines. Upon receipt of the last word, the Programmed Stimulator puts out a trigger pulse and immediately thereafter provides the programmed waveform. The waveform is constructed by sequentially gating D/As for the appropriate durations as measured with a local clock and presettable counters.

Since the Programmed Stimulator's Trigger Output provides a pulse of variable duration that precedes the command potential, it is used to trigger an oscilloscope monitoring the excitable membrane's response. The trigger output is also recorded on FM tape with the analog data signals to provide a trigger pulse for future digitization.

The Shutter Output of the Programmed Stimulator provides a pulse, the duration of which can also be set from the front panels. This pulse is used to close a relay which operates an oscilloscope camera placed in front of the monitoring oscilloscope. It was found convenient to provide this camera relay with a remote arming switch so that the researcher could selectively enable and disable the oscilloscope camera during computer controlled data acquisition.

Since multistep waveforms of the type the Programmed Stimulator provides were previously obtained in our laboratory by the manual operation of the three pulse generators and one ramp generator, we can directly compare the automated and manual methods of stimulation. The Programmed Stimulator is a big improvement in both experiment planning and experiment execution. Under computer control, a table of experiments (each containing a series of families of such wave forms) is displayed on the PLATO IV terminal's screen and can be easily modified or expanded as the researcher decides. In this manner, the experiment can be designed at the scientist's convenience -- not under the pressure that accompanies an experiment on a preparation of short lifetime. Since the execution of the experiment is now under computer control, a protocol is something that will be followed rather than something that should be followed. The hazards of knob-turning are avoided, thus allowing the researcher to spend time on important matters that otherwise would be neglected (for instance, maintenance of the preparation or monitoring the incoming data). Furthermore, computer controlled experiment execution assures voltage clamp reproducibility and hence assists the separation of drug or temperature dependence of the excitable response from its voltage dependence. Perhaps the most obvious improvement that the Programmed Stimulator provides is the maximization of the rate of taking data -- the rate being limited by the recovery time required for the biological preparation rather than by the experimenter's dexterity. Finally, our Programmed Stimulator allows considerable improvement of the realized precision of the stimulus waveform in both time and intensity since each step of the command potential waveform in figure 3B has an eight bit resolution of intensity and a seven bit resolution of duration with one of four decades of clock rate specified.

DESIGN AND OPERATION OF THE TRANSIENT DIGITIZER

Each channel of the Transient Digitizer contains an A/D converter, an averager, a shift register memory, and a D/A converter in the configuration shown in figure 4. A specialized input section with automatic gain control circuitry was also provided to enable the .4% full scale accuracy of the eight bit digital outputs often to be approached. When the Transient Digitizer is in sample mode and receives an external trigger signal, the control section provides appropriately timed pulses to the A/D converter's "take data" input. Within five microseconds after each "take data" pulse, the signal present at the analog input has been digitized. The A/D converter's output bits then accumulate in the averager. This automatic averaging circuitry averages 2, 4, 8, or 16 successive digitizations in order to take advantage of the 200 kHz realized digitizing rate of the selected A/D converter module (Datel model ADC-E HC-1). Thus on slow sampling rates, each of the 256 8-bit "samples" which the averager sends to the shift register memory is actually the average of 16 digitizations performed within an 80 microsecond window. The Transient Digitizer's stored sampling rate can be selected from 12 settings spanning a range from 10 Hz to 100 kHz. In addition, the control section was designed to permit the switching from a first to a second sampling rate at a specified crossover sample

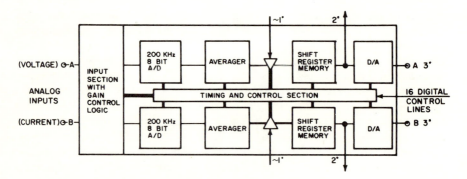

Figure 4. Block diagram of the Transient Digitizer. 1*. 8-bit digital input. 2*. 8-bit digital data output. 3*. Reconstructed analog outputs for scope or pen writer. The Transient Digitizer may be operated in a manual or computer controlled mode.

number. This permits a transient containing two different time constants to be optimally recorded into the 256 sample memory.

When 256 samples have been stored, the control section quits sending "take data" commands to the A/D converters and the 256 8-bit binary numbers representing the sampled transient now reside in the shift register memory. The Transient Digitizer's terminal readout mode will then allow the contents of the memory to be sent to the PLATO IV system via the eight data lines connecting the output of the memory to the auxiliary inputs on the PLATO IV terminal. To minimize the problems created by the slow transfer rate of data onto the present PLATO IV system, we provided a readout feature that allows either all, 1/2, 1/5, or 1/10 of the data in the memory to be sent to PLATO.

It is also possible to put the Transient Digitizer into scope mode, in which the control section rapidly recirculates the shift register memory past the D/A converters. If an oscilloscope is triggered with the Transient Digitizer Trigger Output while displaying the output of the D/A converter, a stable reconstructed analog image of the recorded transient is obtained. Hence the scope mode of the Transient Digitizer effectively converts a standard oscilloscope into a storage oscilloscope.

The Transient Digitizer solves another standard problem, obtaining a quick and inexpensive hard-copy of a transient that is too fast for direct reproduction with mechanical pen writing equipment. This is done by first digitizing and storing the transient in the shift register memory. Then, in readout mode, the memory is slowly cycled while the output of the D/A converter is recorded on a y-t pen writer (an inexpensive slow speed pen writer is then sufficient).

A direct digital input into the shift register memory was provided to allow the Transient Digitizer to act as an interface/buffer between digital systems. For example, this could allow a fast local memory (e.g., a floppy disk cartridge memory) to transfer data through the Transient Digitizer into the PLATO IV system which must acquire the data at a much slower rate than the disk cartridge memory can send it.

Finally, a very important feature of the Transient Digitizer is that it is fully programmable. Thus in "auto", the instrument obtains its control settings in 3

15-bit words which are received over the 16 digital control lines connecting the Transient Digitizer to the auxiliary output of the PLATO terminal. In this manner, the uses of the Transient Digitizer discussed above can be achieved under either manual control from controls on its front panel or computer control from our data acquisition program.

PRESENT OPERATIONS

The present laboratory system is diagrammed in figure 5. Data acquisition software to design experiments, to execute them using the programmed stimulator, to digitize and store the results using the transient digitizer, and to display the stored data is now operational. The data acquisition procedure uses the Programmed Stimulator for automated stimulation and an FM tape recorder to store the analog transient signals. Digitization of these stored signals and transfer of the data to the PLATO system is performed with the Transient Digitizer at a later time. During experiment execution the Transient Digitizer may be used to monitor the tape recorder outputs, switching from sample mode to scope mode under software control. This operation allows stable viewing of the incoming transient signals. Then, when the tape is replayed into the Transient Digitizer, scope mode can again be used after a transient has been digitized to display the contents of the memory, making it possible to recognize false trigger signals and discard spurious data before transmitting the memory to

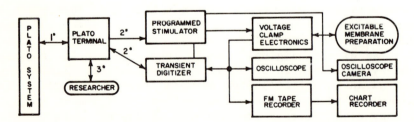

Figure 5. Schematic of present PLATO-interfaced electrophysiology laboratory. 1*. Telephone line. 2*. Digital lines. 3*. Researcher interacts via terminal keyboard and display screen. All other lines indicate coaxial cables.

PLATO. Our present storage allocation is approximately 26,700 x 60-bit words. This can hold records of transients from over 5,000 voltage clamps with 25 points per record or 500 with 256 points per record. A typical screen display of stored data is shown in figure 6. The tracing in the lower left represents current patterns from a series of voltage clamps. Various procedure options are shown to the right. Descriptive information for the last two plotted curves is shown in the upper screen area.

LIMITATIONS

The major limitation of the present laboratory system is the bandwidth of the telephone line connection between our terminal and the central computer. Since data can be generated more rapidly than it can be sent back to the PLATO system, a local memory is needed in the laboratory. The analog tape performs this function at this time but it will be replaced with a "floppy disk" digital storage system. This device will allow us to store the data digitized during an experiment and will alleviate the need to replay the analog tape for digitization. The Transient Digitizer was designed with auxiliary digital inputs and outputs to facilitate the addition of local digital memory.

The effective data transfer rate of the pathway returning to PLATO (see figure 1) is even lower than the 1260-bits per second transferred on a phone line since the site controller multiplexes 32 terminals into each line back to PLATO. The data we send to PLATO is seen as keypresses which are then interpreted in our data acquisition software. During daytime periods of maximal system use, the data transfer rate may fall to 5 data points per second. This may improve to 20 data points per second during nighttime periods of lower system use. Because of the limited data transfer rate, the most effective use of the capabilities of our laboratory system will not be attained until local digital memory allows real-time storage of data.

THE INTELLIGENT TERMINAL

We believe our next step towards improved computer assisted research will be to incorporate an intelligent

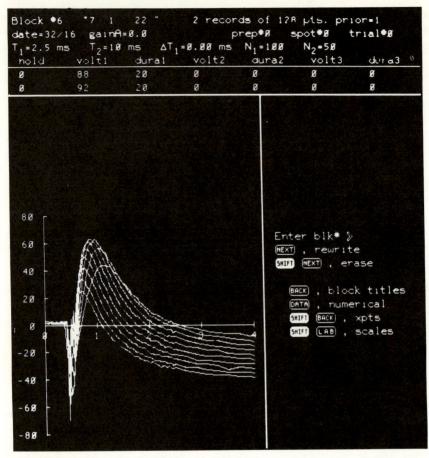

Figure 6. A typical screen display of stored transients. The researcher has many display and analysis options, accessible from the keyboard.

PLATO terminal which is being developed at CERL and the Coordinated Science Laboratory, UIUC, under the direction of R.L. Johnson (1). The intelligent terminal uses a local processor to control and enhance communication with a host system. The local processor used in the PLATO intelligent terminal prototypes is a PDP-11/10. The immediate gain from the intelligent terminal would be the ability to do simple manipulations of stored data locally, only sending a reduced amount of data, we can alleviate the low data rate problem. Presently, assembler language programs for the PDP-11 may be edited and stored in a PLATO lesson. Since a translation program is in preparation to convert from TUTOR into the machine language of the PDP-11, local processing could be achieved without giving up the advantages of programming in TUTOR. Our existing laboratory software would be retained and executed in the normal manner. Additional software written specifically for the intelligent terminal should mesh with the existing laboratory system.

It is interesting to compare the intelligent terminal approach to computer assisted research with other hardware configurations. It is true that the above objectives can be approached by conventional hardware configurations. However, a full-scale computer costs too much for most installations. Furthermore, a terminal linked to a large information system by a telephone line is too slow for on-line research applications requiring high data rates. The most popular approach to this problem has been to use a minicomputer with added peripheral equipment for acquisition, storage, and display. Unfortunately, as languages become more flexible they use more storage space, which leaves less program space. Therefore, there has been a tendency for these installations to require excessive amounts of programming time and to keep increasing in size. As a result of these drawbacks, many minicomputer installations have been restricted to a narrow group of functions.

The intelligent PLATO terminal can have all the advantages of the stand-alone minicomputer without many of the drawbacks. Additionally, an intelligent terminal has the host-system resources to draw from including the computational features, large word size, and a sophisticated program editor. The extensive PLATO network offers inter-terminal communication opportunities not present in any of the other approaches. The prototype intelligent PLATO terminal has been used to mimic an ASCII teletype when communicating with a Burroughs 6700 and a DEC 10 computer (1). Thus, using an intelligent terminal, a researcher could attach to a highly interactive system for on-line activities, or attach to a batch processing system for lengthy computations. Additionally, data could be transferred from one host system through the intelligent terminal to a second host system.

In conclusion, we have developed a laboratory graphics terminal system which uses the PLATO IV system to provide computer assistance in three areas of research; data acquisition, data analysis, and system modeling. The present laboratory system, based on a standard PLATO terminal, would be adequate in research areas not requiring high data rates. We plan to reduce the transfer rate limitation by using an intelligent terminal to provide local processing while retaining the advantages of connection to a large, interactive computer information system.

DISCUSSION

BROWN: What are the prospects for direct hard copy off the plasma screen?

WALTER: There is a prototype which we use quite frequently. It's a wet copy process.

BROWN: For general information, would you describe the basis of operation of the plasma display?

WALTER: The plasma panel is an array of 512 vertical wires and 512 horizontal wires with neon gas in between. They're held at a sustaining voltage; a display is initiated with a pulse, and erased with a pulse of opposite polarity. This gives most of the advantages of a storage or refreshed display on the same screen, since you can selectively write and erase individual stored points.

RANSIL: What does this cost?

WALTER: Terminals are about 5300 dollars. Magnavox is now marketing them, but I'm not sure of their price. Connections can be via microwave or telephone line.

KEHL: It sounds like you're using TUTOR.

WALTER: Yes.

KEHL: That's a computer-aided instruction language, isn't it?

WALTER: Right. So far it's had everything we need. It doesn't do matrix operations, but the system programmers are working on it.

MOISE: What's the cost per connect hour?

WALTER: You're always connected, for some $2800 a year.

HARTLINE: This all sounds about as expensive as a dedicated minicomputer when you add all the peripherals.

WALTER: With the minicomputer-based terminal we envision, you could have the advantages of both a dedicated local processor and hookup to a larger, more powerful system.

MOISE: Do you have trouble with transmission errors?

WALTER: We don't, but some phone lines seem to be noisier than others.

KEHL: You don't do software checking?

WALTER: We do. We run in a hand-shaking mode.

JOHNSON: The terminal does that for you and requests re-transmission whenever there's an error. It tries about five times before giving up.

MCINTYRE: Does everyone at Champaigne-Urbana go through the concentrators?

WALTER: Yes.

REFERENCES

1. Stone, M., Bloemer, R., Fertich, R., and Johnson, R. L. An intelligent graphics terminal with multi-host system compatibility. Proceedings of the IEEE COMPCON 74, Washington, D.C.
2. Wood, Nancy. The PLATO IV Computer-based Education System. Available from PLATO publications, Computer-based Education Research Laboratory, University of Illinois, Urbana, IL.
3. Hodgkin, A. L. and Huxley, A. F. Quantitative description of membrane current and its application to conduction and excitation in nerves. J. Physiol. 116:497-506 (1952).
4. McKown, R. J. and Barr, L. Simulation of excitable membrane experiments. The Physiologist 16:658-668.
5. Connor, J. A., Walter, D., and McKown, R. Neural repetitive firing: modifications of the Hodgkin-Huxley axon suggested by experimental results from crustacean axons. Submitted to J. Neurophysiology.
6. Marquardt, D. W. An algorithm for least-squares estimation of non-linear parameters. J. Soc. Ind. Appl. Math. 11:431-441 (1963).

chapter 13

REAL TIME CLASSIFICATION AND DISPLAY OF POSTSYNAPTIC POTENTIAL AMPLITUDES AND INTERVALS

T. O. CLARKE AND J. H. PEACOCK

Department of Neurology
Stanford University Medical Center
Stanford, California

Intracellular microelectrode recordings from mouse nerve and muscle cell cultures often disclose abundant synaptic activity (1). hand processing of this synaptic data from photographic records or magnetic tape limits the number of cells which can be fully analyzed, must always be carried out after the experiment, and is restrictive in terms of experimental design. In the course of electrophysiologic study of these cultured cells, it has become clear that real time processing of synaptic data before the microelectrode is withdrawn reveals patterns of synaptic interaction which might not otherwise be detected. Certainly many potentially interesting cells would never be analyzed because of time restrictions. Furthermore, experimental advantage may be taken of changes in synaptic relationships as culture conditions are systematically varied by pharmacologic or other agents.

The development of the current program derives from a package by Nelson and Sheriff (2) for collection and analysis of synaptic data. We have restricted our program to excitatory post synaptic postentials (EPSPs)

This work was sponsored in part by NSF GB43526 to JHP and by the KROC foundation.

recorded from cultured muscle cells and extended the real time interactive features. The program provides real time data acquisition of EPSPs from a microelectrode and classifies events by amplitude and interevent interval. Several interactive displays of both input wave form and classified data allow assessment of data quality, and its changes resulting from external perturbation of culture conditions, and are used as an aid in selection of cells for further, more intensive, investigation.

The numerical data are stored in an OS/12 file for subsequent offline data analysis. Program control and option selections are entered from the console device (DECwriter).

PROGRAM OPERATION

The microelectrode amplifier has continuously variable gain, which is often changed in the course of the experiment. Prior to each data collection, a calibration routine is executed and a pulse of known amplitude is inserted in series with the ground return line of the microelectrode channel. The amplitude of this pulse is saved for gain calibration reference. Prior to data storage, the input wave form is displayed on the PDP-12 screen with markers and alphanumeric values for trigger threshold and gain calibration values. At this time, a routine is executed to set the trigger threshold to a level appropriate for the type of data, gain, and noise level.

A data collection sequence is initiated by command from the console. Data collection and storage are set as the foreground tasks and display options as the background tasks. Any event that exceeds the trigger threshold records interevent time at threshold crossing; The input is then scanned for the maximum value of the input voltage until it is below the initial threshold minus a hysteresis value. During the peak detection, the input is sampled at 40 microsecond intervals. This peak value is stored in the data output buffer and the associated amplitude histogram bin and interevent time histogram bin are incremented. Interevent time is saved as an integer value of milliseconds.

The program provides for a foreground display and selection of one of the background displays from the

POTENTIAL AMPLITUDES AND INTERVALS

console. The foreground display is part of the peak detection algorithm. Only the values exceeding the trigger and used in peak detection are displayed. Thus all the baseline activity is removed and only the data waveforms which generate amplitude and time data values are displayed. This is useful as an aid in evaluation of trigger threshold setting compared with the amount of noise present. Each triggering event is aligned at the left margin of the screen so that the waveforms of different classes of post synaptic events may be easily compared.

The first background display is a dual histogram of event peak amplitude and interevent time. The histograms are updated in real time as the data peaks are detected. The foreground display of input data waveform, mentioned above, may be selectively superimposed on the histogram display to allow association of waveform class with peaks in the histograms.

The second background display during data collection is a plot of event peak amplitude vs. interevent time. When there is a diverse population of events, this display is useful in distinguishing different classes of events by regions of clustering of their peak amplitude interevent time.

Any of the background displays may be selected throughout the data collection period, and the foreground waveform display may be superimposed on either background display. Throughout data collection, the peak amplitude and interevent time for each event are saved in a mass storage file for subsequent, more detailed analysis.

Data collection is terminated by a terminal command and the histogram data also are then saved on the output file.

PROGRAM STRUCTURE

The program is written in assembly language and runs under the OS/12 operating system. It consists of two parts: general real time support (SKEL), and task specific routines (MINI). SKEL is a general purpose real time support program, written in assembly language (PAL-8), that provides real time support of all the standard PDP-12 I/O options:

Real time clock (KW-12)
Analog to digital converters
Oscilloscope display (VC12, VR12)
LINCtapes

Console device commands control execution of both foreground and background tasks. SKEL is designed to operate under OS/12 and thus uses the OS/12 file structure and command decoder, but it substitutes its own real time device handlers for the OS/12 handlers, which are not real time handlers.
SKEL is written to provide general purpose real time support with emphasis on support of high speed data acquisition.

In addition to this program for EPSP data collection, task routines for SKEL have been implemented for a peri-event averaged evoked response and unit response histogram, and a cardiac ultrasonic imaging data collection program (3). Task driving routines for PSP data are written in assembly language (PAL-12) (4) and at assembly time are inserted as tasks into SKEL.

These application-specific routines are small compared with the size of the general support package. Thus it will be possible to add several other task routines, and we are developing routines to handle inhibitory PSPS as well as EPSPS, dual microelectrodes (both presynaptic and post synaptic,) and analysis of burst activity.

PROSPECT

We also want to use statistical analysis in the study of neurotransmitter release (5). The results of such computations would be used to determine stimulation and response parameters and to control the processing sequence of real time tasks. Also, a natural step as data collection progresses in automation, is to make task parameters or task selection contingent on processed results from previous operations. However, implementation of these features requires broad mathematical support (floating point operations, exponentiation, logarithms, etc.), and, for versatility, a syntactical command structure. SKEL supports foreground and background tasks, has an elementary command structure, and limited arithmetic, but does not support higher precision arithmetic (floating point, or greater than 12 bit integer), other than addition. Thus, in its current form, SKEL cannot perform the above operations. Development of an interactive, mathematically powerful, fast, real time data acquisition system involves a substantial dedication of resources (time, hardware, software, personnel). However, for PDP-8 and PDP-12 users, several user modifiable supersets of FOCAL (6) (e.g., U/W FOCAL (7)) are available which provide a reasonable method of achieving this objective at a substantial savings of effort. User task specific subroutines are written in assembly language and inserted into the local FOCAL version. If the general purpose real time interupt service, such as provided by SKEL, were combined with a user modified FOCAL, then the interactive language and computational power of FOCAL would be available to perform statistical operations and pass parameters to the various high speed real time subroutines.

These subroutines can be designed so that when high speed data acquisition service is required, the slow interpretive part of FOCAL is turned off, and the high speed processes proceed independently of FOCAL (although the FOCAL internal mathematical routines could still be available for use by the high speed process.) upon completion, or between bursts, of the high speed process, the FOCAL interpreter is re-enabled for data pre-processing, computation of parameters to pass to the high speed subroutines, and process control. Thus the interactive language and computational power of FOCAL would be available for use in conjunction with the real time tasks that have more stringent high speed requirements.

The execution time of several subroutines, particularly if both presynaptic and postsynaptic cells are examined, demands a recording stability of many minutes. In the course of developing the current program, we have maintained stable microelectrode recordings for periods of one to two hours and feel that enough cells in these cultures will meet the time requirements of various interactive subroutines to provide a representative data sample.

These computer routines will significantly expand the electrophysiologic study of cell cultures and the often, rather complex, synaptic phenomena which develop in them.

DISCUSSION

MOISE: One problem with languages like FOCAL, is that interpreters in general are very slow languages, and you very quickly run into time problems.

MCINTYRE: You'd better explain about your brand of FOCAL.

CLARKE: There are several relatively high-speed interpreters now, like BLIP, and you can add user subroutines to FOCAL. The particular version of FOCAL that I use is U/W-FOCAL, from the University of Washington, and available from DECUS. Another version, FOCAL-F, from Georgia Tech, can be used too. FOCAL can have subroutines added which need not be in interpreter language.

MOISE: Sure, but very often you'd like to do continual processing, and the FOCAL interpreter would slow you down too much.

MCINTYRE: Essentially you do the fast stuff in assembler. That part's easy and doesn't require a high-level language. Then, right after you get the data in, you can switch to high-level language.

HARTLINE: The interpreter really does slow you down. The use of compiled high-level programs is much more practical.

CLARKE: My point is that you can have interrupt-driven assembler routines handling data acquisition, and do simultaneous simple analyses in FOCAL, and in most cases, FOCAL will keep up. I didn't mean FOCAL is the best answer.

BROWN: I think incremental compilers are a possible solution to this sort of problem. Processing is set up in a high-level language, and compiled ahead of time.

FRENCH: I have a version of FOCAL which can be compiled or run as interpreter, whichever you prefer. It doesn't require floating point hardware.

CLARKE: The primary point I'm making is that I can do an awful lot with FOCAL, getting online analysis. This is vital for the experiments, and its much easier to do analyses in some language like FOCAL or some other high-level language than in assembler.

REFERENCES

1. Peacock, J. H., Rush, D., and Clarke, T. O., Abs. Subm. Soc. Neurosci., (1975).
2. Nelson, P. G., and Sheriff, W., Personal Communication.
3. McGhie and Dong, DECUS Symposium Proceedings, (in press) spring (1975).
4. Small Computer Group, West Virginia University Medical Center.
5. Katz, B., *The Release of Neural Transmitter Substances*, ed. Charles C. Thomas (1969).
6. Registered Trademark Of Digital Equipment Corporation
7. Digital Equipment Corp. User's Society (DECUS) Library

chapter 14

TWO TIME-VARYING AUTOREGRESSIVE PREDICTORS FOR THE ELECTROENCEPHALOGRAM: A PRELIMINARY STUDY

C. R. PARISOT AND D. O. WALTER
Brain Research Institute
University of California
Los Angeles, California

I. INTRODUCTION

Approximately stationary, approximately Gaussian stochastic processes seem to model the electroencephalogram (EEG) with reasonable success (1, 2). Among the analytic tools suggested by that model is autoregression (AR) in which future values of the EEG are predicted by linear combination of values recently observed (3,4,5,6,7,8). One application of this model has been the detection of possibly epileptic transients (9,10). Although the theory of a time varying AR model for moderately non-stationary processes is not very well developed, it has seemed worthwhile to attempt to consider its application in prediction of the EEG waveform in order to motivate and focus theoretical development. Implications for language design are briefly treated after thorough presentation of the algorithm.

This research was partially supported by NIH grant HD 07524-02 and USPHS grant NS 02501.

II. ESTIMATION OF THE PARAMETERS OF THE AUTOREGRESSIVE MODEL

A. Introduction

Because the EEG signal usually contains non-stationarities and may contain fast changes of state, we reject the concept of simple stationarity, and. generalize our model to be time-varying, in one sense or another.

Our stochastic process $y(u)$, being sampled at a unit rate, is to be the time-varying autoregressive model of the form

$$y(u) = \sum_{i=1}^{p} a_i(u) y(u - i) + e(u) \quad (1)$$

where $a_i(u)$ are coefficients to be estimated, and $e(u)$ is a zero-mean white Gaussian process whose variance will be estimated.

B. Piece-wise Constant Model -- Lag-selecting Estimation

A classical statistical approach can be taken by assuming piece-wise stationarity of the EEG, where the pieces are samples of length $N+p$; in this case, the $a_i(u)$ become step-functions of t, constant over the intervals of length N and denoted by a_i, suppressing notation of their step changes.

B.1. COVARIANCE METHODS. The usual solution of this problem is by a least squares approach. The sample estimate of the fitting error variance over $N+p$ samples is given by:

$$H = \text{Ave}_{u=1,\ldots N} [e(u)]^2 = \text{Ave}_{u=1,\ldots N} [y(u) - \sum_{i=1}^{p} a_i y(u - i)]^2 \quad (2)$$

where Ave is the averaging operator; H will be minimized when

$$\frac{\partial H}{\partial a_j} = 0 \quad \text{for} \quad 1 \le j \le p \tag{3}$$

Applying Eq. (3) to Eq. (2), we get the normal equations

$$\text{Ave}_u[y(u)y(u-j)] = \sum_{i=1}^{p} a_i \text{Ave}_u[y(u-i)y(u-j)]$$

for $1 \le j \le p$ (4)

Solving these equations for a_i involves the sample covariance matrix whose i,jth element is, more explicitly,

$$\text{Ave}_{u=1,\ldots,N}[y(u-i)y(u-j)]$$

$$= \left(\frac{1}{N}\right) \sum_{u=p+1}^{N+p} y(u-i)y(u-j) \tag{5}$$

If we recall our assumption of stationarity over this interval, however, each of the covariances at each particular lag will have the same expected value, that of the (unknown) population autocovariance function, $R(\cdot)$:

$$E_k\{y(u-i)y(u-j)\} = R(i-j) \quad \text{for all} \quad i, j \tag{6}$$

where E_k means the expectation over replications of the experiment. Thus in certain cases it might be preferable to modify our matrix elements Eq. (5) to equal some estimator of unnormalized autocorrelation (11). This makes the elements be a function only of the difference between the indices Eq. (6) which reduces the amount of total computation required. But with N reasonably large, these two matrices are nearly equal, and only when computations based on the matrix are numerically sensitive will it matter which estimators are chosen. The process which we describe later avoids numerically sensitive computations, so the choice of estimator is not critical for us. Various methods sensitive to this choice were applied by Fenwick et al. (4, 7) and Shah (6).

A difficult question in the above time-series analyses is estimation of "the order" of the process. Since the procedures used to "inverse" the autocorrelation matrix are recursive on the order of the model, they suggest one answer to this delicate (and imperfectly defined) estimation question: Gersch suggests increasing the order of the model until no significant reduction of residual variance is observed, and Shah (6) tests the null hypothesis on the partial correlation coefficients, whose computation is incidental to the recursion.

B.2. LAG SELECTING METHOD. Since our eventual goal is to estimate a good predictor for the EEG voltage record, we use a different procedure. Anderson (12) has shown that minimizing the residual sum of squares does not usually minimize the prediction error. This is so for two reasons: prediction error variance always increases with the order of the AR model (13); whereas bias, the other component of error (often ignored in time-series studies) is very little increased by setting to zero any AR coefficient whose standard error of estimate is greater than its absolute value. With such a policy, our AR predictor will have a minimum number of non-zero coefficients, only major coefficients being retained. For example, the lacunary form

$$\hat{y}_u = a_1(u)y(u-1) + a_4(u)y(u-4) \qquad (7)$$
$$+ a_5(u)y(u-5) + a_9(u)y(u-9)$$

may still give a good fit; assuredly this predictor is biased compared to that of the full AR model Eq. (1), but it may well happen that the standard error of our prediction is reduced more by removal of the variance due to poorly-known coefficients, than it is increased by the bias. Anderson (12) derived a statistical test for this situation, which unfortunately requires much computation. But the concept of selecting those lags giving well-estimated contributions can be more simply carried out along with a recursive computation of the coefficients, often called 'stepwise regression'. At each 'step' (these 'steps' are unconnected with the step functions mentioned before), we will introduce that lag which most reduces the (estimated) mean-square prediction error; in unusual steps, we may eliminate one or more lags previously introduced, whose effect is redundant with newly introduced ones (this only occurs in cases of unusual joint distributions). We stop when no more lags can be entered, whose coefficients are larger than their estimated standard errors. This procedure was introduced for multiple regression by Efroymson (14) ; a practical and intuitive description of that algorithm is given by Draper and Smith (15). An implicit result of choosing well-determined coefficients is that our solution of Eq. (4) is equivalent to Gauss-Seidel elimination with pivoting on the largest element; it is well known in computational studies that this elimination method is numerically much more stable than one which pivots in some pre-established way. Nevertheless, the computational load for this method is still of the order of p^2 (being the maximum order examined for inclusion); Appendix 1 gives this algorithm, in the form used by us.

Another feature of stepwise selection, which has both fundamental and practical interest, is that it sidesteps the question (a delicate question for conventional AR estimation) of "the" order of the process. No single stepwise parameter corresponds to that concept, since the number of non-zero coefficients can be (and in practice always is) much less than the index of the maximum lag. The locations of non-zero coefficients resulting from stepwise selection may turn out all to lie within a maximum lag much smaller than p; indeed if some of them are near p, it will be advisable to increase p to check for possibly advantageous longer lags.

In summary, the advantages of using stepwise lag selection are that we obtain a relatively small set of coefficients, while fitting a high-order discrete stochastic process, with low prediction error.

C. Time Recursive Estimation. Kalman Predictor

Consider now the hypothesis that the time dependent AR coefficients obey a linear Markov model:

$$a_i(u+1) = a_i(u) + v_i(u), \quad i=1, 2, \ldots, p \qquad (8)$$

where $v_i(u)$ are zero-mean uncorrelated white Gaussian noises. If all $v = 0$, then $y(u)$ is stationary. The nonstationarities and transitions between stationary segments of the process are reasonably modeled by the effect of non-zero additive noise $v_i(u)$.

C.1 ESTIMATION OF THE AR MODEL BY LINEAR RECURSIVE PREDICTION. This method has been previously used successfully in speech analysis (16, 17) and suggested for EEG analysis by Gueguen. Some results have been reported in (18).

We can restate the time-varying autoregressive model Eq. (1) in a vector form setting:

$$A(u) = [a_1(u), a_2(u), \ldots, a_p(u)]^T \qquad (9)$$

$$V(u) = [v_1(u), v_2(u), \ldots, v_p(u)]^T$$

$$Y(u-1) = [y(u-1) \; y(u-2), \ldots, y(u-p)]^T$$

we get, by using Eq. (8)

$$A(u+1) = A(u) + V(u) \qquad (10a)$$

$$y(u) = Y^T(u-1)A(u) + e(u) \qquad (10b)$$

Thus the stochastic process modeling the EEG is viewed as the output of a linear system whose state vector is the ordered set of autoregressive coefficients. The plant's input is the (unobservable) p-dimensional white-noise vector $V(u)$, whose covariance matrix we call $Q_V(u)$. The observation noise, $e(u)$, has covariance $q^2(u)$.

Our estimation problem can now be stated: Having observed the output, $y(u)$, from a time t_0 up to the "present" time t, we want to predict the output at time $t + 1$; in order to do that, we will (also want to) predict the state vector (ordered set of AR coefficients) for $t + 1$. The goal of our prediction is to minimize the prediction error of $y(t+1)$, in the sense that its expected error variance over replicate runs, with replicate noises and replicate initial state, is minimized.

As we collect more observations of a subject's EEG, we should be able either to improve our knowledge of the state which it represents (if that has not changed), or else adapt our partial knowledge in the direction of any change of state. Since we are dealing only with the observation of one channel, each observation is a single number, yet we want to update a vector estimating the state; if we set

$$\hat{A}_{t+1} = \hat{A}_t + K_t(\hat{A}_t, y(t)) \qquad (11)$$

where $\hat{A}_{t+1}$ is the vector representing the new estimation (or our new knowledge of the state $\hat{A}(t+1)$), A_t is obviously the vector of our previous knowledge, $y(t)$ is the scalar which we observe, and K_t is a correction vector function, which distributes the effect of the new observation onto the various components of the state vector. A very important point now is to specify the correction due to the new information. The prediction of the EEG at t is given by: (giving to $e(t)$ its mean value)

$$\hat{y}_t = Y^T(t-1)\hat{A}_t \qquad (12)$$

If this prediction was exact, $\hat{A}_{t+1}$, the new predictor, would not require any correction. Therefore a reasonable way to correct the predictor is proportionally to its error:

$$\hat{A}_{t+1} = \hat{A}_t + K(t)[y(t) - Y^T(t-1)\hat{A}_t] \qquad (13)$$

The weighting $K(t)$ will have to be determined so that the predictor be optimal in minimizing its own error variance (as stated earlier), taking into account the assumed statistical properties of the noises and initial state.

The Kalman-Bucy predictor (Appendix II) applied to our linear model Eq. (10) will produce this estimate in the form suggested in Eq. (13). (Appendix II gives in some generality an intuitive introduction to this concept of recursive linear prediction, and derives the optimal gain $K(t)$). In Appendix III the theoretical results given in Appendix II are applied to our specific problem, and the computational aspects are considered).

C.2 IMPROVEMENT FOR PREDICTION OF THE TIME RECURSIVE ESTIMATION PROGRAM. The EEG process is in general considered to have an order between 10 and 20 (19, 20). Such a large order requires a rather long computation as shown in Appendix III. We would like not to exceed a computation time equivalent to an order 10. On the other hand, the prediction quality is sensitive to the EEG segment input to the AR model (the last known samples of the signal), and in order not to rely on a short and therefore possibly non-representative segment, we should have some long lags in our AR model. Certainly, as we pointed out in our previous discussion of the AR model

(Lag-selecting Method) increasing the order cannot be accepted without a loss of accuracy in the prediction for statistical reasons. This dilemma was solved for the piece-wise constant model by the lag-selecting procedure. Thus we would like to keep the number of coefficients estimated by the Kalman predictor to a low number, but spread those out so that our AR model will cover more lags and longer lags.

We have not solved the difficult problem of making the time recursive estimation also lag selecting. We propose a partial solution in the form of an original time recursive estimation of a preselected lag-predictor in the sense that the non-zero coefficients of the AR model can be placed at any predetermined lags. This raises the question of the choice of those lags, which could be approached in an experimental way.

The preselection of the lags can be seen as a projection of the p-dimensional vector of coefficients $A(u)$ onto a q-dimensional subspace:

$$\overline{A}(u) = PA(u) \qquad (14)$$

where P is a $q \times p$ projection matrix of 0-1 elements, with exactly one "1" per row and at most one "1" per column. Similarly we define

$$\overline{Y}(u - 1) = PY(u - 1) \qquad (15)$$

Therefore under the arbitrary setting to 0 of $p-q$ coefficients, the "pth order" AR model can now be expressed as

$$y(u) = \overline{Y}^T(u - 1)\overline{A}(u) + e(u) \qquad (16)$$

The previous assumptions Eq. (8) on the evolution of the coefficients certainly remains valid and we obtain:

$$\overline{A}(u + 1) = \overline{A}(u) + U(u) \qquad (17a)$$

$$y(u) = \overline{Y}^T(u - 1)\overline{A}(u) + e(u). \qquad (17b)$$

This reformulation is very similar to the linear system Eq. (10) to which we previously applied the Kalman predictor. We could therefore proceed exactly as we did in Appendix III, and derive similar recursive relations. The only difference would be in the estimated vector $\hat{A}_u$, the non-zero estimated coefficients, and the vector $Y(u-1)$ of the past observations, coefficients and observations relative to the preselected lags.

By spreading out the lags in this way, we have kept the dimension of our vector of coefficients small, while still allowing the AR predictor to be based on values at substantial time lags before the present time.

III. PREDICTION OF THE EEG SIGNAL USING AR MODELS

By using the lag selecting procedure (II B), performed every N EEG samples on the last $N+p$ observed samples, we fitted the EEG with a piece-wise constant time varying AR model. By using the time recursive procedure (II C) (optionally preselecting the lags), repeated with each incoming EEG sample, we fitted the EEG with an adaptive time varying AR model. Due to the form of these AR models, they are simply one step ahead predictors. As we would like to extrapolate the EEG signal supposedly observed only up to time t, for several samples after t, we will derive a longer term optimal predictor in some sense. We assumed the EEG process to be approximately Gaussian, approximately stationary and reasonably fitted by an AR model. If we strengthen these hypotheses on short EEG segments, we will consider locally the EEG to be stationary, Gaussian with rational spectral density (case of the AR model). The best predictor at time $t+k$ for the EEG signal, $\hat{y}_{t+k}$, will be a function of all past observations $[...y(t-1), y(t)]$ and will minimize the variance of the predictor error.

If z denotes the forward shift operator:

$$y(t) = zy(t - 1) \tag{18}$$

the AR model (with some possibly zero coefficients) can be written:

$$y(t) = \frac{1}{1 - a_1 z^{-1} - \cdots - a_p z^{-p}} e(t) \tag{19}$$

or

$$y(t) = \frac{1}{P(z^{-1})} e(t) \text{ with } P(z) = 1 - a_1 z - \cdots - a_p z^p \tag{20}$$

At time t, the EEG samples $y(t)$, $y(t-1)$, ... have been observed and we want to predict $y(t+k)$. Eq. (20) gives:

$$y(t + k) = \frac{1}{P(z^{-1})} e(t + k) \qquad (21)$$

By performing long divisions of 1 by $P(z)$ at the order k we can determine uniquely the polynomials $F(z)$ and $G(z)$ such that:

$$\frac{1}{P(z)} = F(z) + z^k \frac{G(z)}{P(z)} \qquad (22)$$

Then Eq. (21) can be transformed into:

$$y(t + k) = F(z^{-1})e(t + k) + \frac{G(z^{-1})}{P(z^{-1})} e(t) \qquad (23)$$

The first term, as $F(z)$ is of degree $k-1$, is a linear function of $e(t+k)$, $e(t+k-1)$,...,$e(t+1)$ which are white and independent of the observed EEG samples. The noise $e(t)$ is related to the past observed value as (From Eq. (20)):

$$e(t) = P(z^{-1})y(t)$$

By substitution into Eq. (21) we get

$$y(t + k) = F(z^{-1})e(t + k) + G(z^{-1})y(t) \qquad (24)$$

We now express the variance of the prediction error:

$$\mathrm{Var}[y(t + k) - \hat{y}_{t+k}] = \mathrm{Var}[F(z^{-1})e(t + k)]$$
$$+ \mathrm{Var}[\hat{y}_{t+k} - G(z^{-1})y(t)] \qquad (25)$$

The cross product term is null due to the independence of $e(t+1)$, $e(t+2)$, ... $e(t+k)$ from the available EEG observations. The error prediction variance will be minimized when:

$$\hat{y}_{t+k} = G(z^{-1})y(t) \qquad (26)$$

and will have the estimated value:

$$\text{Var}[y(t + k) - \hat{y}_{t+k}] = \text{Var}[F(z^{-1})e(t + k)] \quad (27)$$

If we express $G(z)$ and $P(z)$ explicitly by

$$G(z) = g_0 + g_1 + \cdots + g_{p-1}z^{p-1}$$

$$F(z) = 1 + f_1 z + \cdots + f_{k-1}z^{k-1}$$

we obtain

$$\hat{y}_{t+k} = g_0 y(t) + g_1 y(t - 1) + \cdots + g_{p-1} y(t - p + 1)$$

and

$$\text{Var}[y(t + k) - \hat{y}_{t+k}] = \text{Var}[e(\cdot)] \sum_{i=0}^{k-1} f_i^2 \quad (28)$$

As the prediction is carried further (increase in k) it is reasonable and clear from Eq. (28) that the prediction error variance always increases. Under the hypothesis of a stable filter (Appendix III) we have shown (but do not mention here the rather complicated proof), that the series in Eq. (28) converges and therefore the prediction error variance is bounded.

If we are interested in computing the error variance along with the predictions, then the long division will have to be programmed and the expression Eq. (26) used to compute the predictions (k=1, 2,...). A simpler way to obtain only this same prediction is to move along the AR model by letting the noise $e(t+k)$ for $k>0$ have its zero mean value (eq. 26 could be considered as being derived from eq. 24 in that way). The predictions are in this way fed back into the AR model to produce further estimates. Both methods were examined: the first was used more for prediction performance testing and the second only for pure prediction.

IV. LINGUISTIC CONSIDERATIONS

For the purposes of defining a language for neuroscience, we propose that the architecture of AR is a valuable example. In its utopian form, the piece-wise constant model would begin with a phase of data collection, conceptually followed by a phase of moderately heavy number-crunching (covariances--perhaps better done by double FFT--and partial matrix inversion) followed by a prediction phase involving products of vectors. If all this is to be regularly updated, on more than one EEG channel, much linguistic efficiency is demanded in the tools available to the programmer.

The continually updated model using Kalman filtering or some modification of it, has a more integrated problem architecture, but requires a more powerful machine. But we doubt that the neuroscience community today has the conceptual resources to conceive, develop, and deliver a 'family machine', which could be efficiently organized for the piece-wise model one day, and efficiently reconfigured for the Kalman model tomorrow. Such a machine would need to be armed with a family of languages, each well documented and easily learnable at several levels of effectiveness, and each of which generates rather efficient code for each member of the family of machines. In the absence of such resources, we return to the more familiar position of needing a general (read 'non-specific') language and machine of greater power than would be required for our specific problems.

DISCUSSION

MCINTYRE: The initial segment problem looks to me very similar to the sort of problem you run into with moving average filters. Typically it's just a matter of moving along and rejecting data until the system settles down.

D.O WALTER: The intended application was to include substitution for data which was blocked out by excessive muscle potentials. You don't know when they are going to come in, so you have to back up instead of going forward. It's no great problem but we hadn't thought of it until we were faced by these plots.

KEHL: The weighting function, how is this different from any other sweeping filter technique? Indeed, you could easily develop a comb filter which would do it, or even build a resonant filter bank. I think Grey Walter did it in the late 50s, and it worked in real time.

D. O. WALTER: If I'm thinking of the same gadget, it was used for frequency analysis. Our purpose here isn't just a spectrum, which is just a by-product, but an extrapolation or replacement capability for data which is unavailable because of artifacts. That's really only one application. The unexpected part about a filter scheme which includes all of the weights over a given past period of time is that the inclusion of some of them reduces your prediction accuracy. I think filters developed in terms of infinite time duration can lead to degradation of accuracy in extrapolation.

MCCORMICK: We use a similar technique for two dimensional scene anlysis, where we want to segment the scene, considered as a video signal, into different textured portions of the picture. There you have the same problem: if you use time-series analysis, you can only apply it over the homogeneous texture area. We're trying to define textural boundaries, which I think corresponds very closely to what you're doing.

D. O. WALTER: Yes, that's certainly similar. In the picture terms we would by looking for an unexpected object: the golf ball in the grass, or something like that, or the boundary between sandtrap and green. The point I was making to Ted was that sometimes adding another input may degrade your prediction because your ability to estimate the proper coefficient may be so poor that it's better to leave it out. We think we can develop a stepwise common filter in order to get the correct order of autoregression. The spectra derived from that set of coefficients may be some guide to the right resolution. But we still have a lot to do here.

BROWN: What applications do you envision besides detecting unusual spots and replacing them with "good" data?

WALTER: Well, we could use it for things like detecting changes in sleep stages, and other types of patient monitoring. Also, substituting for messy records can be helpful for the neurologist trying to record EEGs from infants or other subjects who can't be persuaded to relax. It could even obviate the necessity of bringing the patient back for a second recording session.

REFERENCES

1. Walter, D. O., ed. Processing of bioelectrical phenomena. In *Handbook of Electroencephalography and Clinical Neurophysiology*, 4: Evaluation of Bioelectrical Data from Brain, Nerve and Muscle, I, M.A.B. Brazier, ed. Part B., Elsevier, Amsterdam (1972).
2. Matousek, M. ed., Frequency and correlation

analysis. In *Handbook of Electroencephalography and Clinical Neurophysiology*, A. Remond, Editor-in-Chief, 5: Evaluation of Bioelectrical Data from Brain, Nerve and Muscle, II, M. A. B. Brazier, and D. O. Walter eds. Part A. Elsevier, Amsterdam (1973).
3. Makhoul, J. Linear prediction: a tutorial review, Proc. IEEE 63:561-580 (1975).
4. Fenwick, P. B., Mitchie, P., Dollimore, J. and Fenton, G. W. Application of the autoregressive model to EEG analysis. Agressologie 10:553-564 (1969).
5. Gersch, W. Spectral analysis of EEGs by autoregressive decomposition of time series, Math. Biosci., 7:205-222 (1970).
6. Shah, A., Lai, D. C., Anliker, J. E. Prediction of EEG alpha wave form by using an autoregressive model. *San Diego Biomedical Symposium*, In press (1975).
7. Fenwick, P. B., Mitchie, P., Dollimore, J., and Fenton, G. W. Autoregressive series analysis of EEG. Computers for analysis and control in medical and biological research. IEEE Conf. Publ. 79:231-241 (1971).
8. Mathieu, M., Tirsch, W., and Pöppl, S. J. Multichannel on-line EEG analysis by means of an autoregressive model with application. 2nd Symposium: *Quatification of EEG*, In press (1975).
9. Lopes da Silva, F. H., Dijk, A., and Smits, H. Detection of non-stationarities in EEG using the autoregressive model - an application to EEG of epileptics. *Institute of Medical Physics TNO*, Utrecht, The Netherlands (1974).
10. Herolf, M. A predictive detector applied to EEG-signals. *Technical Report Number 81*. Telecommunication Theory, Royal Inst. of Techn. Stockholm, Sweden. (1974).
11. Box, G. E. P. and Jenkins, G. M. *Time Series Analysis, Forecasting and Control*, Holden Day, San Francisco (1969).
12. Anderson, R. L., Allen, D. M., and Cady, F. B. Selection of predictor variables in linear multiple regression. *Statistical Papers in Honor of George W. Snedecor*. ed. T. A. Bancroft· Iowa State University Press (1972).

13. Walls, R. C. and Weeks, D. L. A note on the variance of a predicted response in regression. Am. Statistician. 23(3):24-26 (1969).
14. Efroymson, M. A. *Mathematical Methods for Digital Computers*. John Wiley and Sons, New York (1962).
15. Draper, N. R. and Smith, H. *Applied Regression Analysis*, John Wiley and Sons, New York (1966).
16. Gueguen, C., Caryannis, G. Analyse de la parole par filtrage optimal de Kalman. Automatisme 18:99-105, (1973).
17. Melsa, J. L. et. al. *Development of a Configuration Concept of a Speech Digitizer Based on Adaptive Estimation Techniques*. Final report: Information and Control Sciences Center. Institute of Technology. Southern Methodist University, Dallas Texas, (1973).
18. Mathieu. M. Analyse de l'EEG par Modelisation Predictive. Project de Stage. Ecole Nationale Superieure des Telecommunications, Paris (1973).
19. Rapplesberger, P., and Petsche, H. Spectral Analysis of the EEG by Means of Autoregression. Neurophysiologisches Institut der Universitat und Institut fur Hirnforschung der Osterreichischen Akademie der Wissenschaften, Wien, Osterreich.
20. Zetterberg, L. H. Means and methods for processing of physiological signals with emphasis on EEG analysis. Technical Report No. 84, Telecommunication Theory, Royal Inst. of Techn., Sweden (1974).
21. Dixon, W. J. *BMD02R. Stepwise Regression. Biomedical Computer Programs. BMD*. University of California Press (1973).
22. Afifi, A. and Azen, S. P. *Statistical Analysis: A Computer Oriented Approach*. Academic Press, New York, (1973).
23. Kalman, R. E. and Bucy, R. S. New results in linear filtering and prediction theory. J. of Basic Eng. 83-D:95-108 (1961).
24. Astrom, K. J. *Introduction to Stochastic Control Theory*. Academic Press, New York (1970).
25. Sage, A. P. and Melsa, J. L. *Estimation Theory with Applications to Communications and Control*. McGraw Hill, New York (1971).

APPENDIX I

STEPWISE REGRESSION ALGORITHM

Multiple regression is often used in data analysis to obtain a (least-squares) best fit to an independent variable, based on a set of independent variables, using a relation of the form

$$y = a_0 + a_1 x_1 + a_2 x_2 \cdots + a_p x_p + e \qquad (11)$$

where y is the dependent variable, $x_i, i=1,\ldots p$ is the set of independent variables, and $a_i, i=0,1,\ldots p$ is the set of coefficients to be determined so as to minimize the variance of e, the error variable. This formulation can be translated to apply to zero-mean time series (sampled at unit intervals), as

$$y_0(u) = a_1 y_1(u), + a_2 y_2(u) + \cdots + a_p y_p(u) + e(u) \qquad (12)$$

where $y_k(u) = y(u-k)$, which are the observed values of our process and it is the mean-square value of $e(u)$ which we wish to minimize.

If we consider using this formula for $u = p+1$, $p+2,\ldots p+N$, there will be much redundancy in the multiple appearances of particular observations as values of the successive independent variables; but this is proper and natural for time series (although it will reduce the degrees of freedom of estimate in the distribution functions for various parameters). We will carry out the computation of best values for a_i on the basis of C, the $(p+1) \times (p+1)$ correlation matrix of these dependent and independent variables; its i,jth element is

$$c_{ij} = \left(\frac{1}{N}\right) \sum_{k=1}^{N} y_i(p+k) y_j(p+k), \text{ for } 0 \leq i, \leq p \qquad (13)$$

We could alternatively use an estimate of the autocorrelation matrix, which incorporates the assumption of stationarity of the EEG over the $p+N$ samples: the two estimates are numerically close, as mentioned before. Estimating the autocorrelation matrix requires less computation and furthermore guarantees the stability of the estimated filter (3). The program we used [BMD02R (21)], incorporates the correlation matrix, and rather than re-program, we used it for this exploratory study.

The Algorithm

In sketching the algorithm used by BMD02R(21), we postpone till later the discussion of the various F-tests encountered along the way, and the choice of F_{in}, F_{rem}, and T.

1) Compute and store (in triangular form) the normalized correlation (or autocorrelation) matrix, C; also compute and store the standard deviations

$$s_i = \frac{\sum_{k=1}^{N} y_i^2(p+k)}{N}, \text{ for } i = 0, 1, \ldots p \qquad (14)$$

Also, set q (the number of lags included in the regression) = 0. Go to the next section.

2) Compute F values for lags not included in the present edition of the regression form; for those lags,

$$F_i = \frac{c_{0i}^2(N-q-1)}{c_{00}c_{ii} - c_{0i}^2} \qquad (15)$$

Find the set of those F_i larger than F_{in}, for which c_{ii} is larger than T. If none go to section 4. Choose the largest F_i, and enter the lag corresponding to it as follows:

FLAG = 1; CALL STEP (see below); $q = q + 1$

If q is equal to p go to section 4; otherwise go to next section.

3) Compute F values for lags included in the present edition of the regression form; for those lags

$$F_i = \frac{c_{0i}^2(N-q)}{(-c_{ii})c_{00}} \qquad (16)$$

If any of these F_i is smaller than F_{rem}, remove that lag, as follows:

FLAG = -1 ; CALL STEP (see below, STEP);
$q = q - 1$; repeat section 3. If no F_i is less than F_{rem}, go to section 2.

4) Terminate stepwise and compute the selected set of coefficients corresponding to the included lags:

$$\hat{a}_i = \frac{s_0 c_{0i}}{s_i} \qquad (17)$$

The subroutine "STEP" performs the inclusion (FLAG = 1) on removal (FLAG = -1) or the ith lag as follows:
1) Set

$$u_\ell = \begin{cases} c_{\ell i} & 0 \leq \ell \leq i \\ - \text{Flag} & \ell = i \\ c_i & i \leq \ell \leq p \end{cases} \qquad (18)$$

and

$$c_{\ell i} = c_{im} = 0 \qquad 0 \leq \ell \leq i \leq m \leq p$$

2) pivot the upper triangular part of C

$$c_{\ell m} = c_{\ell m} - \frac{u_\ell u_m}{c_{ii}} \qquad 0 \leq \ell \leq m \leq p \qquad (19)$$

and return.

The F Tests.

Each quantity named F_i in the algorithm is equal at any steps of the regression to the square of the quotient of the ith regression coefficient divided by its estimated standard deviation:

$$F_i(s) = \left[\frac{\hat{a}_i(s)}{\sigma(\hat{a}_i(s))} \right]^2 \qquad (I10)$$

Since q coefficients are presently included in the regression form, the null hypothesis on the ith coefficient (if it is included) can be performed by testing this quantity. In the case of multiple regression analysis, $F_i(s)$ has an F distribution with 1 and $N-q$ degrees of freedom under the hypothesis of independent residual errors (22). Although in the case of autoregression, we reduce the number of measurements from $N(p+1)$ to $N+p$, the number of degrees of freedom of $F_i(s)$ should, we believe, remain the same, because those degrees of freedom originate from the estimation of the variance of each coefficient based on the residual mean square, which because of the whiteness of the noise on the autoregressive model, has $N-q$ degrees of freedom. However, this point seems murky in the statistical literature, and slight adjustment may be required. For included coefficients, the standard error can be expressed (21) as

$$\sigma(a_i) = \frac{s_0}{s_i} \sqrt{\frac{c_{00}(-c_{ii})}{N-q}} \qquad (I11)$$

Eq. (I.10) can be expressed directly from Eq. (19) and Eq. (I.11) in the form identical to (I.6):

$$F_i = \frac{c_{0i}^2 (N-q)}{(-c_{ii}) c_{00}} \qquad (I12)$$

The minus sign appearing with c_{ii} comes from the fact that the subroutine STEP not only reciprocates the diagonal element but also changes its sign. Thus for the included lag the diagonal element has to be corrected for use in an F test.

But now we need to show that the inclusion test Eq. (I.5) represents the same test of the null hypothesis Eq. (I.10), for non-included lags.

For each lag not included, consider the corrections to its c's if it were brought in next, using a prime to signify these tentative values:

$$c'_{ii} = -\frac{1}{c_{ii}}$$

$$c'_{00} = c_{00} - \frac{c_{0i}^2}{c_{ii}} \qquad (\text{I}13)$$

$$c'_{0i} = \frac{c_{0i}}{c_{ii}}$$

Substituting Eq. (I.13) into Eq. (I.5), we get

$$F_i = \frac{c_{0i}(N-q-1)}{(-c_{ii})c_{00}} \qquad (\text{I}14)$$

But Eq. (I.6) would give this value, if this lag were entered at the next step (giving $q+1$ lags entered). Thus Eq. (I.5) is F-distributed, with degrees of freedom: 1 and $N-q-1$ (without the tentative entry), just as Eq. (I.6) had degrees of freedom 1 and $N-q$ (where q is the number of included lags).

As N should be large compared to p for adequate statistical estimation of all parameters, the distributions of the F_i are approximately independent of q; therefore they are approximately F-distributed with 1 and N degrees of freedom. Thus F_{in} and F_{rem} have approximately the same probablility level for all steps. T should be chosen to avoid basing calculations on instrument noise in the analog-digital converter or the number representation used in the computation.

It is interesting to notice that the test to enter a variable Eq. (I.5) can be in the regression, conditional on the included lags:

$$F_i = \frac{(N-q-1)}{1-(1/r_i)^2} \qquad (\text{I}15)$$

Therefore including the lag of highest F level is equivalent to including the lag best partially correlated with the dependent variable.

APPENDIX II

THE CONCEPTS OF RECURSIVE LINEAR PREDICTION OR KALMAN PREDICTION

We will not give a precise, general, nor theoretical derivation of the Kalman predictor (23, 24, 25); rather we present an intuitive but detailed justification oriented toward our application.

We observe a process $y(u)$ at discrete unit increments of time, and in this concept we consider those values as the output of a linear system, whose further properties are developed in the course of this exposition. The system is thought of as having a time-varying (column) vector of state values, $X(u)$, which cannot be directly observed, but which are related to y by the following matrix equations:

$$X(u + 1) = \Phi X(u) + V(u) , \text{ for } u \geq t_0$$
$$y(u) = \theta^T X(u) + e(u) , \text{ for } u \geq t_0 \quad \text{(II-1)}$$

In these equations, Φ is a known constant matrix, Θ a known constant vector; V and e are (unknown) noise processes. In addition, the initial state vector, $X(t_0)$, is assumed to vary over replications, as an unknown vector-valued Gaussian stochastic variable.

We assume that we have observed the outputs $y(u)$ from t_0 up to t, and we want to predict the next state vector $X(t+1)$ (and then $y(t+1)$), on this basis. We naturally desire this estimator to be unbiased, and to minimize the variance of each component of the error in the state vector. These variances appear on the diagonal of the prediction-error covariance matrix.

$$P_{t+1} = E_j\{[X(t + 1) - \hat{X}_{t+1}][X(t + 1) - \hat{X}_{t+1}]^T\} \quad \text{(II-2)}$$

where E_j means the expectation over replicate runs, and $\hat{X}_{t+1}$ means the estimate of $X(t+1)$ based on data available up to time t.

We assume the basic theorem [Åstrom, (24) ch. 7] that because the noises in expression (II.1) are Gaussian, and the criterion for estimation is minimum mean-square error the optimum estimator will be a linear function of the past observations,

Using the recursive filtering concept of including a function of the previous output in the input to a predictor (Part II. C1), we can expect (because of the linearity in the observations), the estimator to be expressible in the form

$$\hat{X}_{t+1} = \phi \hat{X}_t + K(t)[y(t) - \theta^T \hat{X}_t] \qquad (\text{II-3})$$

where $K(t)$ is an unknown time-varying column vector, to be determined. The two terms in this expression can be understood as follows: the first term would equal the expected next state, if $V(t)$ were to have its mean value (the zero vector); the second term is a correction on that, whose second factor represents the "observed" error on the output created by the last estimate of the state, (if $e(t)$ has its mean value of zero), and whose first factor is a gain vector which distributes the observation error to various components of the state vector. Because of the linear property of the estimator, $K(t)$ can be determined independently of the observations: this will be shown later.

The variances and covariances of the noises are assumed known, as a function of time (they may be constant). We denote the known covariance matrix of $V(u)$ by $Q_V(u)$ (diagonal) and the variance of $e(u)$ by $q^2(u)$. The initial state vector $X(t_0)$ is taken to have mean vector M and covariance matrix Q_X. We assume that $X(t_0)$ is independent of both $e(u)$ and $V(u)$. Let's evaluate the prediction-error covariance matrice, whose diagonal terms have to be minimized. Now, by combining Eq. (II.1) and Eq. (II.3) we have

$$X(t+1) - \hat{X}_{t+1} = \phi[X(t) - \hat{X}_t] - K(t)[y(t) - \theta^T \hat{X}_t] + V(t)$$

$$= [\phi - K(t)\theta^T][X(t) - \hat{X}_t] - K(t)e(t) + V(t)$$

$$(\text{II-4})$$

Consider now expressing P_{t+1} in terms of this expression: there will be three "square" terms and three "cross-product" terms. But because of our assumptions that $e(u)$, $V(u)$, and $X(t_0)$ are mutually independent, and $e(u)$, $V(u)$ white, the expected value of P_{t+1} will have zero values for the cross-product terms. $X(t)$ and $\hat{X}_t$ depend, as it can be seen in Eq. (1), only on $X(t_0)$, and $e(u)$ and $V(u)$ up to time $t-1$; they are consequently uncorrelated with $e(t)$ and $V(t)$. Therefore,

$$P_{t+1} = [\Phi - K(t)\theta^T]P_t[\Phi - K(t)\theta^T]^T$$
$$+ K(t)q^2(t)K^T(t) + Q_V(t) \quad (II-5)$$

Let's define a function $h(t)$ and vector-valued function $T(t)$ by

$$h^2(t) = \theta^T P_t \theta + q^2(t)$$

$$T(t) = \frac{P_t \theta}{h(t)}$$

Because P_t is always symmetric positive, h will always be non-negative real; with these definitions, we can express Eq. (II.5) as

$$P_{t+1} = \Phi P_t \Phi^T + [K(t)h(t) - T(t)][K(t)h(t) - T(t)]^T$$
$$- T(t)T(t)^T + Q_V(t) \quad (II-6)$$

For purposes of minimizing the diagonal elements of P_{t+1}, the only adjustable term in the above expression is the one involving $K(t)$ explicitly. The square elements on the diagonal of this matrix are minimized when:

$$K(t) = \frac{T(t)}{h(t)} = \frac{P_t \theta}{\theta^T P_t \theta + q^2(t)} \quad (II-7)$$

Substituting this optimal value of $K(t)$ into Eq. (II.6)

$$P_{t+1} = \Phi P_t \Phi^T + Q_V(t) - \frac{P_t \theta \theta^T P_t^T}{[\theta^T P_t \theta + q^2(t)]^2} \quad (II-8)$$

INITIAL CONDITIONS AND COMPUTATIONAL CONSIDERATIONS

In this derivation we assumed the a priori knowledge of the matrix Φ and of the vector Θ as well as the statistical quantities: M, Q_X (mean and covariance matrix of the initial state) X and $Q_V(.)$, $q(.)$ (covariance matrix and variance of the noises).

Now let's consider the start-up operations for the prediction equations developed so far; we need an estimate for X_{t_0} and its error-covariance matrix. As we have not yet observed any output at t_0, the best estimates are obtained from the statistics of X_{t_0}:

$$\hat{X}_{t_0} = M, \quad P_{t_0} = Q_X$$

By reexamining the expressions which will be used for computation/estimation in applying this scheme to actual data, we see that neither P_u nor $K(u)$ depends on $y(u)$, for any u. That means that our estimates of these matrices as functions of time can be pre-computed on the bases of our knowledge of the known system parameters Φ, Θ, M, and Q_X, and the known variable matrix $Q_V(u)$ and the known function $q^2(u)$. Of course, if Q_V and q^2 are both constant (which would produce stationary outputs y), this simplifies the computations. It also turns out that the derivation is not altered in case Φ and Θ are time-varying, provided that their values at each time are known. It can also be noted that the assumption of the uncorrelation of the component noises of $V(u)$, implying the diagonal form of $Q_V(u)$, can be dropped, and $Q_V(u)$ be any correlation matrix.

APPENDIX III

APPLICATION OF THE KALMAN FILTER TO IDENTIFYING THE COEFFICIENTS OF A TIME-VARYING AUTOREGRESSIVE MODEL

In this appendix we first show how the Kalman Filter, introduced in the previous appendix, is well suited to obtaining a sequential identification of the coefficients of the time-varying linear AR model. Then we deal with the computational effort and see that this algorithm could be real-time feasible. Finally we discuss tests of stability and means of stabilizing the estimated model.

The AR model is of the order p, meaning that p preceding output values are used to predict the next output. As shown in (Part II C1), the time-varying AR model can be reformulated as:

$$A(u + 1) = A(u) + V(u) \qquad \text{(III-1a)}$$

$$y(u) = Y^T(u - 1)A(u) + e(u) \qquad u \geq t_0 \qquad \text{(III-1b)}$$

where: $A(u) = [a_1(u), a_2(u), \ldots, a_p(u)]^T$ is the column vector of the coefficients at time u; $Y(u-1) = [y(u-1), y[u-2], \ldots, y(u-p)]^T$ is the known column vector of the past observations;

$$V(u) = [v_1(u), v_2(u), \ldots, v_p(u)]^T$$

and $e(u)$ are zero-mean white Gaussian noises of covariance matrix $Q_V(u)$ and variance $q^2(u)$ respectively; moreover $e(t)$ is uncorrelated with any component of $V(t)$, and the components of are assumed to be uncorrelated with each other.

We will now relate this formulation with the results obtained in Appendix II on the Kalman predictor. In the present plant Eq. (III.1), the state vector is $A(u)$, the vector of coefficients. The matrix Φ is the identity. The vector Θ, which, as mentioned in Appendix II, can be taken time-varying, is in this case $y(u-1)$, the vector of the p previous observations. We assume at present that we know a priori the variance and covariance matrix of the zero-mean noises $e(u)$ and $V(u)$, as well as the mean A_0 and the covariance matrix Q_A of the intial state $A(t_0)$. We will discuss this assumption later.

Our goal is to estimate the coefficients of the AR model, in the form of the state vector $A(u)$, so that the error variance of the prediction of the EEG signal $y(u)$ is minimized. If we denote by $\hat{y}_{t+1}$ the prediction made at time t, having observed the EEG signal $y(u)$ from t_0 up to time t, the error variance σ^2 can be defined by:

$$\sigma^2 = E_j[(y(t + 1) - \hat{y}_{t+1})^2] \qquad \text{(III-2)}$$

the expected value E_j being taken over replicate runs of the model. Considering the noise $e(t+1)$ to have its mean value of zero, the predicted signal at time $t+1$ is (from III. 1b):

$$\hat{y}_{t+1} = Y^T(t)\hat{A}_{t+1} \qquad (III-3)$$

where $\hat{A}_{t+1}$ denotes a prediction of the state vector. By combining Eq. (III.3) and Eq. (III.1b) we get:

$$y(t+1) - \hat{y}_{t+1} = Y^T(t)[A(t+1) - \hat{A}_{t+1}] + e(t+1)$$

$$(III-4)$$

Then the error variance can be expressed as (from Eq. (III.2) and Eq. (III.4))

$$\sigma^2 = E_j[Y^T(t)[A(t) - \hat{A}_{t+1}][A(t+1) - \hat{A}_{t+1}]^T Y(t)]$$
$$+ q^2(t+1) = Y^T(t)E_j\{[A(t+1) - \hat{A}_{t+1}][A(t+1)$$
$$- \hat{A}_{t+1}]^T\}Y(t) + q^2(t+1) \qquad (III-5)$$

There is no cross product term because, as justified in Appendix II (derivation of eq. II.5), $A(t+1)$ and $\hat{A}_{t+1}$ are uncorrelated with $e(t+1)$. As the expectation E_j is over replicate fitting runs of the model Eq. (III.1) the known observed vector $Y(t)$ can be taken out of the expectation. We notice that in any case the error variance of the prediction cannot be less than the variance of the observation noise $e(u)$.

Let's now show that the minimization of the variance of each component of the error in the stat vector, which is given by the Kalman predictor, is sufficient to minimize the first term of σ^2 in Eq. (III.5). We set

$$P_{t+1} = E_j\{[A(t+1) - \hat{A}_{t+1}][A(t+1) - \hat{A}_{t+1}]^T\} \qquad (III-6)$$

which is the prediction-error covariance matrix; then we can rewrite Eq. (III.5) as

$$\sigma^2 = Y^T(t)P_{t+1}Y(t) + q^2(t+1) \qquad (III-7)$$

P_{t+1} is a covariance matrix and therefore is positive and symmetric. Therefore as the Kalman prediction minimizes the diagonal elements of P_{t+1}, it minimizes also its given values, since they are non-negative. Then it is clear that the quadratic form $Y^T(t) P_{t+1} Y(t)$, and therefore the prediction error variance, are minimized.

As the Kalman predictor has been proved to be a theoretically well-suited tool, will state the recursive relations of its estimation procedure:

Initializations

$\hat{A}_{t_0} = A_0$ mean value of the intial state (vector of coefficients)

$P_{t_0} = Q_A$ covariance matrix of the initial state

At time $t-1$ we computed the prediction of the state at time t, $\hat{A}_t$ and its covariance matrix P_t. We can then compute

$$K(t) = \frac{P_t Y(t-1)}{Y^T(t-1) P_t Y(t-1) + q^2(t)} \qquad \text{(III-8)}$$

and

$$P_{t+1} = P_t + Q_V(t) - \frac{P_t Y(t-1) Y^T(t-1) P_t}{Y^T(t-1) P_t Y(t-1) + q^2(t)} \qquad \text{(III-9)}$$

At time t we get a new observation sample of EEG, $y(t)$. We can compute a prediction of the state (vector of coefficients), at time $t+1$ using the vector $K(t)$ we just calculated:

$$\hat{A}_{t+1} = \hat{A}_t + K(t)[y(t) - Y^T(t-1)\hat{A}_t] \qquad \text{(III-10)}$$

We can also update $Y(t-1)$ to get $Y(t)$, by sifting the components of $Y(t-1)$ and introducing the last sample $y(t)$. We are now ready to repeat the same sequence: compute $K(t+1), P_{t+2}$, get $y(t+1)$, compute the estimation $\hat{A}_{t+2}$ and update $Y(t)$ into $Y(t+1)$. For the block diagram of this algorithm see figure 1.

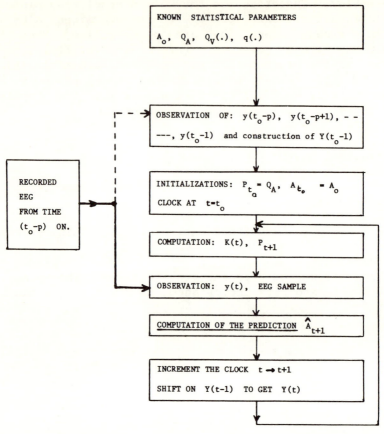

Figure 1. Block diagram of Kalman prediction algorithm.

The particular form of the vector Θ, taken as $Y(t-1)$, whose components are in fact samples of the observed stochastic process, does not cause any complication. Indeed, $Y(t-1)$ is completely known at time t. But the interesting property of some Kalman filtering mentioned at the end of Appendix II -- the possibility of pre-computing the matrices P_u and $K(u)$ -- is no longer true: since the observations constitute $Y(t-1)$, the computation of P_u and $K(u)$ requires knowledge of the signal.

Before we can apply this algorithm to the EEG signal there are four parameters, previously assumed to be known, which have to be selected:

- the value of A_0, the vector of coefficients at the starting time t_0
- the covariance matrix Q_A of this initial state
- the state noise covariance matrix $Q_V(u)$ at each time
- the observation noise variance $q^2(u)$ at each time

We could therefore conduct a parameter study to find reasonable values. It happens that the algorithm is not extremely sensitive to these choices, but they have to be within a certain range.

An interesting aspect of this identification algorithm is its real-time feasibility. Let's examine in some detail the computational requirements of the Kalman predictor. For evaluating the number of multiplications and additions, we give a table, figure 2, where each line reflects the computation of a term. These lines are sequentially ordered and are relative to one sample. The total number of operations required for this recursive estimation algorithm is thus of $2p^2+4p+1$ multiplications and $2p^2+3p$ additions per sample. Running on a PDP-12 with $p = 10$ we would require about 23 ms per sample, which allows us only to analyse an EEG signal sampled at no more than 43 Hz which is not quite fast enough. However we can reach real-time feasibility either by using a slightly faster computer or by using p parallel processors. It can be shown (25) that $2p+4$ sequential multiplications and $3p + 1 + \log_2 p$ sequential additions per sample are required to perform the same computation. For an order $p = 10$, and 53 μs per multiplication and 45 μs per addition (memory--to-memory execution times for the PDP-12), the execution time would be 2.8 ms.

Stabilization of the Estimated Filter

The predictor coefficients obtained by the Kalman filter do not necessarily give a stable model. The transfer function of the filter whose input is the white noise $e(t)$ and output the EEG signal $y(t)$ can be

GAIN	MULTIPLICATIONS OR DIVISIONS	ADDITIONS
$P_t Y(t-1)$	p^2	$p(p-1)$
$Y^T(t-1) P_t Y(t-1)$	p	$p-1$
$[Y^T(t-1) P_t Y(t-1) + q^2(t)]^{-1}$	1	1
$K(t) = P_t Y(t-1)/[Y^T(t-1)P_t Y(t-1) + q^2(t)]$	p	0
ERROR COVARIANCE MATRIX		
$K(t)Y^T(t-1)P_t$	p^2	0
$P_{t+1} = P_t + Q_V(t) - \dfrac{P_t Y(t-1)Y^T(t-1)P_t}{Y^T(t-1)P_t Y(t-1) + q^2(t)}$	0	$p^2 + p$
PREDICTION		
$Y^T(t-1) \hat{A}_t$	p	$p-1$
$K(t)[y(t) - Y^T(t-1) \hat{A}_t]$	p	1
$\hat{A}_{t+1} = \hat{A}_t + K(t)[y(t) - Y^T(t-1) \hat{A}_t]$	0	p
TOTAL	$2p^2 + 4p + 1$	$2p^2 + 3p$

Figure 2. Computation requirements for Kalman predictor per sample.

evaluated by taking the Z-transform of the time-domain AR model:

$$y(t) = a_1 y(t-1) + a_2 y(t-2) + \cdots + a_p y(t-p) + e(t)$$

$$\bar{y}(z) = a_1 \bar{y}(z)z^{-1} + a_2 \bar{y}(z)z^{-2} + \cdots + a_p \bar{y}(z)z^{-p} + \bar{e}(z)$$

(III-11)

then:

$$H(z) = \frac{1}{1 - \sum_{i=1}^{p} a_i z^{-i}} \quad \text{(III-12)}$$

Although the EEG, $y(u)$ may be considered as the output of an occasionally unstable filter, we cannot use such an unstable filter for prediction, because a rather long term predicted signal would be of meaninglessly high amplitude (since we know that EEG amplitude is bounded). Therefore we choose the policy of forcing our estimated filter to be stable. Thus the poles of $H(z)$ (the zeros of the denominator of $H(z)$) have to be inside the unit circle on the Z-plane. This condition has to be checked and some correction performed in case of instability. Unstable estimated filters come mainly from two causes: Roundoff errors in the filter recursive equations, and some low-probability, locally unstable signal segments. In both cases the covariance error matrix becomes ill conditioned. Fortunately the adaptative structure of the filter restabilizes it after a few samples. We encounter these unstable estimates only about 5 percent of the time. We used two different ways to check the stability of the estimated coefficients:

- direct test on the poles
- test on the reflection coefficients

The direct test on the poles requires lengthy computation, to extract the roots of the polynomial

$$z^p - a_1 z^{p-1} - a_2 z^{p-2} - \cdots - a_p = 0 \quad (\text{III-13})$$

We use iterative methods whose convergence can be slow with poor initial guess (Newton-Raphson). However, poles outside the unit circle can be simply moved back inside, or their image inside substituted for the original outside values, which keeps constant the magnitude of the filter frequency response.

Testing the reflection coefficients k_i requires their computation from the AR coefficients (3). This can be done recursively. First the temporary computation variables $b_i^{(i)}$ are initially set:

$$b_j^{(p)} = -\hat{a}_j, 1 \leq j \leq p \quad (\text{III-14})$$

Then for i vaying over p, $p-1$, ...1 in that order, we obtain the k_i by repeating:

$$k_i = b_i(i)$$

$$b_j(i-1) = \frac{b_j(i) - k_i b_{i-j}(i)}{1 - k_i^2} \qquad 1 \leq j \leq i - 1 \qquad \text{(III-15)}$$

A necessary and sufficient condition (3) for the stability of $H(z)$ is that these reflection coefficients satisfy:

$$|k_i| < 1, \ 1 \leq i \leq p \qquad \text{(III-16)}$$

This simple algorithm requires only $(p^2+p-2)/2$ additions and (p^2+p-2) multiplications to check the stability. As for the correction, when the test is negative, we tried two stabilization procedures:

- reduction of the reflection coefficients
- contraction of the pole pattern

The reduction of the reflection coefficients whose value exceeds 1 to some arbitrary value below 1 (for example .98) will certainly yield a stable filter. Then the stabilized filter coefficients can be recomputed by the inverse procedure of (III.15). Certainly only one pass is required to achieve the correction but it is not clear what happens to the poles and therefore, as we experienced it, the spectrum can be rather drastically modified.

The contraction of the pole pattern consists of mapping the Z plane into Z/Δ with $0<\Delta<1$ so that all the poles are brought back inside the unit circle. This is performed iteratively. As soon as a reflection coefficient is found larger than 1 the coefficients are transformed by:

$$a_i = \Delta^i a_i$$

then the reflection coefficients are recomputed. If some poles are still outside the unit circle, a new try is performed with a smaller Δ, and so forth until satisfaction. This method requires certainly more computation than the previous one but the effect on the poles and therefore the spectrum is better behaved. For this reason we chose this approach.

chapter 15

THE ACQUISITION AND GRAPHICAL DISPLAY OF NEUROPHYSIOLOGY DATA WITH AN ONLINE MULTIPROGRAMMING COMPUTER SYSTEM

W. I. WOOD

Department of Biochemistry and Molecular Biology
Harvard University
Cambridge, Massachusetts

A neurophysiology laboratory and several laboratory instruments throughout the National Institute for Dental Research are interfaced for real time data acquisition and control to a central medium sized computer system (Wilson, J., L. Brown, W. Risso, J. Knight, A. Schultz, and F. Brown, 1973). The instruments include a UV-visible spectrophotometer, an X-ray diffractometer, an amino acid analyzer, and four scintillation counters. Computer control of several parameters of a microbial growth experiment and of a behavior training experiment is also performed in real-time. Concurrent data acquisition and control from this variety of instruments and from the neurophysiology laboratory is possible as the instruments have slow or medium data rates (0.001-30

The author wishes to thank Ronald Dubner for whom this neurophysiology system was developed, Frederick Brown who designed and built the necessary interface hardware, Micah Krichevshy who established the NIDR computer system, and John Wilson, chief of the NIDR Computer Branch. Without their effort over the past several years this computer system would not have been possible.

kHz). Interactive data processing and program development are also performed on this computer.

The computer system has two Honeywell 16-bit computers, a DDP-516 (0.96 µsec cycle time) and an H-316 (1.6 µsec cycle time), each with 32k memory (figure 1). The computers are equipped with fixed and moving head disks (3.8M words), card reader/punch, line printer, 9-track magnetic tape, paper tape reader/punch, memory lockout, high-speed integer arithmetic option, and real-time clock. Instrument and laboratory interfaces to the computer system are made through 16 priority interrupts, A/D converter with 16 channel multiplexer, digital I/O device, 20 110-baud teletype interfaces, and 2 1200 baud storage display interfaces. Synchronous telecommunications to the large central computer facilities of the National Institutes of Health (IBM 370 and PDP-10) are also available.

The two computers function independently with dynamically shared access to the two disks (Wood, William I., John J. Wilson, and Frederick Brown, 1975). The other peripherals are dedicated to one computer or the other. The DDP-516 functions on a 24-hour basis for data acquisition and control from the neurophysiology laboratory and all the other instruments. Many small data acquisition programs (500-3500 words) are often

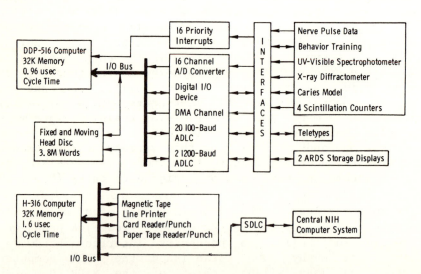

Figure 1. NIDR Computer System.

running concurrently in this computer. These programs run for long periods of time -- some for 24 hours or longer. The DDP-516 computer is also used for remote data processing and display of neurophysiology data. Memory limitations require that these programs be relatively small (maximum of 5k words) with many overlays.

The other computer, the H-316, is used on a sign-up basis for data processing by users, program development and testing, systems development, telecommunications, and hardware maintenance. The computer has 16k of memory available for the large data processing programs and for the FORTRAN compiler.

This dual computer approach to laboratory data acquisition and processing has proved to be a great success in providing: 1. 24-hour trouble-free concurrent data acquisition from several laboratories and 2. hands-on user data processing and program development on a sign-up basis.

The OLERT operating system is used for both computers. OLERT is a real-time, multiprogramming, FORTRAN-oriented system which schedules programs and overlays, performs I/O operations, manages disk program and data storage, and handles interrupts. OLERT also has reentrant format conversion and floating point routines which help conserve memory as they may be called by many FORTRAN programs. An expanded FORTRAN language is used to allow multiple path programming, multiple buffered I/O, and the execution of a special section of a program when an interrupt or timer event occurs. In short OLERT has sufficient capability to allow for real-time multiprogrammed data acquisition and control using FORTRAN programming. The occasional use of assembly language in this computer system is at times convenient and efficient but seldom necessary.

A system as powerful as OLERT is not without its drawbacks. OLERT is large -- occupying half the memory (16k) of each computer. OLERT is slow in some areas -- a minimum of 200 μseconds is required to service an interrupt. OLERT is complex -- bugs in OLERT require expertise and time to correct.

While performing simultaneous data acquisition from several laboratory instruments, this computer system does not have sufficient speed or memory to digitize and analyze, even in a crude manner, analog nerve pulse data. Prior to transmission to the computer system, each nerve pulse is reduced to a single digital event by a special waveform analysis interface (figure 2). An event is a discriminated neurological spike (R1), a stimulus marker -- generally stimulus on (S1), stimulus off (S2) -- or a clock overflow. Each event is transmitted to the computer as two words -- in the first word, a bit designates the event type, in the second word a 15-bit 100 μsec clock is stored. These words are transmitted to the computer memory via DMA to allow a short-term (25-50 msec) maximum data rate of 10 kHz and a long-term maximum data rate of 100-200 Hz.

Nerve pulse data is sensed by a high impedence microelectrode and amplified through a high gain amplifier. The analog wave form is recorded on analog magnetic tape along with the two digital stimulus markers and a voice identification of the experiment. On playback the analog nerve pulse channel is analyzed by a special waveform analysis interface to allow rejection of undesired pulses, i.e., electrical stimulation artifacts or nearby cell pulses. The analysis is made based on two parameters of each pulse: 1. the maximum amplitude and 2. the area under the curve from the start of the pulse to a point where the amplitude has dropped to two-thirds of the maximum amplitude. A user adjustable upper and lower

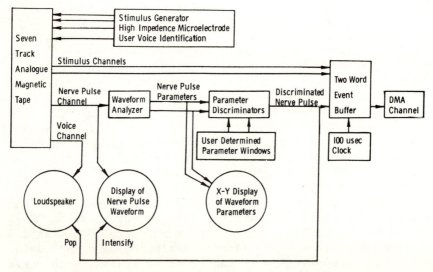

Figure 2. Nerve Pulse Data Interface.

bound or window is provided for each of these two parameters. A discriminated nerve pulse falls within the window for these two parameters and generates an event for which a two-word DMA transfer to the computer is made.

To aid the user, two separate displays are made of the nerve pulse data on storage oscilloscopes: 1. display of each pulse waveform and 2. $X-Y$ display of the two parameters of each pulse. In practice the analog nerve pulse data is played back one or more times for observation of the wave form and of the parameters of each pulse. Based on these observations the parameter windows are adjusted to exclude unwanted pulses and the nerve pulse data is played back again for actual transmission as events to the computer. A discriminated nerve pulse is used to generate an audible pop and to intensify a short section of the waveform display to allow for its easy identification by the user.

Wave form analysis based on the parameters described above has been used very successfully over the past few years. Wave forms have been analysed based on other parameters. 1. The maximum amplitude of a pulse -- when used as the sole parameter, this did not give a sufficient basis for discrimination. 2. The ratio of the maximum rate of rise to the maximum amplitude of a pulse was found to be a good measure of the pulse shape and is insensitive to the maximum amplitude. This has been found to be useful for long bursts of pulses where the amplitude varies greatly.

User communication with the computer is via an ARDS storage display operating at 1200 baud and equipped with a hard-copy unit. This display is located in the neurophysiology laboratory and is used for interaction with the data acquisition program and for display of the processed data. A priority interrupt triggered by a push button is used to terminate a data acquisition run.

A single program with several overlays is used for the acquisition of neurophysiology data. The data which is stored on disk is preceded by a header which contains the following: 1. run identification information, i.e., run number, time, date, user, etc., 2. a list of the variables for the experiment, i.e., the temperature of each stimulus, and 3. a list of pointers to each stimulus in the data -- this will be determined and stored after the data is acquired. Prior to data acquisition, one of several overlays is run, depending on the type of experiment, to set up the header and to allocate the expected disk space requirement. This requires a short dialogue with the user in which the run identification information and experimental variables are entered.

Data is acquired into two 500-word (250-event) buffers via DMA transfer. An interrupt is generated at the end of a buffer to allow the program to switch buffers and to initiate a disk transfer. As the short-term data rate (100 μsec/event) is faster than the system response to an interrupt (minimum of 200 μseconds), some care must be exercised to insure no loss of data during buffer switching. An 8-word overrun at the end of each buffer is allowed for the storage of any data received after the end of buffer interrupt but before the buffer can be switched by the program. A single machine instruction is then used to read the current DMA counter and to start the new buffer. An interchange memory and A-register instruction (IMA) saves the contents of the current DMA buffer pointer for later calculation of the buffer overrun and loads the pointer with the start of the next DMA buffer. Without this IMA instruction, an even more complex buffer switching arrangement would be required for determination of the buffer overrun.

After data acquisition and prior to data processing, a stimulus pointer setup and error detection program is run to examine each data set. This program carries out three functions. 1. It finds stimulus events (S1 and S2) in the data and stores a pointer to each stimulus in the data set header. The stimulus events are used as reference points for user selected time windows during the processing of the data. The stimulus pointers greatly improve the efficiency of the subsequent data processing. 2. The program lists for the user the time of each stimulus, the interval between stimuli, and other stimulus-associated information from the header. This provides the user with a hard-copy record of the important stimulus-associated parameters for each data set. 3. The program verifies that there are no illegal events. A rudimentary check is made of all the event data for erroneous data bits which would indicate a hardware problem.

Use is made of the double precision integer (DPI) to load, store, add, and subtract instructions available as part of the computer's extended integer arithmetic hardware. The stimulus pointers stored in the data set header are in DPI format -- two words (31 usable bits). This allows for pointers beyond the usual one word constraint of 15-bits (32767). The data processing programs also use DPI format in processing the time of events so that the full 100 μsec resolution of the data can be maintained over an entire data set (up to 30 hours). This also allows for much faster processing of the data than could be obtained with floating-point time computation as floating-point arithmetic operations must be carried out by software. DPI format is not supported by OLERT or FORTRAN so that a series of simple assembly language DPI arithmetic routines were written to carry out the DPI instructions. These subroutines are called as needed by the FORTRAN data processing programs.

A common approach is used by all the processing programs to access sections of the data. The stimuli within the data are used as points of reference with each stimulus pair (S1 and S2) being serially numbered and referred to as a trial. A trial generally represents one set of stimulus conditions which may be varied or held constant. Each stimulus is uniquely identified; for example, the notation 5S2 identifies trial 5 stimulus 2. The user specifies a time window which begins at some time before or after a stimulus and has a specified duration. The resolution of the time window is 100 μseconds. This time window can be applied to many user selected trials in various data sets. Thus, it is possible for the user to average or sum over repetitions of an experiment involving several trials and different data sets. For example, the data to be processed could be for the window beginning at S1+5 seconds with a 2.5 second duration averaged or summed over trials 1, 3, and 5 of data set A, trials 1-30 of data set B, and trials 2, 20, 33, and 50 of data set C.

Three graphical display programs are used to present a post stimulus histogram (figure 3), an interspike interval histogram, and a sequential interval histogram of the data. A user dialogue is divided into two sections to allow user specification of 1. the time window and trials to be processed and 2. the parameters of the graph, i.e., bin width, graph size, etc. Suitable defaults are used for the graph parameters whenever possible. After each graph is displayed, the opportunity is given to change the graph parameters while keeping the time window and trials the same. Computation and display of a typical graph requires 15 seconds. In addition to display of the graph itself, also listed are data set identifications, trials processed for each data set, and other input parameters. Thus, rapid and error-free

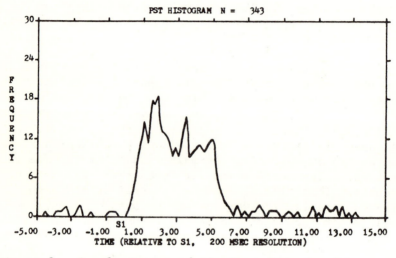

Figure 3. Sample Output of the Post Stimulus Histogram Program.

identification of the output is provided. A non-graphical data summary processing program is used to compute the average and standard deviation of the nerve firing rate, the average and standard deviation of the interspike interval, and other information over the usual time window and series of trials. The data processing programs have 10-15 overlays and use a maximum of 5000 words of memory at any one time.

DISCUSSION

CAPOWSKI: I would love to have a double-precision integer capability. I need that a lot more than floating point.

WOOD: We found it was a much better way to go. A few calls to assembler subroutines do the job very nicely.

CAPOWSKI: How did you produce the output for the figures?

WOOD: I traced it off our hard-copy output.

CAPOWSKI: We've been worrying for years about action potential identification. It's not clear to me how you do it.

WOOD: We use two parameters, the maximum voltage and the area under the curve above two-thirds the peak voltage. This is all done with analog circuitry.

CAPOWSKI: How do you distinguish between monophasic and diphasic spikes?

WOOD: We don't at present. Incidentally, we may go on to using a dedicated mini front end to do this.

FROELICH: Are you using one of your analog channels for timing marks on the tape recorder?

WOOD: No, we just rely on the speed of the tape.

CAPOWSKI: I've seen as much as 10 or 20 msec. worth of "warble" on analog recordings unless I used a separate timing channel.

WOOD: I can't imagine that we have that kind of problem.

KEHL: You use all seven channels at once?

WOOD: Right. Our stimulus timing isn't that accurate anyway.

KEHL: You have to record a timing channel.

BROWN: The only other way is to go through a lot of tape recorders and find one that does it right. We had to do that in Jay Goldberg's lab, and it's not a trivial matter.

KEHL: Which one does it right?

BROWN: The Magnecord was best in our experience, although the Crown isn't bad if you give it time to warm up.

KEHL: On alternate Thursdays, maybe.

L. PUBOLS: I gather you do most of this analysis offline. Do you get any analyses during experiments?

WOOD: You could, but you'd have to know your spike waveform ahead of time.

L. PUBOLS: Have you simultaneously analyzed scintillation data?

WOOD: Yes. We collect from all the instruments at once.

REFERENCES

1. Wilson, J., Brown, L., Risso, W., Knight, J., Schultz, A., and Brown, F. Int. J. Biomed. Comp. 4:135-148 (1973).
2. Wood, William I., Wilson, John J., and Brown, Frederick, ACM/NBS Fourteenth Annual Technical Symposium, In press (1975).

chapter 16

MULTI-CHANNEL NERVE IMPULSE SEPARATION TECHNIQUES

W. M. ROBERTS AND D. K. HARTLINE
Biology Department
University of California at San Diego
La Jolla, California

I. INTRODUCTION

One of the important areas of application of computers in the neurosciences is in the analysis of time-series data, particularly data obtained from recording electrodes. Although the kinds of analysis which can be performed on time-series are virtually limitless, there are a few simple operations, such as signal averaging, spectral analysis, cross- and auto-correlation, convolution, and deconvolution, which are particularly useful. Any computer system which is to be generally useful to neuroscientists must be capable of conveniently acquiring time-series data and performing these operations.

In this chapter we present one application of time-series analysis to the problem of identifying and sorting nerve impulse wave forms in multiple unit recordings (sections II and III), as well as a more general discussion of the problems encountered in handling time-series data (sections IV and V).

This work was supported by grant GJ-43177 from NSF and a grant to UCSD from the Sloan Foundation.

II. MATHEMATICAL DISCUSSION

A variety of nerve impulse sorting techniques have been described in the literature. The choice of method depends on factors such as the number of units to be distinguished, whether the algorithm must operate in real time, whether it is necessary to correctly classify superimposed wave forms, the amount of human intervention allowed, and the costs of equipment and program development. Very simple methods, such as threshold- or window-detectors (4,6,7,14,22) are adequate when only a few units are present or speed is essential. Other simple real-time methods measure a few parameters for each spike, such as the voltage at two times after a threshold crossing (21), the maximum and minimum voltages and their temporal separation (5), or the maximum and minimum voltages and their respective times after the beginning of the spike (15). In many situations, however, it is necessary to resort to more complex procedures (1,8,9,10,11,13,16,17,18,19,20,23,24). These methods can give considerably better results than the simpler ones, but they cannot generally be used for real-time processing without special hardware. They are required when the differences in wave-shape between units are small relative to the background noise, or when the temporal density of impulses becomes so great that there is a significant amount of superposition of wave forms.

One such technique, which requires that the wave form of the nerve impulse from each unit be determined in advance, is to cross correlate a wave form to be identified with the normalized template wave form of each of the units present (10). The unit which is most similar in shape (but not necessarily amplitude) will show the largest peak in the cross correlation. This method cannot distinguish between two units whose nerve impulses have the same shape but different amplitudes since this information is lost when the templates are normalized. The technique is also not applicable to superimposed wave forms.

A better method is to compare the incoming waveforms with the templates for the units by taking the sum of the squared differences in a point-to-point comparison (9,13,16,17). If a histogram of this quantity is plotted for one of the units, it will ofen show a peak near zero, corresponding to spikes from that unit, followed by a trough which can be used as a cut-off for unit assignments (9). Although similar to the normalized cross correlation technique (See below and reference 18), this method gives better results because it utilizes both shape and amplitude information. Again, this method is only applicable to low-density spike trains where superposition events are infrequent.

Correlation techniques can be applied to multi-channel data to distinguish impulses by differences in their conduction velocities (20,23,24). The linear filter technique we will present later also utilizes this type of information.

The problem is complicated further if the wave forms of the units to be identified are not known in advance. One approach is to consider the nerve impulse wave forms as points in an N-dimensional vector space (where N is the number of voltage samples made on each impulse) and to look for clusters of points representing similarly shaped impulses. It is often possible to project this space into a 2- or 3-dimensional subspace with little loss of information(1,8,10,11), making the identification of clusters much easier. It is also possible to construct hardware devices for computing the projections, making real-time analysis feasible (8,10,11).

In situations where the wave form of the nerve impulse from each unit and the statistical properties of the noise are known, it is possible to derive optimal sorting methods. Given a segment of the data record, $v(t)$, sampled at T equally spaced times $(t=1,2,\ldots,T)$, containing at most one nerve impulse wave form superimposed on an independent Gaussian white noise with unit variance, $n(t)$, then:

$$v(t) = w_i(t - \tau) + n(t), \qquad (1)$$

where $w_i(t)$ is the template wave form for the ith unit, with the convention that $w_0(t)=0$, corresponding to the possibility that no nerve impulse has occurred. The problem is to determine which unit is present and its exact time of occurrence, τ. It can be shown (12,18) that the optimal decision procedure, in the sense that it minimizes the probability of making an error, is to choose the values of i and τ which minimize the quantity:

$$\left(\sum_{t=1}^{T} (v(t) - w_i(t - \tau))^2\right) - 2 \cdot \ln(\zeta_i), \qquad (2)$$

where ζ_i is the a priori probability of the ith unit firing, estimated from whatever is known about the firing patterns of the units. In the case that the a priori probabilities are the same for all the units, the optimal procedure is to choose the template which gives the best (least-squares) fit to the data. Several of the implementations described above utilize this type of computation (9,13,16,17). Expression 2 can be stated in a form more familiar to those acquainted with signal detection theory. By expanding the squared term, it can be shown (18) that minimizing 2 is equivalent to maximizing:

$$(w_i(t)*v(t))|_\tau + \left(\frac{1}{T}\right)\ln(\zeta_i) - \frac{P_i}{2}, \qquad (3)$$

where "*" denotes the cross correlation operation and P_i is the average power in the template wave form, $w_i(t)$. Thus, the optimal procedure consists of passing the data through a bank of matched filters, subtracting a constant from each output, and searching for the maximum output. Note that this is similar to the normalized cross correlation method described above, but one subtracts $P_i/2 - (1/T)\ln(\zeta_i)$ from the cross correlation instead of normalizing the template by dividing by $\sqrt{P_i}$. Fairly simple modifications allow for optimal detection in non-white noise, so long as it remains stationary, Gaussian, and independent of the firing of the units. Most sources of electrical noise generated in the recording apparatus will be of this type, but noise caused by small, unresolvable nerve impulses or by variations in the wave form or a spike from a given unit may not fit this requirement.

Other difficulties arise from the assumptions that the wave forms are known in advance and that no superposition occurs. We have chosen to sidestep the problem of initial wave form identification by requiring the experimenter to identify a sufficient number of examples of each unit to allow the extraction of averaged templates. The second problem is more serious, particularly in preparations where the units tend to fire synchronously or in bursts. The optimal method for sorting superimposed wave forms is similar to the method for identifying isolated wave forms, but one must search through all possible combinations of the templates to determine which gives the best fit. This method has been used (13,17) to identify superpositions of two or three units, but the amount of computation becomes prohibitive when the search is carried out for superpositions of more than three units.

In some preparations it is possible to record the activity of a group of cells simultaneously in several electrodes. For example, if the recording is being made from a peripheral nerve, several electrodes can be placed along its length. Although not usually done, the use of multi-channel recordings in other situations (e.g., brain) might, in combination with these techniques, permit determination of activity patterns in a population of simultaneously active cells. This type of multiple electrode data can greatly enhance the accuracy of impulse identification. Furthermore, it makes the problem of identification of superimposed wave forms much more tractable. We have developed a method (18,19) for constructing a set of multi-channel linear filters, each specifically tuned to respond to one of the nerve impulse wave forms and to reject the others. Occurrences of impulses from a given unit are then identified by setting an appropriate threshold on the output of the corresponding filter. Since the filters are linear, the response to the superposition of two impulses will be the superposition of the responses to each separately, so correct identifications of superposition events is also possible. It is important to note that for sorting isolated wave forms this method is not as good as the optimal method described above, and it usually requires the added power of multiple channel data to perform adequately. However, a big advantage in sorting high spike density data is that given a sufficient number of data channels it is fairly undisturbed by the superposition of wave forms. In general, N data channels are sufficient to sort N units; in some cases significantly fewer channels are required, (e.g., using the sequential subtraction technique described below).

The filters are constructed to minimize the sum of the power in the responses to all of the unwanted units and the noise, under the constraint that the peak response to the desired unit is held constant. To generate the filter for the ith unit, we compute $R_i(t)$ using the equation,

$$\hat{R}_i = b\left(\sum_{j \neq i} \hat{W}_j \hat{W}_j{}^a + \hat{C}\right)^{-1} \hat{W}_i, \qquad (4)$$

where $W_i(t)$ is the multiple channel template wave form represented as a column vector, $C(t)$ is the matrix of cross correlation functions of the noise on the different channels, b is a normalization constant, the superscript "a" denotes the adjoint of a matrix, and "^" denotes the Fourier transform operation. $R_i(t)$ is a column vector which characterizes the multiple channel linear filter. The filter output is found by cross correlating $R_i(t)$ with the multiple channel data (i.e., convoluting with $R_i(-t)$) (18,19).

We have found that the accuracy of the identifications can be greatly improved by performing the recognitions sequentially. The data are first passed through the filter corresponding to the most easily identified unit (usually the largest) and its nerve impulses are identified and eliminated from the data by subtracting its template wave form. The simplified data are then passed through the filter for the next unit and the process is repeated until all the nerve impulses have been identified. Thus, each filter can be constructed to take into account only the remaining nerve impulse wave forms.

III. EXPERIMENTAL DATA

We present here one example of the results obtained using data from the cardiac ganglion of the lobster, *Homarus americanus*. More detailed results have been reported elsewhere (18,19). Figure 1A shows a portion of one burst of activity recorded using four bipolar hook electrodes placed along the ganglionic trunk. Seven units (labeled 1-7) are active; templates are shown in figure 2.

The numbers along the top of figure 1A indicate the occurrences of the nerve impulses from the seven units, based on visual identifications. Figure 1B shows the outputs of the seven Filters using the data shown in figure 1a. Outputs were normalized so that the peak response to the corresponding template waveform had an amplitude of 1. Identifications were made by searching for local maxima in the output which exceeded a

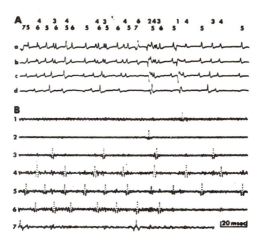

Figure 1. Application of linear filters.
  A. Four channels (a-d) of simultaneously recorded extracellular data from lobster cardiac ganglion. Seven units (1-7) are present and their times of occurrence indicated above the corresponding deflection in the upper channel. Average wave forms for each unit are shown in figure 2.
  B. Outputs from the linear filters based on the templates in figure 2. The output from each filter is shown on a separate line, numbered to correspond to the unit to which it is tuned.

predetermined threshold. A uniform threshold value of .6 was used, but any value between .45 and .74 would correctly identify all the nerve impulses in the data shown.

The sequential recognition scheme described in the preceding section was employed. The output of the filter corresponding to unit 1 was computed first and all nerve impulses from this unit were identified and subtracted from the data. The process was repeated for each of the remaining units in order from 2 to 7.

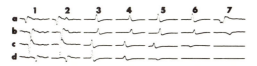

Figure 2. Template wave forms.

In figure 2, the average wave forms of each of the seven units (labeled 1-7 across the top) present in the four pairs of electrodes (a-d) are shown. These are "refined templates" generated from unit firing times obtained by identifying all impulses in eight bursts (except for the first 40 ms. of each burst, where the impulses were unresolvable) from a pass through preliminary filters. There is more noise in the templates for units 1 and 2 than for the others because these two units fired infrequently; they represent the average of only four or five impulses each.

IV. DISCUSSION

This section contains a detailed description of the procedures we used and the problems we encountered in implementing the technique described above. In particular, we want to point out some of the software capabilities we feel are necessary in handling time-series data.

Sampling

Time-series data are usually obtained by sampling a continuous function at equidistant time points. This can be done with no loss of information as long as the sampling rate is greater than twice the highest frequency component of the continuous wave form(2). That is, the original function can be reconstructed exactly from the discrete data, except for errors introduced by jitter in the sampling interval and the finite record length of the computer. It is important that the signal contain no frequencies above one-half the sampling rate because these frequencies will be aliased onto lower frequencies in the sampled data. The most effective way to ensure that no high frequencies are present is to place an anti-aliasing filter immediately in front of the A/D converter input. This filter should have a sharp high-frequency cutoff so that the sampling rate can be made as low as possible while the frequencies of interest are passed unchanged.

Acquisition of multiple channel data presents added problems, particularly if a multiplexed A/D converter is used. Simultaneous sampling of multiple channels is not possible with multiplexed conversion, but this is not generally a problem as long as the delays are fixed and known. At high sampling rates it may be most convenient to make all of the intervals equal; the data shown in figure 1 were obtained in this way. In situations where simultaneous conversion is required, it is possible to digitally reconstruct the simultaneous samples from the delayed samples if the delay is known precisely. This can be done conveniently in the frequency domain by multiplying each frequency by a factor corresponding to the desired time shift(3).

We used sample rates of 5-10 kHz per channel for digitizing spike wave forms. Using a PDP 11/45 computer with an AD001-D analog-to-digital converter we were not able to obtain an overall sample rate of greater than 20kHz, so digitization of eight channels required using tape recorded data played back at 1/4 speed.

The high sampling rates used for digitizing nerve impulse wave forms can quickly generate enormous amounts of data. In some situations the problem can be lessened by discarding the data between nerve impulses, but the high density of spikes in our data required that we sample continuously for the entire duration of the burst. Our sampling program stores the data from each experimental run into a single large disk file consisting of contiguous device blocks, with data from separate channels interleaved. We found that this data format was convenient for speed of storage as well as for the subsequent retrieval of blocks of data for processing or display. Calculated filter outputs were also stored in this format to conserve space and so that the same programs could be used to manipulate the input data and the filter outputs. Shorter data records, such as template wave forms, power spectra, and filter transfer functions were stored in floating-point format with data from different channels in separate records.

Data display

We use three different display programs. A clocked D/A conversion routine allows filming the raw data, using a kymograph camera, for direct comparison with the tape recorded data. We always filmed at least a portion of the data to check for artifacts or noise introduced in the digitization. We also found that this was the most convenient form for rapid visual scanning. The same program was used to display the filter outputs. A second program allows the data to be displayed on an oscilloscope and scanned, either forward or backward, at a rate determined by a knob. Another "fine-tuning" knob adjusts the position of the display when the scanning is stopped. This display capability was essential for the extraction of template wave forms and for verifiction and editing of the computer-identified nerve impulses. The third display program is used for displaying floating-point data files such as templates and filter transfer functions. These display programs have been of general use in displaying other sorts of data such as model outputs.

Signal averaging

We used a general-purpose digital signal averaging program to calculate the averaged template wave forms for the units. The event times could be determined visually from the display or specified in an event table. We sometimes used the latter capability in the following way. Crude templates were found using visual identifications of a few examples of each unit and filters were constructed from these. A segment of data was then passed through these filters to generate a table of the firing times of the identified units. This table could be input to the averaging program, visually checked for errors, edited, and then used to generate refined template wave forms.

To obtain the initial templates, the experimenter scanned through the data to determine the number of units present and to select a representative noise sample for estimating the noise power spectrum. For each unit, he then selected one example to be used as a prototype and set an averaging window on each channel (figure 3). These windows were made as small as possible while still containing the entire nerve impulse wave form; they were necessary because it was often difficult to find temporally isolated wave forms for averaging. The experimenter then chose a sufficient number of examples of each unit to get a good average. To facilitate proper alignment the program displayed the prototype wave form for the unit superimposed on the data display. Discrepancies in alignment were readily detected by the apparent "jumping" of the image with each sweep. Sometimes it was not possible to find a sufficient number of temporally isolated impulses for averaging. In these cases the experimenter identified all of the nerve impulse wave forms in several segments and the program solved for the templates using a deconvolution technique. This method was also used when new templates were generated from the table of firing times obtained from an initial pass through the spike recognition program. In figure 3 a portion of the data from the same experiment as figure 1. is shown. The markers along the top show the times of ocurrence of previously identified nerve impulses, with the number of dots corresponding to the unit number. The intensified portion in the middle of the display shows the prototype wave form for unit three displayed on top of another occurrence of the same unit for alignment. Impulses have been identified without regard for whether they overlap, so a deconvolution method must be used to calculate the templates.

The alignment of sampled wave forms for averaging

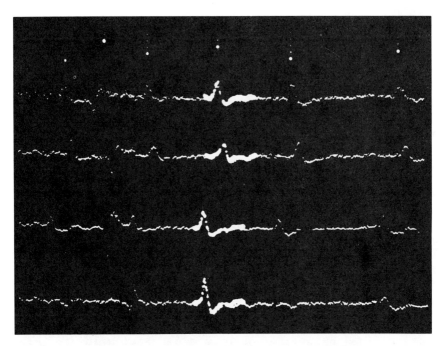

Figure 3. Data and template display.

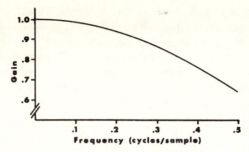

Figure 4. The average transfer function of the linear filter corresponding to rounding event times to the nearest sample point when computing digital signal averages. No phase shift occurs. Frequencies are shown relative to the sampling frequency. In order to obtain a maximum attenuation of 10% one must sample at four times the highest frequency of interest.

presents one important problem: If the event times are rounded to the nearest integral multiple of the sample interval, the resulting jitter will cause the high frequencies in the average to be attenuated, thus blurring it (figure 4). This problem becomes particularly important if one has been careful to use steep anti-aliasing filters and sample at the lowest possible rate. One simple solution is to sample (as we did) at a sufficiently high rate that the frequencies of interest are not distorted. However, since storage space is almost always an important factor in collecting this type of data, this solution is not very satisfactory. Another possibility is to correct the average by dividing (in the frequency domain) by the transfer function shown in figure 4. However, since this represents only the average transfer function, this approach is not applicable when just a few events have been averaged. The remaining alternative is to allow event times which are fractions of the sampling interval and to align the wave forms by digitally reconstructing the samples for fractional time-shifts. This computation can be easily performed on the Fourier-transformed data.

Linear filtering

Any time-invariant linear filter is completely characterized by its response to a δ-function input. The output to any given function is then found by convoluting the input with this impulse response. In practice the data record is likely to be very long, while the impulse response generally contains at most a few hundred points. The most efficient method known for digitally computing such a convolution is to break the data into the largest segments which will fit into core (allowing some overlap), compute the Fourier transform using an FFT algorithm, and perform the computations in the frequency domain (2).

Our linear filtering program uses an FFT subroutine to calculate outputs for multiple channel linear filters, with provisions for searching for threshold crossings and local maxima, subtraction of template wave forms, and outputting event tables. Event times are determined to the nearest discrete time point, but it is also possible to identify event times with greater precision.

V. CONCLUSION

Any computer system intended to handle multiple channel data of the type we work with must have a file structure sufficiently versatile to allow both rapid storage of long data records and subsequent access to shorter segments for manipulation. Since only a small portion of the data can be held in core at any one time, it is usually necessary to perform a large number of read and write operations during the course of the data anlysis. It would be desirable for a programming language to take care of this type of input/output internally, allowing the user to access the data in a much simpler manner. For example, a multi-channel data file could be represented as a two-dimensional array, $F(I,J)$, where I designates the channel number and J the sample number, with all of the input/output made transparent to the user. This would greatly simplify much of the programming, since this type of data transfer constitutes a major portion of most of our programs.

The sampling capabilities must also be quite versatile, allowing precisely clocked multi-channel conversion, either continuously or event-triggered, with the ability to keep track of externally generated event markers. One might also want to perform some simple real-time analysis of the sampled data with the capability of then storing those portions which are of interest. Such a system could be realized with some sort of interrupt-driven sampling and buffering of data running simultaneously with the analysis program.

In addition to the usual plotting routines for the graphical display of data, two other types of display seem essential. The first is a clocked D/A conversion of multiple channel data for filming. We found that this was an easily accessible form of permanent data storage, and often provided the best means of looking at long segments of data. The second requirement is some means of displaying any desired segment of data, in multi-channel format, along with associated event markers.

Once the machinery for accessing the data is established, it is usually fairly easy to perform the desired operations. We found that for ease of computation of Fourier transforms (and hence convolutions and correlations) it was best to first convert the data to floating-point format. In machines without floating-point hardware, care must be used to ensure that a sufficient number of significant bits are preserved when calculating Fourier transforms.

We have tried to limit this discussion to those problems which we have encountered in implementing our spike separation technique. We realize that they represent only one aspect of the data-analysis and experimental control capabilities which would be useful to neuroscientists. However, we believe that they are of sufficiently general interest to merit careful consideration in designing a computer system for neurobiologists.

DISCUSSION

BROWN: How many fibers course through the nerve you are recording from?

ROBERTS: At most ten units per electrode -- the same thing histologically. This one-to-one correspondence between anatomy and physiology is one of the advantages of invertebrates.

BROWN: Can you envision a special-purpose box that would have a mini or micro in it, generating TTL pulses out on separate spike channels?

ROBERTS: Yes. In fact, you can buy similar boxes, but they tend to be pretty expensive.

MCINTYRE: If you're sampling at 10 kHz for a few seconds, you'll need multiple-precision indexes for your virtual arrays.

ROBERTS: We use 32 bits, double-precision.

OLCH: Why don't you throw out the data for epochs where there are no spikes, and just record the length of the empty interval?

ROBERTS: I suppose you could on individual channels, but it's rare to have all channels silent at once.

KEHL: Is your computer fast enough to do all that on the fly?

ROBERTS: No. We can't even sample in real time. With arbitrary slowing down, you could.

KEHL: So your machine just isn't capable of doing it.

ROBERTS: This is a silly thing to do with a computer, in a sense. You're sampling a huge amount of data, most of which you don't want, so you just dump it onto disk.

REFERENCES

1. Abeles, M. Travel into the brain. In *Proceedings of the International Symposium on Signal Analysis and Pattern Recognition*, Technion, Israel (1974).
2. Bergland, G. D. A guided tour of the fast Fourier transform, IEEE Spectrum 6:41-52 (1969).
3. Bracewell, R. *The Fourier Transform and its Applications*, McGraw-Hill, New York (1965).
4. Calvin, W. H. Some simple spike separation techniques for simultaneously recorded neurons. Electroenceph. Clin. Neurophysiol. 34:94-96 (1973).
5. Dill, J. C., Lockemann, P. C. and Naka, K. An attempt to analyze multi-unit recordings. Electroenceph. Clin. Neurophysiol. 28:79-82 (1970).
6. Dyer, R. G., and Beechey, P. A. A window discriminator with visual display, suitable for unit recording. Electroenceph. Clin. Neurophysiol. 31:621-624 (1971).
7. Freeman, J. A. A simple multi-channel spike height discriminator. J. Appl. Physiol. 31:939-941 (1971).
8. Friedman, D. H. *Detection of Signals by Template Matching*, Johns Hopkins Press, Baltimore (1968).
9. Gerstein, G. L., and Clark, W. A. Simultaneous studies of firing patterns in several neurons. Science 143:1325-1327 (1964).
10. Glaser, E. M. Separation of neuronal activity by wave form analysis. Adv. Bio. Med. Engng. 1:77-135 (1970).
11. Glaser, E. M. and Marks, W. B. On-line separation of interleaved neuronal pulse sequences. In

Proceedings of the 1966 Rochester Conference on Data Acquisition in Biology and Medicine, ed. K. Einstein, Pergamon Press, New York (1968).
12. Helstrom, C. W. *Statistical Theory of Signal Detection*, Pergamon Press, Oxford (1968).
13. Keehn, D. G. An iterative spike separation technique. IEEE Trans. Bio. Med. Engng. 13:19-28 (1966).
14. Littauer, R. M. and Walcott, C. Pulse height analyzer for neurophysiological applications. Rev. Sci. Instrument 30:1102-1106 (1959).
15. Mishelevich. J. T. On-line real-time digital computer separation of extracellular neuroelectric signals. IEEE Trans. Bio. Med. Engng. 17:147-150 (1970).
16. Prochazka, V. J., Conrad, B. and Sinderman, F. A neuroelectric signal recognition system. Electroenceph. Clin. Neurophysiol. 32:95-97 (1972).
17. Prochazka, V. J. and Kornhuber, H. H. On-line multi-unit sorting with resolution of superposition potentials. Electroenceph. Clin. Neurophysiol. 74:91-93 (1973).
18. Roberts W. M. Multi-unit nerve impulse sorting techniques. In preparation.
19. Roberts, W. M. and Hartline, D. K. Separation of multi-unit nerve impulse trains by a multi-channel linear filter algorithm. Brain Research 94:141-149 (1975).
20. Schmitt, E. M. Unit activity from peripheral nerve bundles utilizing correlation techniques. Med. and Bio. Engng. 9:665-674 (1971).
21. Simon, W. The real-time sorting of neuro-electric potentials in multiple unit studies. Electroenceph. Clin. Neurophysiol. 18:192-195 (1965).
22. Strong, D. Tatton, W. and Crapper, D. Solid-state amplitude discriminator for neural units. IEEE Trans. Bio. Med. Engng. 18:237-240 (1971)
23. Heetderks, W. J. and Williams, W. J. Partition of gross peripheral nerve activity into single unit responses by correlation techniques. Science 188:373-375 (1973).
24. Sokolove, P. G. and Tatton, W. G. Analysis of postural motoneuron activity in crayfish abdomen. I. coordination by premotoneuron connections. J. Neurophysiol. 38:313-331 (1973).

chapter 17

THE SPIKE PROGRAM: A COMPUTER SYSTEM FOR ANALYSIS OF NEUROPHYSIOLOGICAL ACTION POTENTIALS

J. J. CAPOWSKI

Department of Physiology
University of North Carolina at Chapel Hill
School of Medicine
Chapel Hill, North Carolina

WHAT THE SPIKE PROGRAM DOES

Introduction

In order to extract meaningful quantitative results from neurophysiological experiments, it is necessary to

---

The SPIKE program has been in existence for several years. Many people have invested effort in its preparation. Credit is due to Edward R. Perl, Department of Physiology, University of North Carolina, Chapel Hill, North Carolina, for the original concepts. Louis Schmittroth, Department of Computer Sciences, Montana State University, Bozeman, Montana, was responsible for the original implementation of the program at the University of Utah. Keith Hospers, Department of Physics, University of Utah, implemented the first version at the University of North Carolina. Work on the project has been supported by grants from the Alfred P. Sloan Foundation and the U. S. Public Health Service (NH-11107 and NS-11132).

generate graphs showing the distribution of action potentials (spikes). This is not a theoretically difficult task, but it is quite tedious. The action potentials are recorded versus time; a stimulus might also be recorded versus time. All information necessary to generate the graphs, namely:

(a) spike times
(b) spike shapes
(c) stimulus action

is available, and could be read by a human from strip chart traces. He could then summarize the information in any way. Since this work is quite simple and repetitive, it is possible to program a computer to perform the task. The computer could generate from the recorded information any graphical summary that an experimenter wishes. The SPIKE Program is presently designed to generate the usual statistical summaries of those action potentials which the experimenter selects.

The SPIKE Program has three parts: (1) the digitizing of action potentials and stimulus values, (2) human-computer cooperation to select, according to shape, certain spikes of interest, and (3) the display and plotting of statistical summaries of the distribution of the selected spikes.

Analog to Digital Conversion

Usually, when a neurophysiology experiment is performed, action potential wave forms, read from electrodes, are written onto an analog FM tape recorder. If the experiment is recording a response rather than a background or spontaneous activity, an analog voltage representing stimulus action (temperature, displacement, force, etc.) is simultaneously recorded on a different channel of the analog tape recorder. The purpose of analog to digital conversion is to convert these two tracks of information into digital, quantized form for the computer.

Before beginning an A/D run, the user must be sure that the following five parameters are set correctly.

RUN TIME　　　　　　　　　　　The length of the A/D run

STIMULUS SAMPLE RATE　　　　How frequently the stimulus channel will be sampled

THRESHOLD   A value for the spike
                             channel, which when
                             exceeded, indicates the
                             start of a spike

TWO STIMULUS CALIBRATION   Stimulus values (e.g.,
VALUES   30°C) which the computer
                             will associate with analog
                             input voltage

The user starts an A/D run by typing the RUN command. He must make certain that the information entering the A/D converters at that time is correct. No logic exists for the computer to start sampling when the stimulus reaches a certain value or when the tape recorder reaches a certain time or footage.

Once an A/D run has started, the following events occur. The stimulus channel is sampled at a constant rate determined by the stimulus sample rate parameter. These values are saved in the computer core memory. A clock or counter which keeps the time since the beginning of the run is started. The spike channel is sampled at a rate of about 20 kHz. Only the most recent sixteen samples are saved. When the spike channel sample value exceeds the threshold parameter, a spike is sensed. Then the most recent 16 samples and the next 53 samples are defined to be the shape of a spike. The clock is read to yield the time of occurrence of the spike. The spike shape and spike time are stored in core memory. At the end of a spike, the computer begins sampling the spike channel and checking against the threshold. After 100 spikes have been sensed, the A/D sampling is stopped for about 0.2 seconds while the 100 spike shapes are written from core memory to the computer disk memory. The clock is adjusted accordingly.

The width of a spike which the computer can sample is constrained by the spike time window of

$$\frac{69 \text{ samples}}{2000 \text{ samples/sec}} = 3.4 \text{ milliseconds}$$

This time window limits an A/D run in two ways. First, any spike wider than 3.4 msec cannot be sampled as a single spike. Second, the maximum frequency at which spikes can be sampled is

$$\frac{1}{3.4 \text{ msec}} = 333 \text{ spikes/sec}$$

If either of these constraints is inconsistent with the data coming from the analog tape recorder, the tape recorder speed can be changed to force the tape recorder output to be within the above time limits.

One of the following three conditions terminates an A/D run: (1) The run time parameter is reached, (2) 1000 spikes are taken, or (3) 1000 stimulus samples are taken.

When the A/D run is finished, the number of spikes taken and the number of stimulus values sampled are printed. Then the spikes are displayed on the CRT.

Display and Sorting of Spikes

Once an analog to digital conversion run has been completed, the spike wave forms are displayed on the CRT, ten at a time. Each spike has a displayed number, from one through the number of spikes in the present run. A category number for each spike is also displayed. Initially all the category numbers are set to zero. Figure 1 shows the CRT display.

The user, utilizing some combination of sorting commands and mathematical assistance from the computer, then sorts the spikes according to any criteria he wishes into up to four categories, numbered 0, 1, 2, and 3.

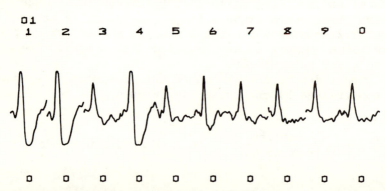

Figure 1. CRT display of spikes, not yet sorted.

Later, summarizing graphs will be generated utilizing only spikes in a certain category. Each time a command to categorize a spike is executed, the category number of that spike is incremented modulo 4. That is, after 3, the next increment sets the category number to zero. The philosophy of the program is that the user is the best judge of whether to include or to exclude a certain spike from a category. The computer simply provides an easy means for executing the user's commands and keeping track of what he has done.

The following commands are available to the user to assist him in sorting spikes.

L (for LEFT)   Cause the CRT to display the 10 spikes to the left (lower in spike number; toward the start of the A/D run) of those currently displayed.

R (for RIGHT) Cause the CRT to display the 10 spikes to the right of those presently displayed.

CA   Increment the category number of all the spikes in the A/D run.

CP   Increment the category number of all the spikes presently displayed on the CRT.

C 24779   Of the 10 spikes presently displayed, increment the category number of the 2nd, 4th, and 9th spike. Increment the category number of the 7th spike two times.

ZA   Zero the category number of all spikes in the A/D run.

An automatic sort feature is also included. To utilize this, the user categorizes up to 10 spikes in each category. He then types "SORT". The program then generates a composite spike, the mean of the spikes which the user has placed in a category. Then every spike in the A/D run is compared to the composite spike. The comparison is performed by summing the absolute value of the error between the present spike and the composite spike for each value in the spike wave form. If the error is less than a sorting tolerance the spike is placed into the category. The sorting tolerance can be changed by the user. Once the automatic sort is done, the user can change any category numbers in order to override automatic sort results.

Experience has shown that with proper adjustment of the sorting tolerance, the automatic sort yields about 95% correct results. The automatic sort has questionable value as a time saver for A/D runs of fewer than 200 spikes. For long runs, an automatic sort followed by the user correcting the sort mistakes seems to be the fastest method of categorizing spikes.

If the user has trouble discriminating among some spikes, a compare option is available. This option displays any two spikes in the A/D run both next to each other and overlaid. This display may aid the user in his decisions. Figure 2 shows the CRT display after a sort.

Statistical Summaries of Spikes

When the user has divided the spikes into desired categories, he may generate on the CRT summarizing graphs

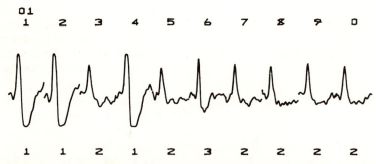

Figure 2. CRT display of spikes after sorting.

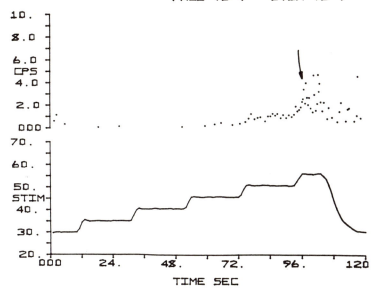

Figure 3. Graphs of instantaneous frequency versus time and stimulus versus time.

of the spikes in a given category. At present, four different graphs are available.

Figure 3 shows a plot of instantaneous frequency versus time and stimulus time. Instantaneous frequency for a given spike is defined as the reciprocal of the time from the previous spike to the given spike. For example, the indicated DOT (figure 3) shows that a spike occurred at 96 seconds after the A/D run started, and 1/4 second after the spike previous to it. The lower graph displays the stimulus action during the A/D run.

The limits on the axes for all the graphs are controlled by the user. It is possible to look at part of a run or more than a run. If the y-axis limits are set too narrow, the graph will "saturate", that is, the data outside the limit range will be displayed at the maximum or minimum range. Stimulus data outside the time range (x-axis) will be clipped. Figure 4 shows a time histogram and a stimulus versus time graph. The times of occurrence of the spikes in a selected category are quantized into bins. The user selects the width of the

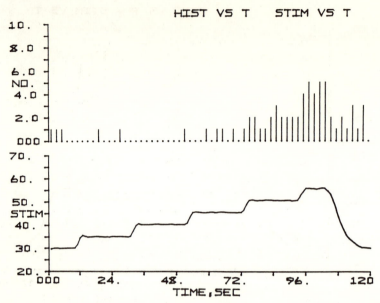

Figure 4. Time histogram and graph of stimulus versus time.

bins and the time limits of the histogram. A maximum of 200 bins is allowed. The count in each bin can also be printed on the teletype.

Figure 5 shows a frequency versus stimulus plot. Here each spike frequency value is displayed at the stimulus value at which the spike occurred. This graph can be of great value for slowly and monotonically changing stimuli (e.g., the application of heat). It must be noted that if a stimulus rises and falls, this graph does not indicate whether a spike occurred when the stimulus passed a given value as the stimulus rose or whether the spike occurred at the stimulus value as the stimulus was falling.

Figure 6 shows an ISI (interspike interval) histogram and a stimulus versus time graph. Here the time difference between spike times of a given category are quantized into bins. The histogram always has 100 bins. The user selects the bin widths. Note that the time scale for the bin values is different from the time scale for the stimulus trace. Only those spikes whose times fall within the time limits shown on the bottom of the graph are used in generating the histogram bin

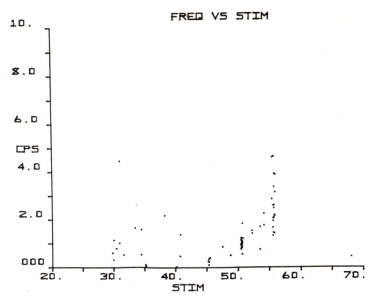

Figure 5. Graph of instantaneous frequency versus stimulus.

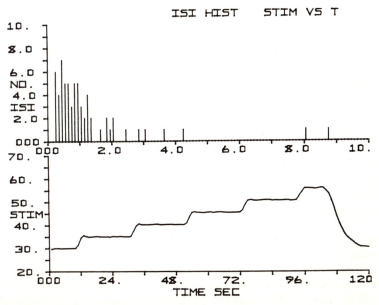

Figure 6. Interspike interval histogram and graph of stimulus versus time.

counts. The histogram bin counts can be printed on the teletype.

The limits on all axes of the graphs can easily be changed; the graphed data are scaled accordingly. A thirty character title can be displayed on any of the graphs. A hard copy of any displayed graph can be made on a Hewlett-Packard plotter by simply typing "PLOT".

Other Features

The most difficult part of using the SPIKE program is performing analog to digital conversion. Therefore the capability of saving the results of the A/D conversion on digital DECtape is designed into the SPIKE program. The user can write the spike and stimulus wave forms onto digital tape and later read them. Thus, more analysis (sorting, graphing) can be done on spike data at a later time without redoing A/D conversion.

The graphs described in the previous section are some of the "usual" graphs used in the field of neurophysiology. The addition of other statistical summaries to the SPIKE program is not a difficult programming task.

THE HARDWARE USED IN THE SPIKE PROGRAM

The PDP-11, LDS-2 Computer System

A block diagram of the computer system on which the SPIKE program runs is shown in figure 7. The heart of the system is a Digital Equipment Corporation (DEC) PDP-11/21 central processing unit (CPU) with 28,000 sixteen-bit words of core memory. Four PDP-11 peripheral devices are shown on the right side of the diagram. The user controls the SPIKE program and receives messages from the SPIKE program through a fast teletype, called a DECwriter. A 256,000 word fixed head disk is used as temporary storage for up to 1000 spike wave forms. DECtapes, small digital magnetic tape reels, are used to store permanent records of spike and stimulus wave forms. The SPIKE computer program is also stored on a DECtape. The SPIKE program uses two channels of a 32 channel

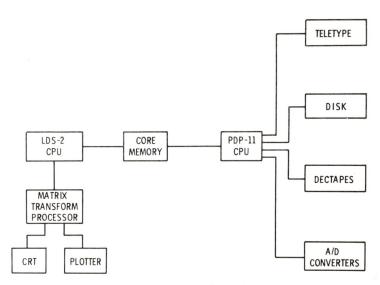

Figure 7. Computer system for the SPIKE program.

analog to digital converter for spike and stimulus analog input.

The graphics display is generated by an Evans and Sutherland line drawing system model 2 (LDS-2). This device is a general purpose central processing unit which also has the capability of executing drawing instructions. The core memory is shared by the two central processing units. The PDP-11 collects and computes data for display, leaves it in core memory, and the LDS-2 passes the data from memory to the matrix transform processor (MTP). The MTP maps the graphic data into the proper coordinate system for display and passes the display data either to the CRT or to the Hewlett-Packard plotter.

This display system has far more power than is used by the SPIKE program. This application could easily be performed on a storage CRT.

The Analog Tape Recorder

A Honeywell Model 7600 seven track analog tape recorder is installed in the computer room to play back analog tapes generated at the neurophysiological experiment. By changing BNC connectors, it is possible to send the output from any channels to the computer A/D converters. The present convention is to use channel one for spike data and channel two for stimulus data. A voice playback track and a time code track (usually channel seven) can be read by the tape recorder. These are not connected to the computer, however.

For any channel, three modes of playback are available. These modes (standard, extended, double extended) allow an increase in frequency response at a given tape recorder speed. It is important that the tape be played back in the same mode at which it is recorded. Since none of the recordings made in present neurophysiology experiments require a great fidelity (frequency response), the mode of recording makes little difference so long as the playback mode is the same. The recording mode should be noted at experiment time, however.

Amplifiers, Oscilloscope, A/D Converters

In order to aid the experimenter when he performs A/D conversion, the electronics system shown in figure 8 has been constructed.

The diagram shows one channel. There are actually two identical channels. The system is designed to give the user complete control over the input to the analog to digital converters. The spike (or stimulus) channel may be inverted to ensure that a positive phase of the spike is first. If the amplifier input is ac coupled, any 60 Hertz noise recorded on the tape recorder is cancelled. The stimulus should be dc coupled as it is usually a slowly varying signal. The oscilloscope monitors the

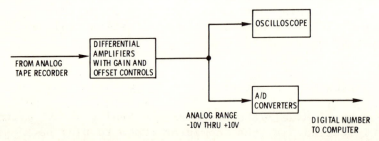

Figure 8. Analog electronics used in the SPIKE program.

input to the A/D converters. It is desirable to use the full range on both channels to maximize the accuracy of the computer graph scaling calculations.

RESULTS

The SPIKE program has been in production use for almost three years. Reaction among the physiologist users has been very good. The ease with which statistical summaries can be generated has allowed more data to be analyzed than would be possible without the SPIKE program. The quality of the graphs has eliminated the need to use an artist for generation of publications.

Constant improvements are being made to both the hardware and the software. The human factors of using the program have been made very smooth.

DISCUSSION

BROWN: I'd like to ask Dr. Brownell. an auditory physiologist: How many action potentials do you gather in one experiment?

BROWNELL: Several million.

BROWN: How about vestibular?

BROWNELL: Somewhat less.

BROWN: We're really talking about two types of neurophysiological data. When you have several million, you can't sit there and examine each one, checking your computer algorithm. In those experiments, we generally have to get well-isolated units which we can just discriminate using a voltage threshold. In your case, how many spikes would be typical in one experiment?

CAPOWSKI: For a single stimulus application, possibly two or three hundred.

BROWN: And you can rapidly scan those?

CAPOWSKI: Easily. An experimenter can categorize a page of 10 spikes by hand in approximately three seconds. Your point is good, but you don't need 100% correct sorting of millions of spikes.

BROWN: As a matter of fact, when doing things like studying binaural convergence, looking at phase coding, which the animal can do with an accuracy of, say, 50 μsec, you don't want the wrong unit in there.

CAPOWSKI: I meant it the other way around. A sort that has nothing spurious included and excludes 95% of good spikes, would probably be all right.

BROWN: Yes, if the ones you lost had no correlation to things you were analyzing. Incidentally, it's absolutely essential that a multi-unit sorting algorithm be able to categorize units when their action potentials overlap, which most hardware and software methods don't do.

ROBERTS: Do you speed up the tape recorder during processing?

CAPOWSKI: Speed is adjusted to make the played-back spike last one or two msec. Some are pretty wide, 8-9 msec, some 1-2 msec.

ROBERTS: So you do realize a real-time gain?

CAPOWSKI: We run our data as fast as we can, given the speed of our computer. Setup time for the hardware far exceeds A/D run time anyway.

BROWNELL: Along similar lines to those Paul was developing, what is the future for chronic implants with multiple units whose wave forms change over time? For anyone who has followed a unit for many hours, you know the wave shape changes. Can you accommodate your template?

CAPOWSKI: The human is the template. We don't do run-to-run analysis. Some other programs do it, but they depend on the assumption that the human can tell the machine that material at one time of day comes from the same unit as material from another time. But our experiments never exceed 6-8 hours.

FROELICH: It seems to me that with the machine making your judgements, you're defining your assumptions better. With a human doing it, it may vary with the weather, how bad your breakfast was, or anything.

CAPOWSKI: That's a valid criticism, but my response is that if you can't define to me how you sort them, you can't define your criteria to the machine.

FRENCH: I'm sypathetic to using human visual parallel processing machinery for pattern recognition. But you can show people your raw data and they can make some terrible changes in the process. So however hard it is, once you've defined the criteria to the machine, you know that what you're getting is consistent. But in your case you just can't tell someone else how you did it.

CAPOWSKI: Fine, but I don't have to tell someone else, because I can show him by example.

FRENCH: Then you can't expect the other person to believe your criterion is invariant.

CAPOWSKI: You're right.

BROWN: This is an old debate that has not originated today, by any means.

chapter 18

NEURAL UNIT DATA ANALYSIS SYSTEM

W. S. RHODE AND V. SONI
Laboratory Computer Facility
University of Wisconsin
Madison, Wisconsin

GENERAL DESCRIPTION

A system (1) has been developed for the analysis of neural unit data collected under a paradigm where two experimental parameters have been systematically varied in a two-dimensional stimulus space. The data are collected assuming that the neural unit will be analyzed only in terms of the times of occurrence of the all-or-none spike potentials. This amounts to neglecting the spike waveshape and the sub-threshold variations in unit potential. Given this assumption, points in the stimulus space are presented and the resulting behavior of a neural unit is recorded as a sequence of event times at which the recorded potential passed a threshold value. This event sequence constitutes a stochastic point process which can then be analyzed using a variety of graphical and statistical analysis techniques. These techniques include post-stimulus time histograms, latency

---

This development benefited substantially from the work of M. Mansfield of the Department of Neurophysiology in developing the plotting programs for the Versatec plotter (7).

This research is supported in part by National Institutes of Health Grant RR00249 from the Biotechnology Resources Branch of the Division of Research Resurces and in part by grant NS 06225 from the National Institute of Neurological and Communicative Disorders and Stroke.

histograms, latency dot displays, joint interval histograms, interval histograms, phase histograms and an assortment of calculations based on the number of neural unit discharges which occur in a specified period.

The analysis system is operated under the supervision of a program monitor which interprets commands and initiates analysis programs. The monitor is also responsible for the transfer of data from a file to the individual analysis programs. The data format of the file is described by a data definition language; the description is called a schema. A compiler translates the schema into an object format file description. A subschema describes the data by name and type for each analysis program. The translation of data from its storage format to the format required by analysis programs provides a flexibility in system development which would otherwise be difficult to achieve.

The principal goals which guided the development of the system were:

a) To provide a general approach to data analysis and thereby eliminate redundant efforts by our investigative staff.
b) To provide a measure of independence between analysis programs and the data structure.
c) To provide the capability to extend the system by the inclusion of additional analyses to meet unanticipated needs.
d) To provide flexible data structuring which allows data to be structured in a manner best suited to each application.
e) To provide a system which would be useful in other areas of biomedical research.
f) To provide a high level language to facilitate interaction between the user and the data.

Since the system is used to analyze neural unit data in an auditory neurophysiology laboratory, the examples of usage given will be in terms of two important parameters in audition: frequency and intensity; however, they could just as well be any two parameters. Both parameters do not have to be varied; one or both could be fixed. Therefore, this is a general analysis system applicable to a variety of experiments. The results of each analysis can be displayed, printed, plotted, or written into a data file upon request. The details of the unit analysis programs will not be described. Rather the general approach to implementing the data analysis system will be emphasized.

ANALYSIS

The data consists of the times of occurrence of discrete events or neural discharges called 'spikes', and is a stochastic point process. A diagram of such a neuroelectric potential is shown in figure 1. Whenever the magnitude of this recorded potential becomes greater than some arbitrary value called the detection threshold, the time is recorded as a data value. A particular combination of stimulus parameters can be repeated several times, permitting an averaging of the response. The parameters can be systematically or randomly varied in the process of presenting points or a particular combination of parameters, in the stimulus space.

Figure 2 illustrates this process for a stimulus space of four points. The extent of the stimulus space is defined by four parameters, low and high SPL (sound pressure level) and low and high frequency. The number of points in the stimulus space is then determined by the parameters 'delta SPL' and 'delta frequency'. The stimulus space shown above has 35 points. These points can be presented to the subject in any order. In general

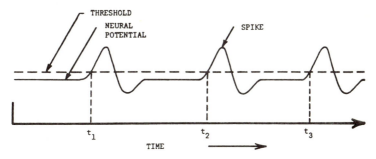

Figure 1. The sequence of times $(t_1, t_2, t_3, ...)$ corresponding to the time at which the electrical potential measured with a microelectrode from a neural unit becomes greater than a predetermined value called the detection threshold which is used to differentiate the desired neural discharges from noise and other (smaller) spikes. This sequence of times is a stochastic point process.

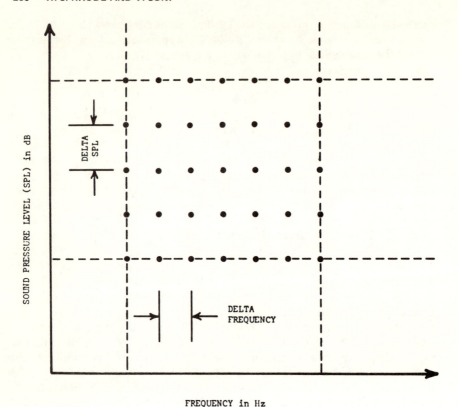

Figure 2. A stimulus space specification.

as far as the analysis program is concerned, the x and y variables of the stimulus space can be any parameters with a range and increment specified for each.

At each point in the stimulus space a record is made of the time of occurrence of each neural discharge for each repetition of the stimulus. These records can then be analyzed with the following types of analyses: 1) post-stimulus time histogram; 2) interspike interval histogram; 3) joint interval histogram; 4) latency histogram; 5) period (or phase or cycle) histogram; 6) latency dot pattern; and 7) autocorrelogram. The algorithms that are used are specified in the Appendix. A discussion of cycle histograms can be found in Stein (2) while other analyses are discussed by Kiang, et al. (3). The results of each of these analyses can be displayed for the entire stimulus space or for any part of it. A sample plot is shown in figure 3 for post-stimulus time histograms where a part of the stimulus space, 3700 Hz to 4900 Hz and 20 dB to 70 dB SPL, was analyzed. One can select any part of the complete stimulus space for analysis. The identification of the data and the parameters for the analysis are printed in the legend. Five trials for each stimulus combination were given. The stimulus duration was 500 msec and the display window is set to 1000 msec with 100 bins. A similar output could be obtained for interspike interval histograms, phase histograms, joint-interval

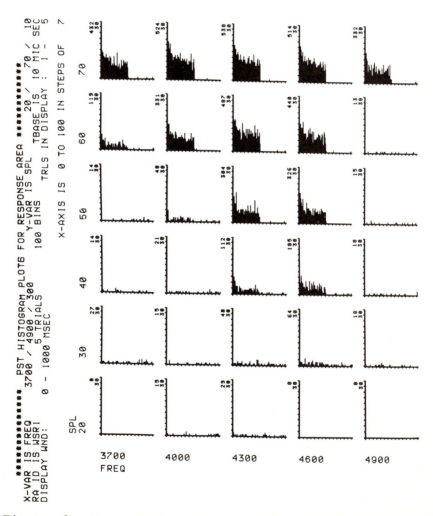

Figure 3. Poststimulus time histograms are shown for a part of the complete stimulus space which had been specified.

histograms, latency histograms, autocorrelogram, and the latency-dot display.

By respecifying the extents of stimulus space or starting the display at a particular point in the stimulus space, single histograms can be plotted as shown in figure 4A. An interspike interval histogram is shown for 1300 Hz at 90 dB SPL with the x-axis extending to 50 milliseconds. At low frequencies neural discharges in the peripheral auditory system tend to be phase-locked to the cycle of a sinusoidal stimulus and therefore certain interspike intervals are dominant, that is, intervals which are a multiple of the stimulus period. By changing the horizontal scale to 10 milliseconds these modes appear in the interspike interval histogram, as shown in figure 4B. An autocorrelogram of this data shows the same frequency dependency. The display parameters which can be varied will be discussed in the next section.

There are a number of measures of unit response which can be explored, such as mean latency and its variance, phase (for units that phase-lock) and synchronization coefficients, curves of cumulative phase shift vs. frequency, and isorate curves. The mean latency is derived from the latency histograms along with the variance. The phase and synchronization coefficients are derived from phase histograms and are the mean and a measure inversely proportional to the variance of the distribution. A period histogram is shown in figure 5 for the same stimulus conditions shown in figure 4. Marked stimulus locking of peripheral auditory nerve fibers is the rule at low frequencies. The range of the sync coefficient is 0 to 1, with 1 implying perfect locking or all discharges in a single bin of the histogram. The period histogram has been used extensively to explore the behavior of the auditory system when two tones, usually phase-locked, are the stimulus. Several modes will appear in the phase histogram. The amplitude and phase of the response are then usually recovered either by Fourier techniques or by fitting of sinusoids to the histogram (4).

Parametric curves can be generated to explore the unit's response characteristics. Some measure of the unit's responsivity to each stimulus point can be tabulated in an array for examination. These measures can be spike counts, mean and variance of latency or phase and synchronization coefficient. One of the most frequently used analyses is the isointensity curve, which is a plot of the number of spikes vs frequency at

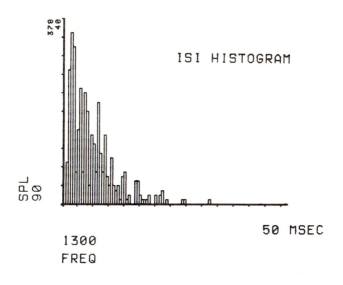

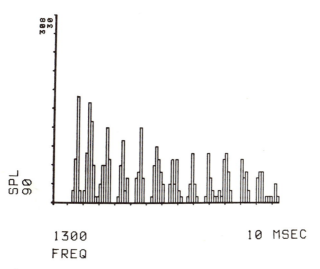

Figure 4. An interspike interval histogram for the same neural unit shown in figure 3. Frequency = 1300 Hz and SPL = 90 dB. A display window of 0 to 50 msec is used in A. There appears to be some large variability in the number of spikes/bins which is explored by increasing the temporal resolution of the histogram in figure 4B where the display window is set to 0 to 10 msec. The multimode histogram is an expression of phase locking which occurs at low frequencies in the peripheral auditory system.

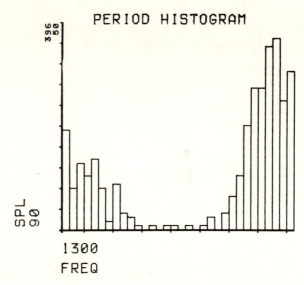

Figure 5. A period histogram for a stimulus of 90 dB SPL at 1300 Hz showing prominent phase locking.

constant sound pressure level. A set of isointensity curves is plotted in figure 6 for SPLs of 20 to 90 dB in increments of 10 dB. Five trials (repeats of the stimulus) were given. Various data file identifiers are also given. A different symbol was used in plotting the curve corresponding to each intensity. Similar displays can be made of the mean and variance of latency, mean and variance of the period histogram, isorate curves and cumulative phase curves. One can readily observe the spread in effective frequency range for the nerve fiber in figure 3 as SPL is increased. These curves define the 'response area' of a neural unit. These curves could also have been plotted with normalized rate as the ordinate, i.e., with vertical scale expressed as a percentage of the maximum observed discharge rate. The time window, a period relative to the synchronization or start of a stimulus trial for which the spikes are counted, can be specified. This allows the elimination of onset transient effects or, in general, the exploration of selected temporal periods of unit behavior.

Other analyses can be performed using measures of unit behavior. Isorate curves are derived from the spike array; they are a plot of the intensity necessary to effect a specified rate of discharge vs frequency.

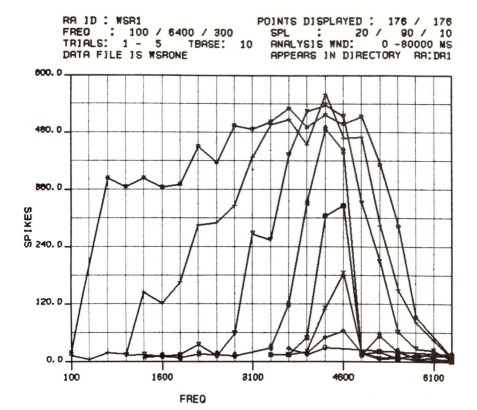

Figure 6. The spike counts/stimulus point in a 100-6400 Hz, 20-90 dB SPL stimulus space.

Similarly, the cumulative phase curves are derived from the array of phases. They are a measure of the rate of change of phase as a function of frequency. More importantly, these measures illustrate how analysis can be performed using measures derived from the point process rather than using only primary data or the point process itself.

MONITOR OPERATION

Each analysis routine is an independent program which interacts with the data through a monitor program. The monitor is responsible for the generation of a table of pointers which identify program variables and data every time a program is to be executed. The storage structure of the data is specified by a schema the description which specifies data names and types, while the variables needed by an analysis program are described by a subschema. The concept of a schema and sub-schema have been borrowed from database management system technology (5).

The monitor has three principal functions: command entry, parsing and execution of analysis. Once initialized the monitor waits until a line of characters is entered. The line of code is then parsed into a SPECIFY, CREATE, OUTPUT, or EXIT command. If an invalid command is entered, an error message is given. The basic structure of the monitor program is shown in figure 7. The system is programmed for a Harris 6024/5 computer (formerly Datacraft). The system program loader supports multiphase, multisegment overlays. This feature was used to minimize the memory size requirements of the program. Only one segment of the program is core resident at any one time. The monitor requires 4000 words while each analysis program requires between 2k and 5k words. Each program exists as a disc load module, which facilitates

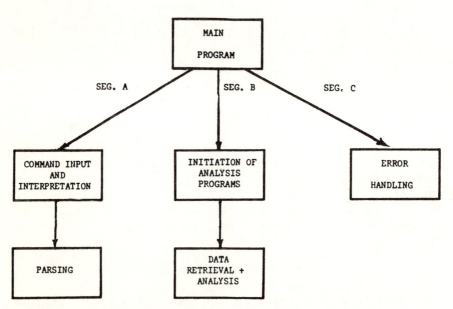

Figure 7. The organization of the monitor program as a multisegment program. Only one segment is core resident at any one time. The maximum size of the monitor is 4k while the maximum size of any analysis program is less than 5k due to the use of overlay techniques.

rapid execution. The computer system configuration is described in a companion chapter (6).

The specify commands are used to furnish to the system the following parameters: data file name, the range of parameters for the stimulus space, the identity of the abscissa parameter, time base, number of bins, output devices, range of trials to be analyzed, time windows for both display and analysis purposes, whether histograms are to be shaded, whether the default parameters are to be used. The default parameters are preset and can be substituted for the current set of parameters upon request. Only the first two letters of a command need be given for each command. For example: ODEVICE PLOTTER or OD PL are equivalent and state that the output from any analysis that will subsequently be performed is to be directed to the plotter. Any of the possible output devices can be specified as current output devices. The printer, plotter and display could simultaneously be output devices.

The output command causes the specified analysis to be performed with the results of the analysis directed to the specified device(s). Analysis can take the form of isorate curves, spike count curves (see figure 6), phase, synchronization coefficients and cumulative phase (when period histograms are available), and the mean and variance of the latency. The command to perform each of the above analyses can have a parameter appended to it to specify what the dependent parameter should be for analysis. For example in the following command:

OU(tput) SP(ikes)(SPL = 90)

the spike count vs frequency curve at 90 dB SPL will be 'displayed' on the output device. Parentheses in the command indicate optional code.

The histograms which can be computed are post-stimulus time histograms, interspike interval histograms, latency histograms, joint interval histograms, period histograms and autocorrelograms. The command to execute any of these analyses can optionally have one or two parameters appended to specify where in the current stimulus space the analysis should begin. Other output commands include the 'display' of latency dot patterns, the current range of parameters for the stimulus space, the current parameters, and the directory entries. The results of an output parameter command are given below.

OUT PARA

```
DIRECTORY FILE NAME : RA:DR1
DATA FILE NAME : WSRONE                    RA ID : WSR1
     TRIALS :      1 - 10                  NBINS :   100
     TBASE :       10 MICRO SEC            NTH SPK : 1
     DISPLAY WINDOW :       0 - 1000 MSEC
     ANALYSIS WINDOW :      0 -80000 MSEC
     NHCOL = 16
     LDROW = 4
     % RATES FOR ISORATE CURVES : 50
     OUTPUT VARIABLE FOR RA CURVES IS X-VARIABLE
     PLOT SYMBOL SIZE = .12 INCHES
```

A third type of monitor command is CREATE. This command instructs the system to compute the arrays for spikes, phase coefficients, synchronization coefficients, cumulative phase, mean of the Nth spike latency and variance for the Nth spike latency. These arrays are sorted and used whenever the appropriate output is requested. Their existence facilitates further analysis.

The last command is merely an exit from the system, EXIT, which returns control to the Harris Disc Monitor System.

THE DATA DESCRIPTION LANGUAGE

The schema DDL, or Data Description Language, is used for describing a Data Base which may be accessed by many programs. The schema is a description of the names and characteristics of the Data Items and Data Aggregates of the data base.

Definitions:

    A Data Item is an occurrence of the smallest unit of named data. It is represented in a data base by a value.

    A Data Aggregate is an occurrence of a named collection of data items within a record. There are two kinds: vectors and repeating groups. A vector is a 1-dimensional sequence of data items having identical characteristics. A repeating group is a collection of data that occurs a number of times within a record occurrence. The collection may consist of data items, vectors, and repeating groups.

The sub-schema is a description of the data items and data aggregates (arrays) which are of interest to a program. This allows an individual programmer to be concerned only with that portion of the data base relevant to the program he is writing. The rest of the data base can be changed without modification of the programs which do not use any of the changed part of the data base.

The description of the data is straightforward. There are five data types: integer, real, alphanumeric (string), vector and repeating group. An occurrence clause may be appended (see figure 8) to a data type to indicate the number of times the data item will occur in the record. A level number is also appended to the description to indicate the order in which the variables change. The schema for the response area program is listed in figure 8. It was found to be advantageous to impose certain requirements in order to facilitate rapid analysis. A number of data items must be present along with a variable called STATB or status table. This array contains an address pointer for each point in the stimulus space which indicates the start of the data in the file. The variable length data incorporate information about the length of the vectors into the data file. For a stimulus that is repeated N times the first N locations of the data file contain pointers to the end of a sequence of values corresponding to each repetition of the stimulus. This format allows quick access to the response to any individual stimulus and is equivalent to storing the length of the sequence. The advantage of the status table for response area data is that the points in the stimulus space can be present in any order and the data stored in the most appropriate order. The status table tells where to find the data and whether points in the stimulus space were skipped. The status table permits data to be collected with different experimental paradigms, yet to be analyzed in the same manner using the same programs.

```
           01   TIME TYPE INTEGER
           01   UNSN
           01   ANIMAL
           01   FREQ TYPE RG
                     02   LOW
                     02   HIGH
                     02   INC
                     02   NFREQ

           01   SPL TYPE RG
                     02   LOW
                     02   HIGH
                     02   INC
                     02   NSPL

           01   NREPS
           01   NSPON
           01   BINPHT
           01   NUMPHT
           01   REPINT
           01   STMDUR
           01   STATTB TYPE RG
                     02   FREQ TYPE RG OCCURS NFREQ TIMES
                          03   SPL TYPE RG OCCURS NSPL TIMES
                               04   ADDRPT

           01   SPONTB OCCURS NSPON TIMES
           01   DATA TYPE RG OCCURS NFREQ TIMES
                     02   DATARG TYPE RG OCCURS NSPL TIMES
                          03   TSDATA TYPE VECTOR INTEGER OCCURS NREPS TIMES
                          03   PHDATA TYPE VECTOR

           00
```

Figure 8. Source schema or description of data file WSRONE, the file used for all the illustrations in this chapter. A level number is used to keep track of indices for arrays or repeating groups. Type integer was made implicit since every variable in file is integer. Other data types must be explicitly stated. OCCURS clause specifies number of times a repeating group occurs. A separate schema must be written for every database.

The subschema gives each variable name and type used by an analysis program and is straightforward; it consists of a series of 3-word entries in a table. The first two words are the FORTRAN variable name in ASCII while the third is the data type. The monitor is responsible for obtaining the data from the data file for each analysis program after the schema and subschema have been checked.

The present use of the terms schema, subschema and data base take a great deal of liberty with the meaning of these terms as given by Codasyl. Nevertheless, the system implemented using these concepts has proved beneficial to the development of the neural unit data analysis system. The inclusion of new analyses into the system has been greatly facilitated.

SUMMARY

A neural analysis system has been described which borrows from data base management system technology. It has resulted in a flexible and easily modified and expandable system. The use of a description of the data file which can be read by a monitor eases the development of new data files while permitting use of the same analysis programs for many experimental paradigms.

DISCUSSION

MCINTYRE: Is this your descriptor structure, or is it part of the software that came with the machine?

RHODE: What I haven't mentioned is that in another group we're working on a data-base management system, best described in (5); we really abstracted a minimal part of it because the whole thing would be too cumbersome for a laboratory environment.

REFERENCES

1. Rhode, W. S. and Soni, V., Response Area Analysis Program, Monograph No. 1, Laboratory Computer Facility, Madison, Wisconsin (1975).
2. Stein, R. B. The stochastic properties of spike trains recorded from nerve cells. In: *Stochastic Point Processes: Statistical Analysis, Theory and Applications.* ed. P. A. Lewis, Wiley - Interscience, New York (1972).
3. Kiang, N. Y., Watanabe, T., Thomas, E. C. and Clark, L. G. *Discharge Patterns of Single Fibers in the Cat's Auditory Nerve.* M.I.T. Res. Monograph, No. 35. Cambridge: M.I.T. Press (1965).
4. Rose, J. E., Kitzes, L. M., Gibson, M. M. and Hind, J. E. Observations on phase-sensitive neurons of the anteroventral cochlear nucleus of the cat: nonlinearity of the cochlear output. J. Neurophysiol., 37:218-253 (1974).
5. Codasyl. Data Base Task Report, April (1971).
6. Rhode, W. S. A digital system for auditory neurophysiology research. In: *Computer Technology in Neuroscience.* Washingtom: Hemisphere (1975).
7. Mansfield. Graphpac, Monograph No. 3, Laboratory Computer Facility, Madison Wisconsin (1975).

APPENDIX

ALGORITHMS FOR UNIT ANALYSIS

Definitions

BIN(J): the contents of the Jth bin of a histogram.
INT(X): the integer portion of the argument X. e.g., INT (3.2) = 3.
t_i: the time of occurrence of the ith spike.
Display Window: the minimum and maximum times for the abscissa of a display.
BW = the Display Window width/number of Bins. $(t_i)_r$: the time relative to the start of the display window.

1. PST: Post Stimulus Time Histogram

    The histogram bin contents are computed as,
    $$BIN(J) = BIN(J) + 1$$

    $$\text{if } J - 1 < INT\left(\frac{(t_i)_r}{BW}\right) \leq J$$

and if t_i < Display Window, that is t_i must be between limits specified by the Display Window.

   The number of bins and display window is specifiable at analysis time, through the monitor.

2. IST: Inter-spike Interval Histogram

    The bin contents are computed as,
    $$BIN(J) = BIN(J) + 1$$

    $$\text{if } J - 1 < INT\left(\frac{(t_i - t_{i-1})_r}{BW}\right) \leq , \forall i > 1$$

and if $t_i - t_{i-1}$ < Display Window

3. JIH: Joint Interval Histogram

$$BIN(J,K) = BIN(J,K) + 1$$

$$\text{if } J - 1 < INT\left(\frac{t_{i+1} - t_i}{BW}\right) \leq J$$

$$\text{if } K - 1 < INT\left(\frac{t_i - t_{i-1}}{BW}\right) \leq K$$

4. LH: The Nth spike Latency Histogram

$$BIN(J) = BIN(J) + 1$$

$$\text{if } J - 1 < INT\left(\frac{(t_n)_r}{BW}\right) \leq J$$

The computation of the mean and variance of the Nth spike latency obeys the following formulae:

$$\text{Mean} = \frac{\sum_{}^{SPK} t_i^2}{SPK}$$

$$\text{Variance} = \frac{SPK^* \sum_{}^{SPK} t_i^2 - \left(\sum_{}^{SPK} t_i\right)^2}{SPK^*(SPK - 1)}$$

where SPK is the number of spikes in the latency histogram and t_i is the time of occurrence of the Nth spike from each stimulus trial included in the latency histogram.

5. PH: Period or Cycle Histogram

$$BIN(J) = BIN(J) + 1$$

$$\text{if } J - 1 < INT\left(\frac{t_i \text{ modulo } (T)}{BW}\right) \le J$$

where T is the period of the stimulus.

t_i modulo (T) is the remainder of the division of t_i by T. For example 18 modulo (8) = 2. The phase and synchronization coefficient, θ and R, are calculated using a resultant vector calculation in a polar coordinate system:

$$Re^{j\theta} = \sum_{i}^{N} r_i\, e^{j\theta_i}$$

where the r_is are the number of spikes in each of the N bins of the period histogram and $\theta_i = 2\pi i/N + 2\pi/N/2$ which is the phase corresponding to the center of each bin. The calculations are actually performed in a cartesian coordinate system with the following formulae:

$$X = \sum_{i}^{N} r_i \cos\theta_i$$

$$Y = \sum_{i}^{N} r_i \sin\theta_i$$

$$R = (X^2 + Y^2)^{1/2}$$

$$\theta = \tan^{-1} Y/X$$

6. AC: Autocorrelogram

$$BIN(J) = BIN(J) + 1$$

$$\text{if } J - 1 < INT\left(\frac{t_j - t_i}{BW}\right) \le J$$

for all $i, < j$

chapter 19

SEQUENTIAL INTERVAL HISTOGRAM ANALYSIS OF NON-STATIONARY SPIKE TRAIN DATA

A. C. SANDERSON AND B. KOBLER

Biotechnology Program
Carnegie Mellon University
Pittsburgh, Pennsylvania

INTRODUCTION

Although stationarity is often discussed as a problem inherent in physiological signal analysis, remarkably little attention has been paid to either the stationarity requirements for particular types of a..alysis, or to the quantitative characterization of non-stationary signals with the goal of relating non-stationarity properties to physiological correlates. We are interested in both of these problems with specific application to neuronal spike train analysis. The sequential interval histogram (SQIH) technique is a first step toward this description of non-stationarity and offers much more information than is given by the mean rate alone. An important attribute of the technique is that it can be easily implemented on a minicomputer system and is therefore compatible with facilities in many neurophysiology laboratories. In this chapter we present briefly the theoretical basis for the SQIH procedure, then describe in some detail the implementation of this technique on our PDP-8L system, and finally, present an example of its application.

BACKGROUND

A neuronal spike train may be described mathematically as a stochastic point process where each spike is represented by the time at which it occurs. The set of occurrence times $\{t_i\}, i = 0, 1, 2, \ldots$, or equivalently the set of interval lengths, $\{X_i\}$ where $X_i = t_{i+1} - t_i$, completely describes the process.

As defined by Cox and Lewis (1965), in the strict sense stationarity of the process requires that the nth order joint distribution of the number of events in fixed intervals $(t_1, t_1'], (t_2, t_2'], \ldots (t_n, t_n']$, is invariant under translation, that is, the joint distribution is the same for any other set of intervals $(t_1+h, t_1'+h]$, $(t_2+h, t_2'+h], \ldots (t_n+h, t_n'+h]$, for all h and for all n. This requirement is extremely restrictive and less stringent criteria are sufficient for most practical purposes, for example, invariance of the first and second order statistics, $n = 1, 2$. In the practical application to spike train statistics, however, it is common to test only for mean rate stationarity. This corresponds to requiring that only the expectation of the first order distribution be invariant with translation. This is a very weak requirement and is not sufficient to ensure reliable estimation of even first order distributions such as the interval histogram.

While tests for stationarity may be based on higher order moments of the counting or interval distributions, we have chosen to describe a process in terms of its generalized, or non-stationary, probability density function (pdf). The inter-event interval density function, $f(x t)$, of a point process may be defined as the conditional probability:

$$f(x|t) = \text{prob } \{\text{event in } (t + x, t + x + dx) \text{ given an event in } (t, t + dt), \text{ with no events in between}\}, \quad (1)$$

where,

$$\int_0^\infty f(x'|t)dx' = 1, \text{ for all } t. \quad (2)$$

For a stationary process, the pdf $f(x|t) = f(x)$ is invariant with respect to translation of t. This is a statement of first order ($n=1$) stationarity. Our objective is to estimate the function, $f(x|t)$, as a means to characterize the non-stationarity of the process.

ESTIMATION

The stationary pdf, $f(x)$, is most commonly estimated by compiling a histogram of the relative frequencies of occurrence of interval lengths. Given a stochastic point process defined by the sequence $\{X_i\}$, $i = 1, 2, \ldots, N$, this estimate takes the form:

$$\hat{f}(2kh) = \frac{n_k(2h)}{N}, \quad k = 0,1,2,\ldots M \qquad (3)$$

where

n_k = no. of X_i such that $2kh \leq X < 2(k+1)h$,
$2h$ = bin width,
M = number of bins.

Parzen (1962) has discussed the problem of estimation pdf's and has provided an analytical formalism for describing a certain class of estimators. The generalized Parzen estimator takes the following form:

$$\hat{f}(x) = \frac{1}{N} \int_{-\infty}^{\infty} \left(\frac{1}{h(N)}\right) K\left[\frac{x-u}{h(N)}\right] d\hat{F}(u), \qquad (4)$$

where

$K(x)$ = weighting function
$h(N)$ = generalized bin width,
$\hat{F}(x)$ = estimate of the distribution function.

Parzen discusses the desirable mathematical properties of $K(x)$ and $h(N)$ for consistent, unbiased, and efficient estimators. The choice of $K(x)$ and $h(N)$ depends on the type of pdf to be estimated.

The conventional histogram estimate of $f(x)$ defined by (3) above is a special case of (4) for

$$K(x) = \frac{1}{2} \quad -h \leq x \leq h$$
$$ 0 \quad \text{otherwise,}$$

which has been sampled at points $x = 2kh$, $k = 0, 1, 2, \ldots, M$. The method used by Levick and Zacks (1970) to smooth single post-stimulus responses of retinal ganglion cells is also a special case of (4).

In order to choose a weighting function, $K(x)$, we carried out a series of Monte Carlo simulations which evaluated the mean square error in the estimation of a Gaussian and an exponential pdf. The weighting functions used were rectangular, Gaussian, and one-sided exponential. The conclusion from these simulations was that the Gaussian weighting function provided the lowest error for all values of h. Choosing the Gaussian weighting function, we then computed the function $h(N)$ for optimum estimation of each pdf type. The results of these calculations suggest that a function

$$h(N) = 1.0 \nu N^{-0.23},$$

where ν is the standard deviation of the actual pdf, provides a good estimate for a whole class of monomodal pdf's. $K(x)$ and $h(N)$ thus define an optimum estimation procedure for a pdf based on N samples and a priori knowledge of the variance. An example of the convolution of a Gaussian weighting function with a set of data points is shown in figure 1. The value of the estimated function at z is the sum of the contributions from each of the Gaussian functions.

We utilize the Parzen estimation procedure in conjunction with a moving window technique (Bendat and Piersol, 1971) to estimate the non-stationary pdf,

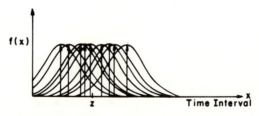

Figure 1. Illustration of the Parzen estimation process. Events which occur at time intervals indicated by the arrows are convolved with Gaussian weighting functions. The estimate of the pdf at point z is obtained by summing the contributions of all the Gaussian functions at that point.

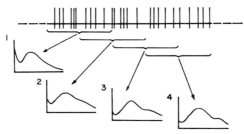

Figure 2. Illustration of the use of the moving window technique to estimate the non-stationary interval pdf. The inter-event intervals contained in each of the windows, 1, 2, 3, and 4, are used to obtain four estimates of pdf's using Parzen estimation. The windows shown are overlapping to obtain greater continuity in the sequential estimates.

$f(x|t)$. The principle of this approach is illustrated in figure 2. The process is assumed to be quasi-stationqry over a range $(t-T/2, t+T/2)$, and the estimate is compiled from intervals contained in the segment T. The selection of T is critical since a small T reduces N, the number of samples available for the estimate, while a large T decreases the temporal resolution of non-stationarity. The compelling reason for introducing the Parzen estimation procedure is to improve the temporal resolution by decreasing the number of samples, N, necessary for an estimate of a given accuracy. The temporal resolution may be increased in this manner by a factor of 5-10 over conventional histogram techniques.

In the implementation of this technique we have chosen to define a segment length in terms of N, a fixed number of samples, rather than T a fixed time period. This allows us to advance segments automatically using the same h for the whole record. We refer to the series of histograms generated in this fashion as sequential interval histograms (SQIH).

IMPLEMENTATION

A hardware and software package has been implemented on the PDP-8L computer to digitize spike train signals and compute estimates of the sequential interval histogram. The basic computer system shown in figure 3 consists of a PDP-8L processor with 8k of core memory and dual DECtape units. Primary I/O devices are teletype and plotter. The spike train analyzer system consists of a hardware Time Interval Digitizer system including a programmable clock, a clock interface module, and a four channel data interface module, a software Time Interval Digitizer Program (TIDIG), and a Parzen Estimation Program (PARZ). These three parts of the system will be described below.

Time Interval Digitizer Hardware

The time interval digitizer hardware consists of a clock interface module with device code 32 (see figure 4), and four data interface modules with device codes 14, 15, 16, and 17. The purpose of these modules is to interface external pulses to the interrupt line, maintain the interrupt flag registers, and assert the SKIP line when necessary. These devices are basically all identical, except for slight differences at the input stage.

The input of the clock module is the output of a programmable crystal controlled clock, capable of generating pulses with a period ranging from 20 μsec. to 4096 msec. This crystal clock is tied to the PDP-8L input-output bus as device 33 and is controlled with the

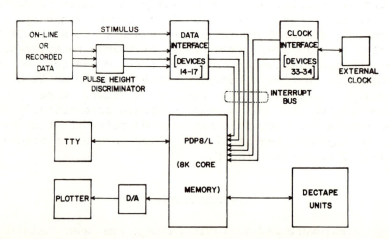

Figure 3. Block diagram of the computer system and peripherals used.

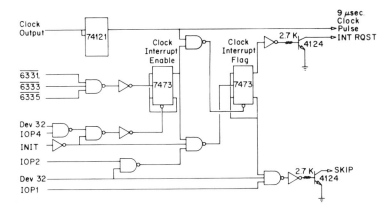

Figure 4. Clock interface module circuitry. Operation of the circuit is discussed in the text. Similar modules are used for the four data interface channels.

IOT instructions 6331, 6333, and 6335. These instructions cause internal clock pulses of 1000, 100, and 10 μsec. respectively, which are used to decrement an internal register set to the value of the accumulator when the instruction was issued. When the register has been decremented to zero, the next pulse will reset it to its original value, and will also be generated as an output pulse. Pulses with periods ranging from 2^1 to 2^{12}, or 4096, times the internal clock period can thus be generated. The width of the output pulse is, however, always either 1000, 100, or 10 μsec., depending only on the internal clock rate. At the clock interface module, the leading edge of this pulse is used to trigger a 74121 monostable which generates a 10 μsec pulse. If the clock interrupt enable register is set, then this 10 μsec pulse is used to set the clock flag register, to indicate that a clock interrupt has occurred. A short pulse is necessary for this purpose in order to ensure that no conflict will arise when the interrupt service routine later tries to clear the flag register.

The interrupt flag register is tied to the INTERRUPT line with an inverter stage consisting of an open collector transistor. This is necessary because standard TTL circuitry has internal active pull up, and would therefore not allow the wired ORing of several devices onto the INTERRUPT line. A similar circuit is used to tie the clock module to the SKIP line. If the interrupt flag register is set, then the IOT instruction 6321 will assert the SKIP line by pulling it down to ground. The rest of the circuitry is fairly straightforward. The 6322 instruction will clear the interrupt flag register, and the 6324 instruction will disable the clock interrupt enable register. This register is also automatically turned off at machine turn on, and automatically turned on when the clock is started with either 6331, 6333, or 6335.

The circuitry for the four pulse train modules is virtually identical to that for the clock module, except that the inputs are derived from external sources and are therefore voltage protected. The two types of input to the data interface are (a) standard digital pulses output from the window discriminator, and (b) stimulus pulses generated either on-line or recorded simultaneously with the spike train data. For the clock interface module, the IOP1 instruction will assert the SKIP line if the appropriate device selection code is present and if the interrupt flag is set. The IOP2 pulse will in turn clear the appropriate flag register. The IOP4 pulse, however, is shared for several operations. The instruction 6144 will enable both device 14 and device 15 interrupt enable registers, while 6154 will turn both registers off. Similarly, 6164 will enable both device 16 and 17, while 6174 will disable both devices.

Time Interval Digital Software (TIDIG)

The Time Interval Digitizer is a hardware/software package on the PDP-8L computer which ties programmable clock pulses and neuronal spike train signals to the PDP-8L interrupt line, and then transfers time interval data to DECtape. The hardware described above provides for up to four different pulse train inputs, though the software to be described below handles only the first two of these channels. In double channel mode, pulses on channel 1 are interpreted as stimulus pulses, and pulses on channel 2 as response pulses. The time intervals transferred in this case are those between the channel 1 and channel 2 pulses. In order to indicate the presence of the stimulus pulse, partitioning zeroes are transferred ahead of all channel 1 pulses. The two types of data transfers handled by TIDIG are shown in figure 5.

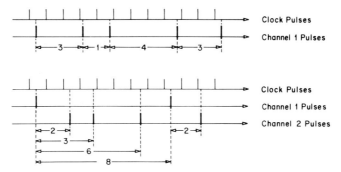

Figure 5. Illustration of the operation of TIDIG in both single and double channel mode. In single channel mode the program counts the number of clock pulses which occur between data spikes (see figure 6). In double channel mode, the program counts the number of clock pulses between each channel 2 (spike train) pulse and the last channel 1 (stimulus) pulse.

The operation of TIDIG is shown schematically in figure 6. The program has a background program, B, which waits for either a clock interrupt or a data interrupt. A clock interrupt causes the program to jump to a clock

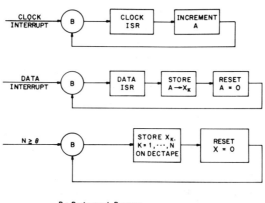

Figure 6. Block diagram of the TIDIG software program.

interrupt service routine (ISR). The clock ISR causes a memory word A to be incremented by 1. A serves as a register which keeps time (counts clock pulses) between spikes. After incrementing A the program jumps back to B. If a data interrupt occurs, the program jumps from B to the data ISR, stores word A (number of clock pulses since the last data interrupt) in a memory buffer as X_k, then resets A zero and jumps back to B. The program periodically checks to see if the number of words in the memory buffer are stored on DECtape, and the buffer is reset to zero in preparation for more intervals. This diagram applies only to the single channel mode operation. The double channel mode is similar except that zeroes are stored as markers preceding each stimulus pulse. Data transfers will start on DECtape unit 1, but will continue on unit 2 when unit 1 is filled, and will then return to unit 1 when unit 2 is filled. Data transfers, once initiated by the user, will continue until a preassigned number of transfers has occurred. In single channel mode this number refers to the total number of data transfers, while in double channel mode it refers to the number of stimulus pulses. In either case it is possible to place the program in a very large loop (2^{24}) which will last until the user halts data transfers.

Because time intervals due to hardware interrupts are stored in a temporary buffer, and then transferred to DECtape in large blocks, it is important to ensure that the buffer will not overflow. This is done with a threshold variable, Θ, which determines when data transfers should begin. It is set by the user, and should contain the expected number of data points to be transferred per second. In double channel mode this has to include both channels as well as the partitioning zeros. Since the threshold is limited to 2^{12}, the fastest pulse rate which can be safely digitized is 4096 pulses per second.

The time resolution can be specified in 10 μsec. intervals, up to a maximum time resolution of 100 μsec. Since the time duration of a single spike is about 1 msec. this is adequate for spike train analysis. Up to 2^{12}, or 4096, different time interval values can be transferred and stored on DECtape, although those larger than 2^{11} will be interpreted by programs expecting 2's complement numbers, as negative integers.

NON-STATIONARY SPIKE TRAIN DATA   281

TIDIG Program Interaction

1. TTY will respond with:
    TIME INTERVAL DIGITIZER
    BLOCK (OCT) = 0000
    ALTER (OR CR)

2. If you wish the time interval data transfers to start at DECtape block number indicated, then type carriage return, otherwise type the octal block number you wish, followed by carriage return.

3. TTY will respond with:

    # CHAN'S:

4. Type either 1 or 2, depending on whether you wish to operate in single or double mode. Single channel mode will expect time interval pulses on channel #1, while double channel mode will expect stimulus pulses on channel #1, and post stimulus response pulses on channel #2.

5. TTY will respond with:

    CLOCK PERIOD:

6. Type the desired clock period as the number of 10 μsec intervals.

7. TTY will respond with:

    # SAMPLES (OR CR)

8. If you wish to record only a certain number of channel #1 interrrupts, then enter that number, otherwise type carriage return.

9. TTY will respond with:

    INPUT RATE:

10. Type the approximate number of data points per second that you expect to transfer. If you are operating in double channel mode, then this should

include both channels as well as the partitioning zeros. This will set an internal threshold so that DECtape motion will be initiated on time, so that the data buffer will not overflow. The largest threshold value which can be entered is 4096.

11. TTY will respond with:

    CONTROL P TO GO

12. Type Control P to start data transfers.

13. Data transfers will stop when the number of channel #1 pulses equals the number set in step 11, or when any TTY key is struck.

14. TTY will then respond with:

    BLOCK (OCT) = n
    CONTROL C TO QUIT
    BLOCK (OCT) = $n+1$
    ALTER OR CR

15. If you wish to continue with more data transfers, then go to step 5, otherwise type Control C to get monitor.

The Parzen Estimation Program (PARZ)

The Parzen Estimation Program is a FORTRAN program on the PDP-8L computer which uses the data stored by TIDIG to estimate sequential time interval probability density functions. The function of PARZ is shown schematically in figure 7. Because of core limitations, the full complement of 4096 different time resolutions cannot be used, so that these are collapsed into a maximum of 100 rectangular bins, which are then Parzen processed. The number of bins and the number of bits per bin can be set by the user. The first and second moments and the variance are calculated for the data, the pre-Parzen estimate, and then printed. If desired, the entire pre-Parzen and post-Parzen estimate can then be displayed on teletype.

The time window was chosen as a rectangular weighting function, and as a function of index number

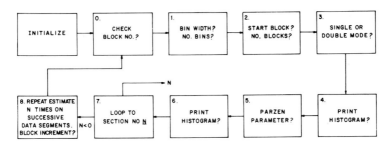

Figure 7. Block diagram of the PARZ software program. The sequence of operations can best be followed by studying the section, PARZ Program Interaction in the text.

rather than time. The Monte Carlo results for $h(N)$ are therefore applicable. The window can be programmed to begin at any block boundary, and to extend an arbitrary number of data points. There is an option which allows the user to automatically estimate sequential sections of the spike train, so that analysis of non-stationarity is simplified.

PARZ Program Interaction

Section 0:

1. TTY will respond with:

    SECT 0. CHECK OCTAL BLOCK #? =1 TO CONTINUE

2. If you wish to list a particular data block on the TTY, then type that block number, in octal, otherwise type -1 and control will pass to section 1. If a block number is specified then that data block will be listed 12 lines of 10 words each, and control will pass back to section 0.

Section 1:

1. TTY will respond with:

    SECT 1. # OF BITS/BIN?

2. Enter the number of bits you wish to accumulate per bin, in the pre-Parzen histogram estimate.

3. TTY will respond with:

    # of BINS?

4. Enter the number of bins you wish to consider in the histogram. Because of core limitations this is limited to 100. In this section, as in all subsequent sections, a response of −1 will result in control passing to the previous section; all other illegal responses will result in the question being repeated.

Section 2:

1. TTY will respond with:

    SECT 2. START BLOCK, OCTAL

2. Enter the DECtape block number, in octal, where histogram estimation is to begin. The program will compute and print the decimal equivalent.

3. TTY will respond with:

    # OF BLOCKS, OCTAL

4. Enter the number of data blocks, in octal, that you wish included in the estimation of the histogram.

5. If above response is 0, then TTY will respond with:

    # OF DATA POINTS, DECIMAL

6. Enter the number of data points, in decimal, that you wish included in the estimation of the histogram.

Section 3:

1. TTY will respond with:

    SECT 3. TYPE 1 OR 2 FOR SINGLE OR DOUBLE MODE

2. Enter a 1 or a 2, in order to compute either a simple or a post stimulus time interval histogram respectively. For the post stimulus case, the number of data points specified in section 2 will refer to the number of stimulus pulses, not the total number of pulses.

Section 4:

1. TTY will respond with:

    SECT 4. PRINT HISTOGRAM? TYPE 1 FOR YES

2. Enter 1, if you wish to print the pre-Parzen histogram.

Section 5:

1. TTY will respond with:

    SECT 5. ST. DEV. DESIRED FOR PARZEN ESTIMATION = ?

2. Enter the desired value of h for the Parzen Estimation Procedure. This can be a floating point number. If it is 0, then the Parzen Estimation will be skipped.

Further Program Flow:

1. The Parzen Estimation Program will begin the histogram estimation procedure, print the results, and then respond with:

    LOOP TO SECTION ? IF -N THEN CYCLE THROUGH N TIMES

2. Enter the section number you wish to branch to, and control will pass there. A negative number allows one to set up a loop counter which will repeat the estimation procedure just specified over successive sections of the data.

3. In the case of a negative number , the TTY will respond with:

BLOCK INCREMENT

4. Enter the desired block increment between successive loops. If each loop consists of two or more blocks of data points, then overlapping sections are easily made possible.

EXAMPLES

Examples of the application of the SQIH system will be discussed for single unit neuronal spike train data recorded from the lateral geniculate nucleus of the cat under Urethan anesthesia. Retinal illumination was uniform in all cases shown.

Figure 8 shows the dependence of the pdf estimate on the parameter h for a Gaussian weighting function and $N = 120$ points. The multimodal nature of the pdf suggests that $h(N)$ should be estimated based on average single mode characteristics.

Figure 9 shows histograms computed for a set of 2400 intervals. Notice that the Parzen estimate is still an improved estimate of the pdf. In order to investigate the stationarity of the spike train from which this histogram was generated, we computed the mean and standard deviation of interval lengths for segments of the spike train. These results are plotted in figure 10. Each segment was 240 intervals long and successive segments overlapped by 120 intervals. Both the mean and standard deviation are reasonably constant for this process.

Figure 11 shows the result of SQIH analysis for the same 20 segments of data. Each segment overlaps the next by 120 intervals. These results suggest that while the mean rate and the standard deviation (figure 10) are relatively constant for this series, the pattern of spike firing can change considerably during the course of spontaneous discharge. Note, for example, the increase in burst firing activity (short interval peak) between the first five histograms and the rest of the record. Also note the appearance and disappearance of one, two, or three distinct modes in the histograms. All histograms are normalized to the highest value. Full scale horizontal is 100 msec. The detailed pattern of multimodal peaks changes significantly during this process. The process does not appear to be strictly stationary in terms of the pattern of firing although the average rate of firing is quite constant.

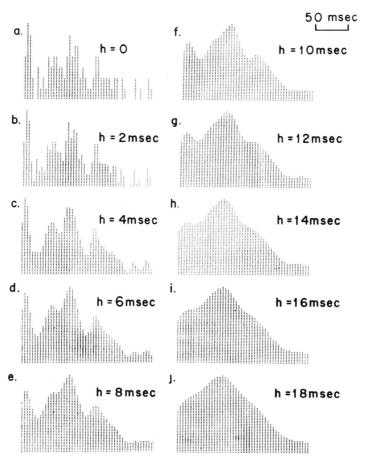

Figure 8. Application of the PARZ program to neuronal spike train data. Ten different values of h were used. N = 120 intervals. This sequence of histograms shows the effects of smoothing on real data. An optimum estimator occurs for this set at about h = 6 msec. When is small the estimate is very noisy. All histograms shown are normalized to the largest value.

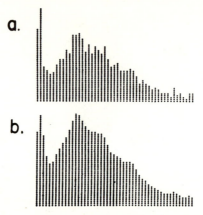

Figure 9. Two estimates compiled from 2400 intervals from the data described in figure 8. (a) Simle rectangular bin histogram estimate. (b) Parzen estimate with h = 4 msec. Full scale horizontal is 100 msec. Both histograms are normalized to the largest value.

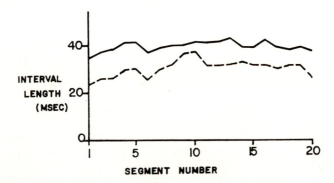

Figure 10. Mean interval length and standard deviation of the interval length as a function of segment number for a series of sequential interval histogram estimates (see figure 11).

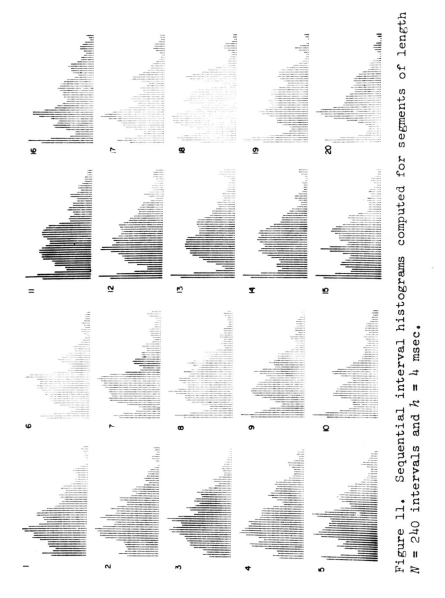

Figure 11. Sequential interval histograms computed for segments of length $N = 240$ intervals and $h = 4$ msec.

CONCLUSION

We have described a technique for computing sequential interval histograms from neuronal spike train data. This chapter has emphasized the practical implementation of the data acquisition and processing system on a PDP-8L computer system. These techniques are compatible on any minimum configuration minicomputer system with tape or disc capability, though the details of hardware design and interfacing procedures will differ. The system is currently in use in the Neurophysiology Laboratory of the Biotechnology Program, Carnegie-Mellon University. The system is used either on-line, in which case TIDIG may be used to store data directly on DECtape, or via an analog magnetic tape recorder. In addition to the TIDIG system and PARZ programs, we commonly utilize a small special purpose computer (Victoreen PIP-400, modified for neurophysiological applications) which allows us to compute conventional histogram estimates as data is being stored by TIDIG. These rough estimates are photographed and provide a useful guide for further analysis using PARZ.

It is hoped that sufficient information has been provided here to make these techniques useful to others as well as to encourage interest in the analysis of neuronal spike train stationarity.

DISCUSSION

D.O. WALTER: I suppose that a series of smoothed histograms suggests the presence of change, but do you have any formal criterion of significant difference?

SANDERSON: That's what we're working on now. We haven't devised these criteria yet. We're doing decomposition analysis and peak-following analysis to follow changes with time, trying to distinguish trends from noise.

D.O. WALTER: Have you written up the considerations that led you to choose Parzen windowing?

SANDERSON: Not yet.

REFERENCES

1. Bendat, J. S. and Piersol A. G., *Random Data: Analysis and Measurement Procedures*, John Wiley, New York (1971).

2. Cox, D. R. and P. A. W. Lewis, *The Statistical Analysis of Series of Events*, John Wiley, New York (1966).
3. Levick W. R. and Zacks J. E., Responses of cat retinal ganglion cells to brief flashes of light. J. Physiol. 206:677 (1970).
4. Parzen, E. On estimation of probability density function and mode. Ann. Math. Stat. 33:1065 (1962).

chapter 20

A GENERAL-PURPOSE ALGORITHM FOR HISTOGRAM GENERATION

P. B. BROWN, J. FROELICH, C. ROBY, AND J. MARLER

Department of Physiology and Biophysics
West Virginia University Medical Center
Morgantown, West Virginia

In writing histogram software, the traditional approach is to write a subroutine for each type of histogram desired. We have decided to write a single general-purpose algorithm for all types of histograms in order to (1) simplify programming of the software package; (2) minimize core requirements; (3) provide maximum flexibility, in that (a) a larger number of histogram types is made available, and (b) new, previously unheard-of histograms can be devised and used without re-programming.

Our package could operate in a minimum environment of 8k of core (not including device handlers and monitor) on a PDP-8 or PDP-12, with any mass storage medium on which output and input files can be maintained simultaneously. Our system provides greater speed and flexibility: we have a 32k disk operating system, 2M disk space, extended arithmetic and floating point processor, two types of graphics terminals, and several alternative forms of hard copy output.

A portable graphics terminal/hard copy unit allows online use in any of the laboratories wired to the departmental computer.

This work was supported by PHS grants NS 09970 and NS 12061, and by NSF grant DCR74-24765.

The "language", or set of commands available, is concise and powerful, and can be learned in an afternoon by anyone familiar with histograms. In order to allow even more concise instruction format and the use of English-like commands, we are also preparing a macro-defining procedure.

We have broken the task of histogram generation into three parts (not all of which are completed):

(1) Selection of data for inclusion in the analysis. This may range from simple selection of individual stimulus presentations to quite complex data-selection paradigms. Input consists of inter- event interval data generated by a Pulse Interval Timer, or equivalent.

(2) Generation of histogram-format interval data. The output consists of a file of histogram bin-incrementing x-axis values. This data can be used to generate histogram displays, for statistical analysis, or in histogram-manipulation procedures (computation of difference histograms, correlation histograms, etc.).

(3) Generation of graphics-format display files, for storage, numerical procedures, or display. These consist of fixed-length strings of successive x-values. other graphics format displays with (x,y) scattergram data. numerical analyses, and histogram manipulation are envisioned, but are not yet completed.

I. INTER-EVENT INTERVAL FORMAT DATA

The input data used for steps (1) and (2) consists of a variable-length string of 24-bit (2 PDP-8 words) numbers, of which the first 6 bits are a channel code (indicating which channel(s) received an input) and the last 18 are an interval code (length of interval initiated by previous event, terminated by current event). The data word format is illustrated in figure 1. Data from our Pulse Interval Timer (PIT) has a variable resolution, depending on clock rate, with a maximum resolution of 10 μsec. A PIT overflow causes generation of a zero output word, signifying no channel input and an

```
                                              MOST SIGNIFICANT 6
                               CHANNEL CODE   BITS OF INTERVAL CODE
                              ⎧            ⎫ ⎧                    ⎫
       FIRST PDP - 12 WORD    | M                                L |
                              | S                                S |
                              | B                                B |
BIT NUMBERS IN 12-BIT WORD:     0  1  2  3  4  5  6  7  8  9 10 11
BIT NUMBERS IN 24-BIT WORD:     0  1  2  3  4  5  6  7  8  9 10 11
BIT NUMBERS IN CHANNEL CODE:    0  1  2  3  4  5  ------------------
BIT NUMBERS IN INTERVAL CODE:   ----------------  0  1  2  3  4  5


       SECOND PDP - 12 WORD   | M                                L |
                              | S                                S |
                              | B                                B |
BIT NUMBERS IN 12-BIT WORD:     0  1  2  3  4  5  6  7  8  9 10 11
BIT NUMBERS IN 24-BIT WORD:    12 13 14 15 16 17 18 19 20 21 22 23
BIT NUMBERS IN INTERVAL CODE:   6  7  8  9 10 11 12 13 14 15 16 17
```

Figure 1. Format of inter-event interval data. Channel code represents input channel active, interval code indicates length of interval terminated by input event.

interval of 1,000,000(8). End of file is end of data, at present. Subsequent versions of our programs will have the capability of handling sub-files with identifying header sections. Our PIT data file names consist of FILENAME.EX, where FILENAME= 6-digit date, and .EX= .file number for that date, from 00-99. An automatic permute feature in the data acquisition program automatically increments the extensions of successive files.

II. GENERATION OF HISTOGRAM-FORMAT FILES FROM INTERVAL DATA

This is step (2) of the three-step analysis sequence outlined above. We will describe it first, because it is simpler than step (1) and establishes certain conventions for command-line format. Input data may be original PIT data: FILENAME.EX, or a data-edited file: FILENAME.Dx. they both are in interval format. The extension Dx allows permuting from 1 through 0 (D then permutes to E, etc.).

The command structure consists of a single command line, of whatever length is compatible with user needs, terminated by a carriage return. The usual line editing is possible before the carriage return, and execution of a command string begins at the terminator. The carriage return locks out all but an abort command during execution. More advanced systems should have a batch or pseudobatch type of operation, where commands could be entered at any time and executed in order as prior command execution is completed. One output file is generated per line, with extension .Hx appended automatically. Hx can be permuted, similar to Dx. The line consists of a file-manipulation command similar to that used on Digital Equipment Corporation operating systems, followed by a slash and code specifying histogram format:

DEV1:HISTONAME.Hx < DEV2:DATANAME.EX (or .Dx)/CODE.

Spaces embedded in the command line are ignored. The above command line means: "Process file DATANAME.EX (or .Dx) on DEV2 as specified in /CODE, and write resulting output as file HISTONAME.Hx on DEV1." Omitted device names imply default devices. If no output file name is given, file TEMP.HS is assumed. If no input file is specified, the implicit specification is taken from the last histogram or data selection command, whichever is most recent, such that it specifies: (a) the input file specified by a histogram command, if most recent, or (b) the output file specified by a data selection command, if most recent. Other types of commands, such as histogram-manipulation or graphic output commands, are ignored in determining file specification. Thus, in the following sequence,

 (1) /(data-selection CODE)
 (2) /(histogram generation CODE)
 (3) /(histogram manipulation)
 (4) /(histogram display)
 (5) /(histogram generation CODE)

line (1) generates output file TEMP.HS, line (2) uses TEMP.HS as output file, and line (4) uses TEMP.HS as input.

When ".$\pm n$" is used as input file specification, it is interpreted as follows: "." refers to the most recently referenced interval-data file, and "$\pm n$" is an increment applied to the extension. Thus, in the sequence

 (1) DATANAME.13/(data selection CODE)
 (2) /(histogram generation CODE)
 (3) .-3/(data selection CODE),

line (3) refers to DATANAME.10.

The CODE portion of the command consists of a three-digit number, specifying the type of histogram, and a pair of numbers indicating origin and termination of x-axis, i.e.,

/ ABC, X TO Y,
where
   A=x-axis origin event (initiator)
   B=bin-incrementing event (incrementor)
   C=x-axis reset event. C=0 implies no reset (resetter).

In our system, A, B, and C refer to PIT input channels 1-6. The reset event forces a pause in histogram incrementing until the next initiator is encountered.

   X=x-axis lower limit, in units of time relative to initiator
   Y=x-axis upper limit, in units of time relative to initiator. Y need not be greater than X, and the histogram may be in first or second quadrant, or both.

Units of time are U (microseconds), M (milliseconds), or S (seconds). Note that the x-axis boundaries do not specify display boundaries: they specify boundaries for analysis, in order to limit the amount of histogram data to be generated. For example, some limit would usually be placed on cross-correlation histograms, which would otherwise grow as the product of the number of events in the two unit potential channels. Data from a mammalian system could easily include over 1,000 spikes per unit in one file, or more than 1,000,000 histogram events -- clearly an unwieldy number. By setting reasonable limits on number of events or length of time covered relative to each initiator, this can be reduced to a few thousand.

Examples   (DATANAME.EX/ABC, X TO Y):

(a) PST histogram:
   A = stimulus synch channel,
   B = single unit action potential channel,
   C = 0;
   -X TO -Y prestimulus time histogram,
   -X TO +Y peristimulus time histogram,

+X TO +Y poststimulus time histogram,
Y>X.

(b) Interspike interval histogram:
A=B=C=single-unit action potential channel

(c) Autocorrelogram:
A=B=single-unit action potential channel
C=0.

(d) Cross correlogram:
A = action potentials, unit A
B = action potentials, unit B
C = 0.

III. DATA SELECTION

The command format for data selection is similar to that for histogram generation:

(file manipulation) / (processing CODE).

However, both file manipulation and processing may be more complicated. In file manipulation, format is:

DEV0:OUTNAME.D < DEV1:INNAME1.EX
(or .D_x),DEV2:INNAME2.EX (or .D_x),DEV3:INNAME3.EX (or .D_x),DEV4:INNAME4.EX (or .D_x).

Under most circumstances, only one input device is used, but for special processing described later, two, three or four input devices may be specified. We will consider use of single input files first.

Default and . device and file assignments are the same as for histogram generation. Thus in the following sequence,

(1)   .+1 / (histogram code)
(2)   / (display code)
(3)   / (data editing code)
(4)   / (data editing code)
(5)   / (histogram code)
(6)   ./ (histogram code)

line (1) refers to the previous interval data file .EX (or D_x) + 1,
(2) refers to the output file from (1)

(3) refers to the same file as (1),
(4) refers to the output file from (3),
(5) refers to the output file from (4),

and

(6) refers to the output file from (4).

The /CODE format is:

/(A = X, Y, ... Z) AND/OR (A' = X', Y', ... Z') ...

where A specifies data to be selected, and X, Y ... Z specifies a set of acceptable tolerance ranges. The right side of each specification constitutes a set of inclusive intervals, either in units of time or number of events, e.g.,

(1) A = -100 U TO 10 M;
(2) A = 1 TO 50;
(3) A = 10, 25, 33 TO 36.

Single-entry ranges are interpreted as a two-entry single-valued range:

A = ..., 10, ... ≡ A = ..., 10 TO 10, ...

Thus, examples (1) to (3) specify the following ranges:

(1) : -100 microseconds to 10 milliseconds, inclusive;
(2) : events 1 TO 50, inclusive;
(3) : events 10, 25, 33 TO 36, inclusive.

Ranges in a statement may overlap; ranges need not be in any orderly sequence; limits within a range need not be ordered from low to high; negative numbers of events are interpreted as positive numbers; time units are ignored in event-number ranges; missing time units are assumed to be in milliseconds; spaces are ignored. Thus

(1) A = -100 U TO 0, 0 TO 10;
(2) A = 50 TO 1;
(3) A = 33 TO 35, 10, 25, 33 TO 36,

are equivalent to lines (1) to (3) of the original example.

The commands available for the left side of the command line are of the following forms:

NA = "channel A event numbers, starting from the beginning of data, must fall in the following tolerance range(s)"

T = "cumulative time, starting from beginning of data, must fall in the following tolerance range(s)"

NABC = "select for all events B (incrementors) such that they are Nth events following the initiator A, where N falls in the histogram tolerance ranges for numbers of events following the initiator."

TABC = "select for event B (incrementor) falling in histogram ABC, such that it occurs in the following histogram-time tolerance ranges."

Data selection may proceed in reverse time, starting at end of data, by preceding the commands above with R:

RNA, RT, RNABC, or RTABC.

Examples:

(1) T = 100 S TO 600 S selects data in range 100 seconds to 600 seconds after start of data.
(2) RT = 0S TO 5S: selects last 5 seconds of data
(3) N1 = 10 TO 100: selects the 10th through 100th occurrence of events in channel 1, e.g., trials 10 TO 100, if channel 1 = stimulus synch. All such events retain a 1 bit in channel 1, and all channel 1 events outside this range (e.g., stimuli 1-9 and 101 onward) have bit 1 (channel 1) cleared. Other channel bits are left unchanged.
(4) RN1 = 1 TO 90. Select last 90 events in ch. 1. Thus for example, if there are 100 such events this is equivalent to N = 10 TO 100.
(5) N120 = 2: select for the second channel 2 event following each channel 1 event. If channel 1 is a stimulus synch and channel 2 is a unit spike, this selects the second post-stimulus spike in each trial, and could be used to construct a latency histogram for the second spike.

(6) RN120 = 2: select for the second channel 2 event preceding each channel 1 event.
(7) T120 = 10M TO 100M. Select events on channel 2 occurring from 10 TO 100 msec after events on channel 1, with no reset. In the preceding example, this selects post-stimulus time data between 10 and 100 msec after the stimulus synch.
(8) RT120 = 10M TO 100M. In the preceding example, this would select spikes occurring 10-100 msec before stimulus synchs.

In data selection, the output file contains the same number of data words as the input file, with rejected input events modified so they become "transparent." This is accomplished by clearing the channel bit in the data word under examination, and setting it again if it occurs within any of the specified tolerance ranges. If it does not occur inside a tolerance range, it is not set, and hence remains zero. As an example, consider the following data selection (channel code in octal, interval in decimal):

/T120 = 10M TO 100M

Before editing, the data string is:

... 01(8)450(10), 07(8)1040(10), 60(8)320(10), 00(8)0(10), 06(8)930(10), 23(8)221(10), 22(8)640(10), 32(8)740(10), ...
[00(8)0(10) = overflow = 2.62144 sec.]

After editing:

... 01(8)450(10), 07(8)1040(10), 60(8)320(10), 00(8)0(10), 04(8)930(10), 21(8)221(10), 20(8)640(10), 32(8)740(10), ...

The operator NOT may precede CODE:

/ NOT CODE

In this case, data in the tolerance ranges is discarded, and other data is retained. Thus, in a data file with 100 trials (stimulus synch = channel 1),

NOT N1 = 1 TO 9 = N1 = 10 TO 100.

Operators OR and AND are used to combine code expressions in parentheses:

/ (CODE1) AND (CODE2),
/ (CODE1) OR (CODE2).

NOT operators should occur within the parentheses. In execution, each parenthetical portion produces a temporary output file, and the files are then ANDed or ORed on a bit-by-bit basis. All parenthetical CODEs refer to the same input file, and AND and OR conjunctions are performed sequentially. Order is important in some cases; for example,

/ (CODE1) AND (CODE2) OR (CODE3)

may produce an output different from

/ (CODE3) OR (CODE2) AND (CODE1),

just as

0 AND 0 OR 1 = 1

and

1 OR 0 AND 0 = 0

are different, if evaluated left-to-right.

In order to combine input files derived from different input files, more than one input file is specified:

FILE1, FILE2, FILE3/ (CODE1) AND (CODE2) OR (CODE3).

CODE 1 uses FILE 1, CODE 2 uses FILE 2, and CODE 3 uses FILE 3. Extra input files are ignored; missing input files are assumed identical to the last specified input file:

FILE 1, FILE 2, FILE 3, FILE 4 / (CODE1) AND (CODE2) AND (CODE3)

is equivalent to

FILE 1, FILE 2, FILE 3 / (CODE1), AND (CODE2), AND (CODE3),

and

FILE 1, FILE 2, / (CODE1) OR (CODE2) OR (CODE3)

is equivalent to

FILE 1, FILE 2, FILE 2 / (CODE1) OR (CODE2) OR (CODE3).

Programming example:

Generate a PST histogram file using the first 100 trials other than trials 5, 26, 31, and 46-70, which have an action potential occurring in the interval 0.6-1.2 msec after stimulus. Synch = 1, Unit = 2. The data is from the most recently analyzed file plus one:

.+1 / (NOT N1 = 5, 26, 31, 46 TO 70) AND (RT210 = 600 U TO 1200 U) AND (N1 = 1 TO 100).

/ 120, X TO Y.

Consider the following example:

Select all trials in which there are no interspike intervals less than 10 msec during PST time less than 100 msec. From these trials, generate PSTs over the interval 0-1 sec. The first steps are simple:

A < B / (T120 = 0 TO 100 M)
C < / (RT222 = 0 TO 10 M) AND (T222 = 0 TO 10 M)
D < / NOT RT210 = 10 TO 100 M.

However, we cannot go back and do PSTs over the interval 0-1 sec, because there is no file containing all the interval data over this range in which the synch channel has been edited. The solution is to use one file for synch channel data, and to use another for unit channel data. This is done by specifying multiple input files for a single data-selecting code:

FILE 1, FILE 2, FILE 3, FILE 4 / CODE.

Here, there must be only one CODE statement, not in parentheses, to distinguish from

FILE1,FILE2,FILE3,FILE4/(CODE1)*(CODE2)*(CODE3)*(CODE 4),

where * is a logical operator. With multiple parenthetical codes, the file specifications serve as sources for each data selection procedure. With one non-parenthetical code, the file specifications serve as sources for the initiator bit (FILE 1), increment bit (FILE 2), reset bit (FILE 3), and all other bits (FILE 4). If less than four files are specified, the remainder are assumed to be the same as the last. Hence, in the ridiculous extreme:

FILE1/CODE=FILE1,FILE1,FILE1,FILE1/CODE.

Using the files specified in the above example, the designated histogram file can be generated using the most recent data-selection output file, for synch channel data, and using the original input file for unit data:

E<D,A/120,0 TO 15.

In such a procedure, a pseudo-input file is created from the four input files, and it is this file which is processed. The pseudo-input file is always PSEUDO.DE.

IV. DISPLAY GENERATION

Histogram files can be used for statistical analyses, histogram manipulations (difference histograms, correlation histograms, etc.), or display file generation. We will only describe display file generation here. The display command line structure is:

DEV1:DISPNAME.P< DEV2:HISTONAME.Hx/CODE

The display may be stored as a graphic-format file, displayed on a video device, or output on a hard-copy device. The portable laboratory terminal is the default device for our system. CODE selects display parameters and the following CODE is only one of many possible:

CODE = XLIMITS, YLIMITS, BINWIDTH
     = A TO B, C TO D

where A and B are units of histogram time relative to initiator (origin); C and D are number of events setting upper and lower y-axis limits. Either C or D can be automatically scaled by simply not entering them. In this case it is scaled to $10^m \times 1$, 2, or 5, m an integer, such that it gives the largest sensitivity on the y-scale without overflowing or underflowing. Frozen (specified) y-scale specification results in truncation of overflowed or underflowed bins. The bin width is the number of display points per bin. Thus, if there are 1000 display points and a bin width of 25, there will be 40 bins in the display, each consisting of 25 adjacent display points of equal y-value.

V. MACRO DEFINITION

We have not implemented this feature, but it might be done as follows:

```
OPEN DEFINE CROSS (B) RE (A),    (C), FROM (D)
[D] / (RT [A][B]0 = 0 TO 10M) AND (RT[B][A]0 = 0 TO
10M)
/ [A][B]0, 0 TO 1100M
CLOSE DEFINE
```

Then

```
CROSS (3) RE (2),    (50M), FROM (< .+ 10)
```

might be a handier form for the operator than

```
<.+10 / (RT210 = 0 TO 10M) AND (RT310 = 0 TO 10M)
/ 230, -50M TO 50M.
```

If succinctness were desired,

```
OPEN DEFINE (B) (A) (C) (D)
[D] / (RT[A]10 = 0 TO 10M) AND (RT[B]10 = 0 TO 10M)
/ [A][B]0, 0 TO 1100M
CLOSE DEFINE
```

The operator could then achieve the same procedure with

```
(3) (2) (50M) (< . + 10)
```

Ultimately, the defining of macros allows the individual laboratory to define its own language, and still take advantage of a) the use of a single general-purpose algorithm, b) the ability to generate new histograms; and c) the opportunity to modify its idiolingua. Clearly, the ability to write implicit programs for automatic processing of whole batches of data will also be a useful feature, and will eventually be incorporated.

VI. THE ALGORITHM

We will consider here a general-purpose algorithm to be used in histogram-format processing, including commands of the form N..., RN..., T..., RT..., TABC..., RTABC..., NABC..., RNABC... (data selection), and ABC... (histogram generation).

Input data in interval format is read into a circular double-buffered vector called the I/O VECTOR (figure 2), where data is read into buffer A or B, in alternation. The scan pointer points to the word currently being analyzed, and the end-of-vector pointer points to the word ending the buffer currently being scanned. Whenever the scan pointer passes the end-of-vector pointer, it moves to the beginning of the next buffer, the new buffer is refreshed with more data, and the end-of-vector pointer is moved to the end of the new buffer. As data is processed, the end of completely processed data is also tracked in the analysis routine, and whenever it crosses the end of its buffer, the buffer is written onto the output file (in data selection mode).

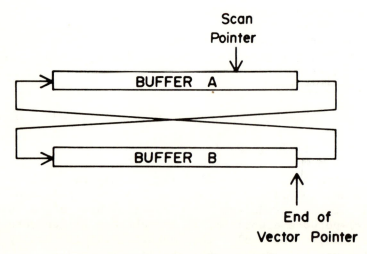

Figure 2. The I/O VECTOR.

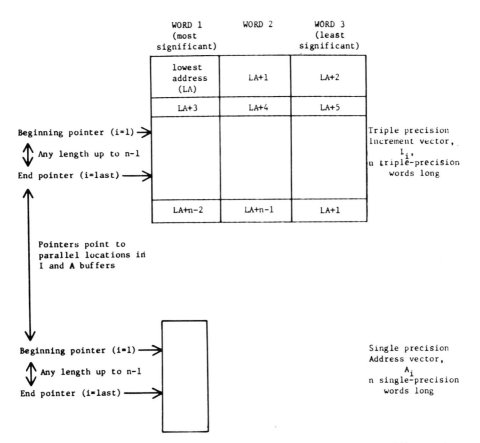

Figure 3. The INCREMENT and ADDRESS vectors, I(i), and A(i), with their common pointers, i = 1, and I = last.

A second pair of vectors consists of two parallel vectors, one triple precision and one single precision (figure 3). The triple-precision vector, the INCREMENT VECTOR or I(i), and the single-precision vector, the ADDRESS VECTOR or A(i), contain the cumulative interval or number of events since successive increment events and the addresses of those events in the I/O VECTOR, respectively. Both are circular vectors, with common beginning-of-vector (i = 1) and end-of-vector (i = last) pointers. The use of two pointers permits variable-length work areas. If i(1) = i(last) + 1, the vector is of zero length (empty), by convention.

A third vector, the INITIATOR VECTOR S(j) (figure 4), is identical to the INCREMENT VECTOR: it is a

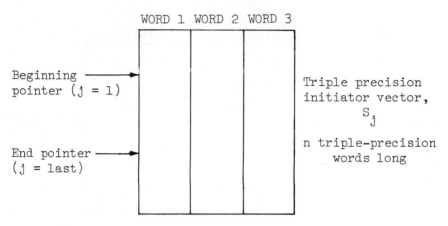

Figure 4. The INITIATOR VECTOR, S(j), with its pointers, j = 1, and j = last.

triple-precision circular vector with beginning-of-vector pointer (j = 1) and end-of-vector pointer (j = last). If j(1) = j(last) + 1, the vector is empty. Initiator vector entries constitute the cumulative intervals or number of events since successive initiating events.

A final vector, the HISTOGRAM VECTOR, H(h), is a double-buffered vector used to generate the histogram output file. It is a triple-precision vector.

At the beginning of data analysis, buffer A is filled with the first buffer-full of data from the input or pseudo-input file. The scan pointer is set to word 1 of buffer A, the end-of-vector pointer is set to the end of buffer A. The pointers i(1) and j(1) point to the first words in A(i), I(i), and S(j), i(last) and j(last) point to the last words in A(i), I(i), and S(j), indicating empty vectors.

Table 1 summarizes the analysis procedures for T and N commands, in histogram and data selection modes, with and without NOT, in normal and reverse time. The implementation of these procedures is flow-diagrammed in figures 5-8. In figure 5, the mainstream program is outlined. Figures 6, 7, and 8 outline processing of increment, reset, and initiate events, respectively.

Briefly, the I table codes negative intervals in histogram time relative to the current initiator events; they are updated and processed on initiator events. New, zero-valued entries are added as new increments occur. They thus represent negative numbers of events or

Table 1. Processing procedures for increment and initiate events under different instructions.

Instruction	Input event	Processing procedure
ABC = (note 1)	Increment	$T = S(j)$. Set pointer $j(1)$ to first $s(j)$ such that $s(1) <$ Upper -axis limit. Add a zero-value event to $I(last + 1)$, making this new $I(last)$. Generate $H(h) = S(j)$ for all $S(j)$ falling within -axis limits.
	Initiate	$T = -I(j)$. Set pointer $i(1)$ to first $I(i)$ such that $I(1) >$ lower -axis limit. Add a zero-value event to $S(last + 1)$, making this the new $S(last)$. Generate $H(h) = -I(i)$ for all $I(i)$ falling within -axis limits.
N = (note 2)	Increment	$N = S(j)$. Set pointer $j(1)$ to first $s(j)$ such that $S(1) <$ highest upper tolerance limit. Add a zero-valued event to $S(last + 1)$. Search for first $S(j)$ within any tolerance range. If any found, set increment bit of current IO word if not NOT command, clear increment bit if NOT command.
	Initiate	$N = -I(j)$. Set pointer $j(1)$ to first $I(i)$ such that $-I(1) >$ lowest lower tolerance limit. Add zero-valued event to $I(last + 1)$. Search for all $I(i)$ within any tolerance range. For each found, set increment bit of corresponding IO word it not NOT command, clear if NOT command.

Table 1. (*continued*) Processing procedures for increment and initiate events under different instructions.

RN = (note 2)	Increment	On second pass, RN = I(cumulative) -S(j). Process RN as though it were N in second pass.
	Initiate	No action on first pass. On second pass, add zero word to S(last + 1), making it new S(last).
T = (note 3)	Increment	T = I(cumulative). If I(cumulative) is within any tolerance range, set increment bit of current IO word if not NOT, clear if NOT.
	Initiate	No action.
NABC = (note 4)	Increment	N = S(j). Set pointer to first S(j) such that S(1) < highest upper tolerance limit. If any S(j) in any tolerance range, set increment bit os current IO word if not NOT, clear if NOT.
	Initiate	N = -I(j). Set pointer to first I(i) such that -I(1) > lowest lower tolerance limit. For each I(i) falling within any tolerance range, set increment bit of corresponding IO word if not NOT, clear if NOT.
RNABC =	Increment	N = -S(j). Set S(1) pointer such that -S(1) > lowest lower tolerance limit. If any S(j) falls within any tolerance range, set current

Table 1. (*continued*) Processing procedures for increment and initiate events under different instructions.

		increment bit if not NOT, clear if NOT.
	Initiate	$N = I(i)$. Set $I(1)$ pointer for $I(i) <$ highest upper limit. For each $I(i)$ in a tolerance range, set or clear increment bit according to absence or presence of NOT command.
TABC =	Increment	$T = S(j)$. Process as in NABC.
	Initiate	$T = -I(j)$. Process as in NABC.
RTABC =	Increment	$T = -S(j)$. Process as in RNABC.
	Initiate	$T = I(j)$. Process as in RNABC.

intervals in normal time or positive numbers of events or intervals in reverse time. As such values become larger than the largest tolerance limits of appropriate sign they are eliminated.

In cumulative mode, a single counter counts time or number of events since the beginning or end of data.

The S table codes positive intervals in histogram time, relative to current increment events. They are updated and processed on increment events. New, zero valued entries are added as new initiate events occur. They thus represent positive intervals in normal time or negative intervals in negative time. As such intervals become longer than the largest tolerance limits of appropriate sign they are eliminated.

In reverse time, no output is generated on the first pass for cumulative mode. Instead, on a first pass the total cumulative time or number of events is obtained, and then a second pass is taken through the data, taking the difference between the cumulative sum of events or time obtained in the first pass and the current sums, to

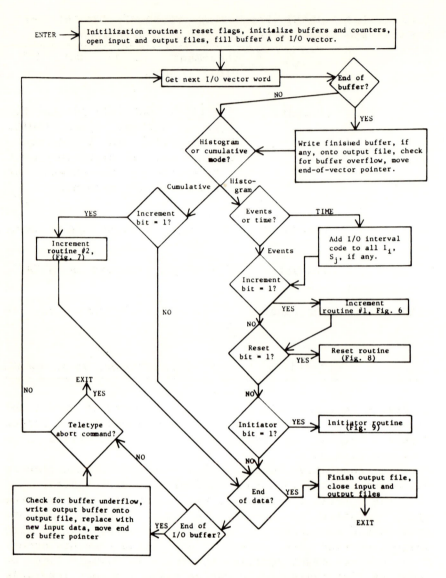

Figure 5. Flow diagram of mainstream of algorithm. Subroutines are diagrammed in Figs. 6,7,8. *Note: I/O vector buffers are filled from an original data file, data-selected output file, or a pseudo-input file generated from multiple files, as specified by file specification. ** See Figure 9 for RN = and RT = instructions.

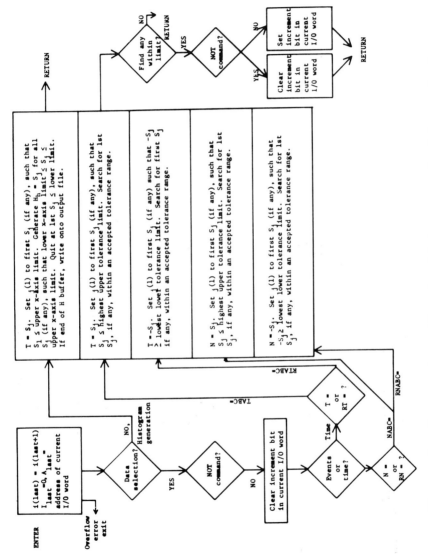

Figure 6. Increment routine for histogram-mode data analysis.

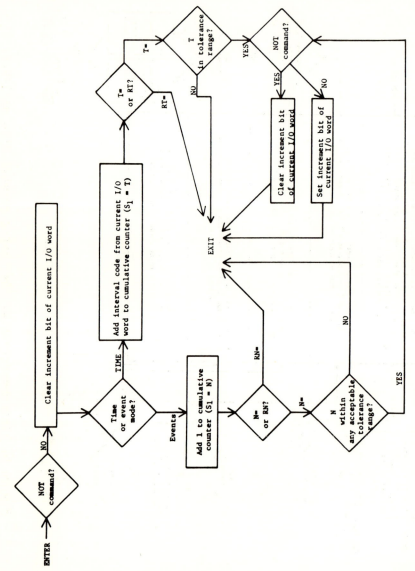

Figure 7. Increment routine for cumulative-mode data selection.

ALGORITHM FOR HISTOGRAM GENERATION

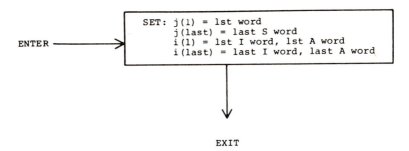

Figure 8. Reset routine.

determine whether data are within accepted tolerance limits. It might be appropriate in future versions of the system to include cumulative counts of time, and events in all six channels in a file header.

As valid histogram intervals are encountered (in histogram generation), they are added to the histogram file as triple-precision histogram-time words. As valid event numbers or times are encountered (in data selection), the corresponding increment bits of the original data string are cleared or set, according to presence or absence of the NOT instruction, for output onto the .Dx file.

There is sufficient similarity between the processing of increment events and initiating events so they use many common subroutines. With regard to requirements for buffer lengths, it is necessary that both beginning and end pointers for the I/O VECTOR be in the same buffer when the end of the buffer is reached, so the other buffer may be refreshed wihout loss of data. In histogram generation, the processed buffer is simply written over by new data; in data editing, the processed buffer is first written onto the output file, and then written over.

In the last buffer of the file or sub-file, some means must be used to determine end-of-data. We have chosen to fill out the end of the last buffer of a file or sub-file with words in which the channel code is 7(8), and the interval code is 000001(8). The second such word encountered signals end-of-data, because the PIT is incapable of generating two of that value in a row.

A rule of thumb for selecting length of the IO VECTOR is therefore: regardless of histogram type or whether analysis or editing is taking place, the I/O

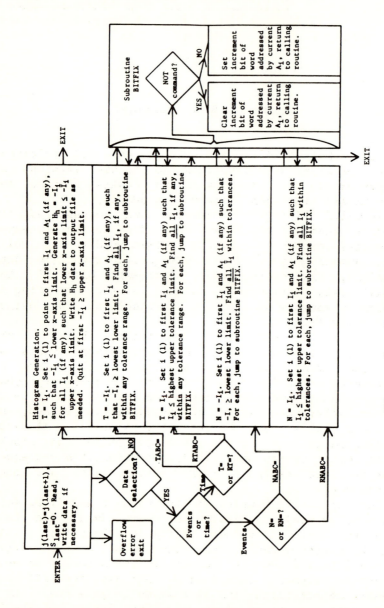

Figure 9. Initiate routine.

VECTOR must be at least twice as long as the longest active vector area: i.e., twice as long as the most dense data string in the longest histogram time interval ever analyzed. We never analyze more than 1 sec. of data relative to the histogram origin, and our data never exceeds a mean density of 100 histogram events /sec over such a long period, so the IO VECTOR is 2 sec x 100/sec x 2 words/event x 2, or no more than 800 words long.

The HISTOGRAM VECTOR need only be twice as long as the shortest buffer practical for output: In our system, 128 x 3 (output words are triple precision) = 384 words / buffer x 2 buffers = 768 words.

The INCREMENT and INITIATOR VECTORs should be half as long as the IO VECTOR. Using the criteria above, INCREMENT VECTOR (both I and A) = 2 sec x 100/sec x 4 words / event = 800 words. INITIATOR VECTOR = 2 sec x 100/sec x 3 words / event = 600 words. Thus, the major vectors come to a total of 800 + 786 + 800 + 600 = 2968 words for our applications. Clearly, core can be traded off against more limited analysis ranges. Use of double precision vectors instead of triple precision can also save space, but then the largest interval which can be represented is some 168 sec. rather than 688,000 seconds. This is undoubtedly acceptable in many situations.

VII. GENERATION OF GRAPHIC FORMAT DATA

Histogram data is converted very easily to graphic-format data. The input file, consisting of triple precision histogram data, is scanned from start to end, converting each entry to an (x,y) coordinate as shown in figure 10. Assume there are 1000 display points. Our histogram table then consists of 1000 successive y-values. We first increment bins, one increment per data word, distributing increments to bins according to the value of the data word. Off-scale events are discarded here, but they could be used to increment additional underflow or overflow words.

For each data word we compute

I = (XVAL-XMIN) / (XMAX-XMIN) * BW,

where XVAL = histogram-mode data word,
       XMAX = -axis upper limit,

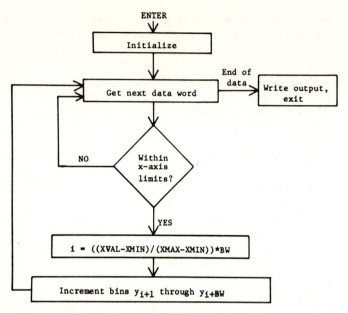

Figure 10. Generation of histogram y-coordinates from histogram-mode data.

XMIN = x-axis lower limit,
BW = number of display points per bin,
using triple-precision integer arithmetic.

If I and I + BW fall within the range of allowable y-values (i.e., 1 TO 1000, for 1000 display points), we increment bins (I + 1) through (I + BW), and go on to the next word, until all data words have been processed. The output file is filled out and closed after appropriate scaling, axis labelling and title routines.

VIII. CONCLUSION

All of the material described above is still in the process of realization and laboratory debugging. We hope to add features for well-defined Macros, as well as a number of statistical procedures. Sorting and shuffling routines will be added in order to use the data as input for modelling. We feel such a package should be a part of a general neurobiology laboratory computing package.

DISCUSSION

J. GIBSON: This algorithm seems idealy suited to a hardware implementation. Have you thought about that?

BROWN: My philosophy is that we want a general-purpose computer in the lab or tied to the lab to do analyses, and we want to hook powerful interfaces to it which will allow the CPU to control many processes at one time during an experiment. Once one has the minicomputer one might as well use it to do the analyses. A hardwired device could probably be constructed for almost as little as a minicomputer, but it would do a lot less things for the same money.

CLARKE: What's the final output? You still have to go through a histogram generation.

BROWN: The histogram file consists of a string of numbers representing the histogram times of increment events. We can generate a histogram by binning these times. We could use the same data for dot patterns, joint interval displays, or statistical analysis. You might want to do statistics at this stage, because you still have maximum resolution. You obviously don't want to do statistics on the data after binning, because by then you have lost resolution.

L. PUBOLS: Were the numbers you showed octal or decimal?

BROWN: Decimal in the range statements, and the channel code is 0-6. If you have more than ten channels, you might want to go to hexadecimal or alphabetic representation, to specify each channel by one character.

MCINTYRE: The 18-bit count is binary.

BROWN: Thanks, Tom. In our system the pulse interval timer has internal resolution of 10 µsec., but longer timer units can be selected if the user wants to use an external clock.

CLARKE: This syntax is under your program rather than OS-8?

BROWN: We can make use of OS-8 file handling capabilities, using the OS-8 USR subroutines. These essentially permit programs to use OS-8 file management the same way the utility programs do. This system also provides command line decoding functions.

chapter 21

A CONTOUR MAPPING ALGORITHM SUITABLE FOR SMALL COMPUTERS

L. S. DAVIDOW

Department of Biology
Massachusetts Institute of Technology
Cambridge, Massachusetts

P. B. BROWN

Department of Physiology and Biophysics
West Virginia University Medical Center
Morgantown, West Virginia

We envision a contour-mapping algorithm of the type presented here as an example of the kind of general-purpose routine which could be applied to a large variety of neurobiological applications, including:

single-cell receptive field response-magnitude maps;
single stimulus locus, neural space response-magnitude maps;
cell density maps;
current source-sink density maps; and simulation of activity in neural networks.

For the usual laboratory facility, such an algorithm should take little core space, should be fast, should use

---

This work was supported by PHS grants NS 07505, NS 40636, NS 09970, and NS 12061, and NSF grant DCR74-24765. Figure 6 was reproduced from Brown et al. (1973) by permission of Journal of Neurophysiology.

any mass-storage available for scratch files, and should operate on data obtained from irregularly spaced sampling loci in the map field.

Our algorithm was written in FORTRAN, making it highly portable. Using overlays, the program and vectors occupy a maximum of 5600 core locations in a PDP-15 time-sharing system -- this could be reduced substantially with machine language programming.

To save both core space and operator's time, the program is divided into three logical and physical parts. In the conversational portion the user is interrogated for all parameters. In the second section, this information is used to manipulate experimental data into a form that can be used by the final contour mapping section. The large blocks of data involved are passed between successive links in the overlay system via mass storage (tape or disk).

In our system (see figure 1) the first stage of processing consisted of obtaining post-stimulus time response histograms for a single unit during repetitive stimulation at many different cutaneous receptors in its receptive field (RF). A separate histogram was generated for each receptor stimulated. Each point on the histogram represented the average number of spikes recorded per "bin width" of time, for successive intervals following the stimulus. We adopted the convention that reception of stimuli can be represented as continous over the skin throughout the RF, rather than merely at discrete receptor loci. The question could then be asked: "Is there an organized relation between the PST activity of the unit and the spatial location of the stimulus?" The construction of an iso-response contour map of the RF for each interval was chosen to graphically examine the relation between response magnitude and receptor location. Our algorithm could be used, with minor modifications, to provide elevation calculations for other types of displays.

To arrange the data for calculation of such a contour map, the activity elicited by stimulation of each receptor during the interval in question must be stored in a single array. The first section of the program stores the data on disks as blocks of successive PST histograms. We represent this as: Block 1 = A_1, A_2, A_3,....,A_m; Block 1 = B_1, B_2, B_3.....B_m etc.; where A_1 represents the activity seen during the first interval following stimulation of receptor A alone and B_1 is the

```
┌─────────────────────────────────────────────────────────────┐
│ Interrogate operator for catalogue #, x and y coordinates,   │
│ and data file for any new receptors. Interrogate for the     │
│ selection of receptors, contour values, intervals of the     │
│ PST desired, and format of the final map.                    │
└─────────────────────────────────────────────────────────────┘
                              │
                              ▼
┌─────────────────────────────────────────────────────────────┐
│ Calculate and store all PSTs for newly added receptors.      │
│ Scan through the updated PST library, normalize and modify   │
│ the PSTs of the receptors chosen for this run. Store in a    │
│ scratch area on tape or disk.                                │
└─────────────────────────────────────────────────────────────┘
                              │
                              ▼
┌─────────────────────────────────────────────────────────────┐
│ Read in the modified PSTs as row vectors, and the columns    │
│ now represent activities during a given interval of time.    │
│ Store these column vectors as sequential intervals.          │
└─────────────────────────────────────────────────────────────┘
                              │
                              ▼
┌─────────────────────────────────────────────────────────────┐
│ Using the stored activities, operator-chosen parameters,     │
│ and spatial coordinates of the receptors, compute and draw   │
│ the desired series of contour maps for sequential intervals. │
│ The drawing I-O time and the CPU time for computing the maps │
│ is shared. See figs. 2,3,4, and 5.                           │
└─────────────────────────────────────────────────────────────┘
                              │
                              ▼
                          ┌──────┐
                          │ EXIT │
                          └──────┘
```

Figure 1. Flow diagram of program operation. Each section is a separate core overlay.

activity in that same interval when only receptor B is stimulated. To make a contour map of the response characteristics for the first interval, we want a block consisting of A_1, B_1, C_1, D_1 ... Therefore, the data is read into a rectangular array from successive PSTs, where each PST is regarded as a row vector. Then the desired output blocks of "activities" are obtained by storing the successive columns.

The spatial location of each receptor (a tactile dome or *Haarscheibe*), obtained in the first phase, is also arranged on disk, the x and y coordinates on two separate blocks. Also stored from the conversational program portion are the total number of receptors involved, the number of PST intervals under investigation the specific value of each iso-response level to be displayed and the total number of such contour lines desired.

For the calculation of the contour map (see figure 2) the large blocks of data are read into two vectors called X and Y in a point storage table. The X and Y vectors contain the spatial coordinates of the receptors and the Z vector contains the order number of that receptor, i.e., for the first receptor the value is 1, for the second it is 2 etc. The Z vector will later contain the PST activities of the corresponding receptors. To find the points in the receptive field which would give the activity equal to a desired contour line value, we must interpolate between the data points.

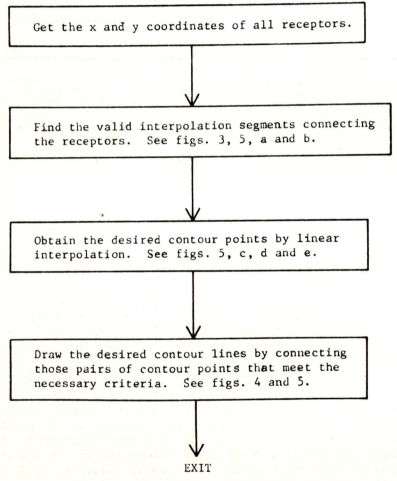

Figure 2. Procedure for generation of contour map.

The problem of just how to do this interpolation was solved using only two assumptions: (1) The shorter the distance between two points, the more valid the interpolation. (2) In the absence of evidence for any other mode, a linear interpolation between data points should be made.

The enumeration of the interpolation-lines proved to be the most CPU-time consuming aspect of this program. Simplified conceptually, the method is as follows (see figure 3): First generate all possible interpolation

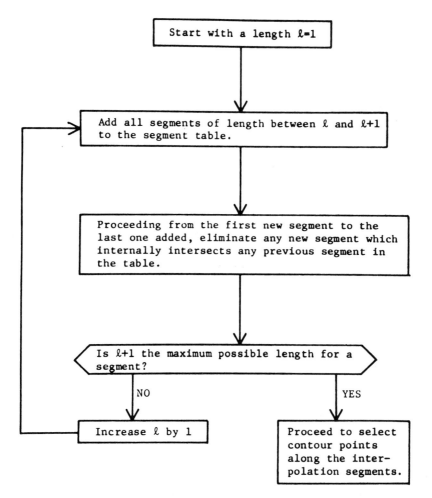

Figure 3. Generation of interpolation segments. This procedure is performed only once for a given RF, regardless of the number of contour maps required.

lines by making all possible pairwise combinations between receptor points. For 50 receptors, this would be 50 C2 = (50·49)/2 = 1225 line segments. Then order the segments by length, shortest to longest. Finally, starting with the second segment, eliminate any line which intersects the interior of any previous (i.e., smaller) line segment. The end result forms a set of non-overlapping triangles, whose vertices are the receptors, which cover the entire RF.

In practice, however, our program only has enough space to hold 256 segments. Also, since the coordinates of the receptors are confined to $-150 \leq X \leq 150$ and $-100 \leq Y \leq +100$, the longest segment must be less than 361 units long. This coordinate system is chosen to allow all arithmetic to be done using 15-bit integers, which is more rapid than floating point arithmetic. Therefore, the set of interpolation lines is generated in 361 passes through a subroutine as follows: Let N = the number of the pass that we are in. L is the square of the length of a proposed segment:

$$L = \Delta X^2 + \Delta Y^2$$

If $(N-1)^2 < L \leq N^2$, then the segment is added to the segment table as follows: The x, y, z values of the endpoint whose x value is smaller is inserted first into the three segment vectors and is then followed by the other endpoint. This convention is used to simplify procedures later on. Since N increases by 1 on each pass, the segments are automatically generated in order of length to an accuracy of 1 unit within a pass.

After a batch of segments is generated, the first segment of the new batch is tested for intersection with the previous segment. If it does not intersect, it is tested against the one before that, etc., all the way back to the first accepted segment. If a new segment does not intersect with any previous segment, it is left in the table, and the next one is then tested against all previous segments, beginning with the one which was just accepted. If a new segment does cross a (shorter) previous segment, it is eliminated by moving all later segments back one position in the segment table. In this manner, the segment table is consolidated and condensed before the next pass begins, generating more and longer segments. The order in which a new segment is tested against previously accepted segments is important, because the new segment is more likely to intersect the longer ones just below in the table than the shortest ones at the beginning. Clearly the fate of most segments is elimination and the sooner one is eliminated, the more efficiently the program operates.

The method for determining whether two segments intersect utilized the single parametric form for determining a line segment:

$$\binom{X}{Y} = P = t \cdot P_1 + (1 - t)P_2$$

where $-\infty \leq t \leq \infty$
describes the line determined by the two points P_1 and P_2. If we wish to restrict P to the locus of points on the segment $\overline{P_1 P_2}$ then we restrict the parameter t as follows: $0 \leq t \leq 1$. Similarly for another segment, with endpoints P_3 and P_4 we have

$$P = u \cdot P_3 + (1 - u)P_4,$$

assuming that the two infinite lines do intersect for some P_1 we get

$$t \cdot P_1 + (1 - t)P_2 = uP_3 + (1 - u)P_4$$
$$tP_1 + P_2 - tP_2 = uP_3 + P_4 - uP_4$$
$$t(P_1 - P_2) + u(P_4 - P_3) = (P_4 - P_2).$$

We are only interested in determining whether both t and u are between 0 and 1 or not, which tells us whether the two segments intersect.

$$P_1 = \binom{X_1}{Y_1}$$

Let

$$a_1 = X_1 - X_2$$
$$a_2 = Y_1 - Y_2$$
$$a_3 = X_4 - X_3$$

$$a_4 = Y_4 - Y_3$$
$$a_5 = X_4 - X_2$$
$$a_6 = Y_4 - Y_2$$

We can set up the extended matrix in thought, without actually using any matrix subroutines in the program, since its size is so small:

$$\begin{array}{ccc} t & u & \text{scalar} \end{array}$$
$$\begin{vmatrix} a_1 & a_3 & a_5 \\ a_2 & a_4 & a_6 \end{vmatrix}$$

We cannot actually perform the divisions involved because truncation errors might make an intersection near the endpoint (not allowed) seem like one at an endpoint (allowed) or vice versa. In addition, we wish to keep all arithmetic in the integer mode. Most cases can be handled using the usual approach: multiply row 1 by a_2 and row 2 by a_1 to get

$$\begin{vmatrix} a_1 a_2 & a_3 a_2 & a_5 a_2 \\ a_1 a_2 & a_1 a_4 & a_1 a_6 \end{vmatrix}$$

Subtract row 2 from row 1:

$$\begin{vmatrix} 0 & a_3 a_2 - a_1 a_4 & a_5 a_2 - a_1 a_6 \\ a_1 a_2 & a_1 a_4 & a_1 a_6 \end{vmatrix}$$

This means that $u = (a_5 a_2 - a_1 a_6) / (a_3 a_2 - a_1 a_4)$. However, since we are only interested in determining whether $0 < u < 1$, we test the absolute values and signs of the two parenthetical quantities.

If the absolute value of $(a_5 a_2 - a_1 a_6)$ is greater than or equal to the absolute value of $(a_3 a_2 - a_1 a_4)$ then $u \geq 1$ and the segments don't intersect internally. If the signs of the two quantities are different, the segments don't intersect. If $(a_5 a_2 - a_1 a_6) = 0$, any intersection is at an endpoint and is allowed.

By reversing the first two columns of the original matrix and proceeding through the same arithmetic operations, we can see if t is also in the range $0 \le t \le 1$. In order for the two segments to intersect in a disallowed manner, both t and u must be between 0 and 1.

The special cases we must take into account in the program all involve avoiding multiplication or division by zero. For example, if a_1 or $a_2 = 0$ (the first or second segment is vertical) we skip the intermediate multiplication and go directly to comparing signs and absolute values to see if falls in the desired range. Similarly, when the columns are shifted in the second pass to look at the parameter t, the same problem arises if either segment is horizontal. The last problem corresponds to the lines determined by the segments being parallel or identical. If the lines are parallel, but not identical, we will have the matrix

$$\begin{vmatrix} 0 & 0 & \text{nonzero} \\ a_1 a_2 & a_1 a_4 & a_1 a_6 \end{vmatrix}$$

after the subtraction step, which corresponds to the nonsense equation $0 \ne 0$.

This is not surprising, since we set up the matrix with the incorrect assumption that the two lines involved intersected.

If the two lines on which the segments fall are identical, we will have the matrix

$$\begin{vmatrix} 0 & 0 & 0 \\ a_1 a_2 & a_1 a_4 & a_1 a_6 \end{vmatrix}.$$

We can check for intersection of the segments which are part of the same line by checking for overlap in the x range. This is relatively easy since by convention the larger x-valued point of each segment is the even numbered point in the segment table. In fact, we routinely pre-screen each pair of segments for overlap in their x-ranges to detect obvious non-intersection.

After the final pass through the subroutine creating new segments and eliminating the larger ones which intersect previous ones, we have a set of acceptable

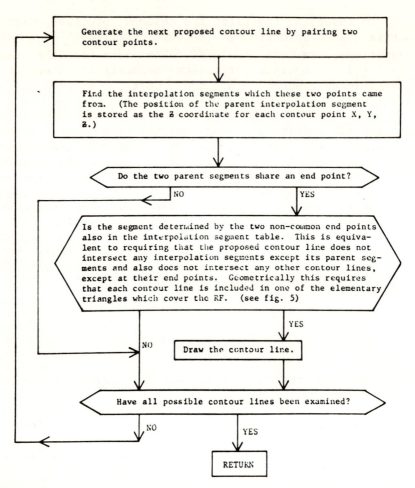

Figure 4. Generation of the contour map from contour points and their z values.

segments. This set of interpolation segments forms a large jigsaw puzzle-like covering of the RF where each of the pieces is a triangle whose vertices are receptors. We will call these "elementary triangles" (see figure 5b).

The covering is a convex set, in that a segment drawn between any two points in the set lies wholly within the set. If the RF in reality has a concave shape, or has "forbidden areas" inside which you do not want any interpolation axes, you must set up segments which delineate these areas and include them in the segment table before even the shortest proposed interpolation line. In this way, no interpolation line will cross into a forbidden area and no contour points will be found there. This possibility of a restricted geometry of the field was not included in our project.

The set of interpolation segments obtained is wholly a property of the location of the receptors. It depends only on the x and y values and not the z values. Therefore, the segment table and a variable indicating how many segments are in the table are stored on a bulk storage device and recalled every time a new contour map

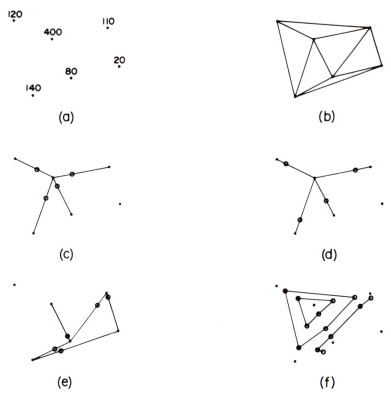

Figure 5. Steps in the process: (a) Start with RF of 6 receptors, and their normalized activities over some PST epoch. (b) Interpolation segments: of the 15 possible, 11 are rejected due to intersections with shorter segments. (c) Interpolated contour points, $z = 300$. (d) Contour points, $z = 200$. (e) $z = 100$. (f) The completed contour map.

is desired which uses the same receptors, but involves these activities during a different interval. Since we studied the changes in receptor activity during many sequential intervals, this feature saved much execution time. The x, y and z values of the segments were recalled before each run. Remember that the z values stored were the positions in the standardized tables of activities rather than the activities themselves. Then the table of activities for the interval desired was found and read into core. Finally the position numbers in the vector were replacd by the activity numbers from the table.

Beginning with the lowest contour value, each interpolation axis from the segment table was screened to see if its endpoints "straddled" the contour value. If they did, we used our hypothesis that activity varied linearly with distance and interpolated to find the coordinates of the contour point. The coordinates of the contour points generated were stored in the x and y vectors from the receptor points table, since we no longer needed the receptor points in core, and we know that the z value of the point is the contour value anyway. The point's z vector was used to store the position of the "parent" interpolation segment which gave us the point. This information was used to determine which contour points to connect and draw as contour lines in the final hard-copy map. In this manner, the x, y and z vectors in the table of receptor points and the x, y and z vectors in the table of segments were the only large core data arrays needed.

After all contour points were generated the points table was sorted so that the x values were in ascending numerical order. This was done so that the generation of contour lines would flow evenly from left to right on our final x-y plotter output to minimize pen manipulations and thereby speed up the output substantially.

The rule for deciding which contour lines to draw (see fig. 4) between the contour points follows directly from the geometry of the set of interpolation segments. Two contour points were connected only if the two interpolation axes they spanned formed sides of the same elementary triangle. This condition guaranteed that a contour line could only intersect an interpolation axis at a contour point, and that contour lines only intersected each other at their endpoints. The table of contour points was examined pairwise for prospective contour lines. Since the z vector in the points table showed which interpolation segment each point came from the algorithm was fairly simple. We made the simplifying assumption that contour points are connected by straight line segments rather than curvilinear ones. To see if two contour points should be connected in a line on the final map, we first examined the two parent interpolation segments. They must have a common endpoint; also, the segment determined by the non-common endpoints must also be an interpolation axis. If both conditions were satisfied, the line was drawn. After all contour lines were drawn for a given contour value, the program would go back to the segment table to get all contour points for the next iso-response value until the map was finished. Figure 6, from Brown et al. (1973) shows several contour maps generated from real data, using our program.

Output may be in several alternative forms: for universality, we recommend generation of an output table of line segments, in a file format optimized for this type of data, onto a hard-copy device. This can then be used as input for any of a number of graphics output programs under operator control.

DISCUSSION

L. PUBOLS: Did you stimulate only the domes, and not in between?

DAVIDOW: Right. That's the artificial nature of this representation, in that one tends to think of contour maps as representing continuous surfaces.

L. PUBOLS: Did you try stimulating elsewhere to try responses to other receptors?

DAVIDOW: Tapper's laboratory is setting up to do that now. The point of the project was to examine the responses of postsynaptic cells to precisely identical inputs: single action potentials coursing through single afferent fibers, the only variable being location of the receptor in the receptive field. Next, Tapper will go on to do this with other receptors and with more complex spatiotemporal inputs. The domes were the easiest receptors to start with.

CAPOWSKI: Have you tried higher-order interpolation?

DAVIDOW: No, we had no justification for doing them.

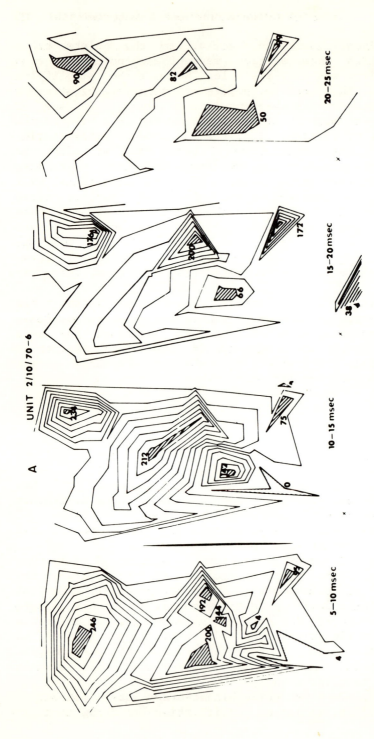

Figure 6. A set of contour maps showing responsiveness of a neuron as a function of receptive field position, for different "windows" of post-stimulus time. Input: single action potentials elicited from single *Haarscheiben* in the receptive field.

REFERENCE

1. Brown, P. B., Moraff, H., and Tapper, D. N. Functional organization of the cat's dorsal horn: spontaneous activity and central cell response to single impulses in single type I fibers. J. Neurophysiol. 36:827-839 (1973).

chapter 22

NEURON POPULATION ANALYSIS WITH THE NVAR SYSTEM

G. W. HARDING, A. L. TOWE, AND T. H. KEHL
Department of Physiology and Biophysics
University of Washington
Seattle, Washington

INTRODUCTION

A population sampling technique is being employed to investigate the responsiveness of neurons in the cerebral cortex to peripheral somatosensory stimulation (4). Several sites are sampled in cytoarchitectonic areas 4, 3a, and 3b, as indicated in figure 1A. Extracellular microelectrode recordings are made within a cylinder of cerebral tissue at each site. The postcruciate cylinder shown in figure 1B has been magnified; it is a cylinder 1 mm in diameter and 2.5 mm long, its long axis being normal to the cortical surface. The set of neurons contained within the cylinder is termed a population. The microelectrode (figure 1C) is driven through the tissue parallel to the long axis of the cylinder to search for neuronal responses evoked by a peripheral hunting stimulus. When the electrode is in close proximity to such a neuron (figure 1D), a standard set of experimental questions is asked of the neuron (figure 1E)

This research was supported by U.S. Public Health Service Grants NS00396 and NS05136 from the National Institute for Neurological Diseases and Stroke, and RR00374 from the Division of Research Resources, Biotechnology Branch. We are grateful to the following people for their assistance in this work: Hanna Atkins, Alan Henkins, Paul Moreau, Christine Moss, and Rose Stogsdill.

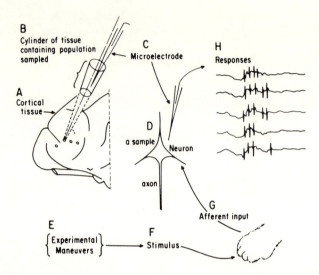

Figure 1. The neuron population sampling technique, showing a cortical region under study (A), a population sampling site (B), the microelectrode (C), an isolated neuron (D), the experiment execution (E), with the stimuli (F) delivered to the periphery, resulting in afferent inputs (G) which cause the cell to actively discharge (H).

and the pattern of answers to these qustions constitutes the object of study. Each of these questions is asked by delivering a known stimulus (figure 1F) to the skin (or to a more central site along the input/ output pathways). The receptor organs (or nerve fibers) in the immediate neighborhood of the stimulus site are activated, and the resulting afferent input (figure 1G) synaptically reaches the extracellularly isolated neuron (figure 1D). A neuron's responses to five sequential stimuli are depicted in figure 1H; these are part of the sample of the neuron's behavior. The neuron's efferent output passes down the axon, perhaps on its way to the execution of a motor response to the sensory input. Pyramidal tract neurons are detected by antidromic stimulation of the medullary pyramid, through which such a neuron's axon passes.

The interface between the microelectrode and the neuron is a critical element in the population sampling technique. The combination of microelectrode properties and the investigator's microelectrode manipulation technique generates a sampling bias -- effectively increasing the probability that certain types of neurons will be sampled from the population (6). This bias can be corrected when the sample is large enough and the precise nature of the bias can be determined.

Currently, the population sample is partitioned into subsets according to the size and location of each neuron's peripheral exitatory field, and whether or not the neuron's axon passes throuh the medullary pyramid. Analysis concentrates upon the response properties of each of these subsets, with particular attention being directed to possible interactions among the members of the different subsets. Detailed neuronal circuitry is derived from the data, through both weak and strong inference, and further experiments are then generated to test and refine the model of the circuitry.

For analysis, the sampled data are prepared as a matrix of column vectors, each vector having a symbolic name. This format is employed as the primary numeric data structure in an interactive, real-time data collection and anlysis system called NVAR (2). NVAR maintains and manages the user's data structures and provides processing modules which are executed on a call-by-name basis, to perform analysis. The data are continuously available in an open file, as one processing module after another is executed. Each module performs a specific function, and leaves the open data file in the form expected by all other modules that the user might request. The NVAR system, in our laboratories for nearly 10 years, is implemented on the Raytheon PB 440 computer*, and is organized as a named-file data structure for both data and processing modules.

HARDWARE AND SYSTEMS PROGRAMMING

It is customary in a software presentation to describe the computer hardware and then proceed with the software. This custom will be followed here. The Raytheon PB 440 is a 24-bit machine with 8 CPU registers, 8k of main memory, 512 words of fast memory (Biaxial core

---

*This machine was originally produced by Packard Bell in the early 1960's. Raytheon subsequently acquired Packard Bell Computer Division and the expanded, repackaged version of the 440 was called the Raytheon 520.

with 200-μsec nondestructive readout), and a 1 microsecond basic cycle time. Peripheral devices include a CDC 852 disk drive, with 74.5k capacity (100 tracks at 745 words each), 2 Datamec 7-track tape drives, a Potter line printer, a Sorban card reader, a 16-channel multiplexed 10-bit A/D system, a 2-channel 10-bit D/A system (used in conjunction with a CRT display), and a Calcomp plotter. Figure 2 shows the organization of the CPU, memories, and I/O. Note that the L, N, and P registers are not full 24-bit registers, and that the Q register is a pseudo register, always containing zeros. The PB 440 is microprogrammed, each 24-bit program word containing 2 12-bit instructions (E register, figure 2), in either an OP R1 R2 or and OP M form. The P register is the program counter. Six bits encode the operators (OP) and 3 bits each specify the origin register, R1, and the destination register, R2. One of the registers is always used to reference memory. One does not normally microprogram large problems, however. The fast memory (Biax) is employed to define a set of programming stored-logic consisting of microprogrammed routines (microroutines). These microroutines each execute an assembly level language instruction, with a control sequence (CNSQ) managing the execution of the assembly language. Figure 3, center, shows the sequence of this

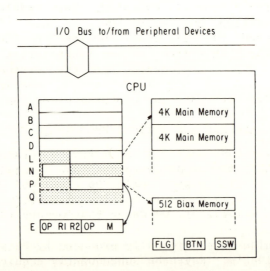

Figure 2. The CPU, memory, and I/O Bus organization in the Raytheon PB440.

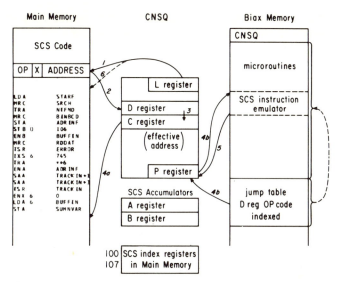

Figure 3. The execution of microprogrammed stored-logic for the assembly language SCS.

execution within CNSQ. The L register (assembly level program counter) points to the assembly level instruction to be executed (figure 3-1), and this instruction is fetched to the D register (figure 3-2). The L register is incremented by 1, the address and index (if any) are assembled into the C register (figure 3-3), and the result is an effective address (figure 3-4a). The op code in D is added to 70300 to point to a jump address, which in turn points to a microutine (figure 3-4b). The microutine executes the assembly level instruction, and returns to CNSQ (figure 3-5), which then fetches the next instruction pointed to by the L register (figure 3-6).

System processing modules are primarily written in the assembly language termed Systems Command Set (SCS), with microcoded subroutines when appropriate (such as for I/O drivers). This language uses the A and B registers for accumulators, and has 8 memory index registers (figure 3, lower center). The first half of Biax is used by SCS, as shown in figure 3, for CNSQ, microutine instruction stored-logic and for a vectored jump table. The last 3/8ths of Biax is used for system microutines, leaving 64 words for module use. Peripheral device input and output is done via the I/O Bus (top of figure 2), which is common to all peripheral devices and is manipulated directly -- passing device select, data and status responses through the CPU registers.

As will be seen below, the operating system resides on the disk. The disk is treated as a virtual memory in which a memory fetch results in an entire track (745 words) being copied to main memory. Usually, the data structure is such that a track contains sequentially processed elements. Some processing modules must fetch in a random fashion inside a file, and a software, single-word oriented, virtual memory package is used to buffer tracks to and from memory and to manage the random access buffers -- fetching from and storing into memory when the correct track is in a buffer, and loading the buffer with the needed track if not. The software selects a buffer to be overlaid on the basis of which one has been inactive the longest, writing the data to be overlaid back out onto the disk track of origin if any location in the buffer has been changed.

The systems programmer has access to the machine's stored logic, and can change parts of it, or add to it, when a processing module is written, provided that the original logic is restored before the module exists to the operating system (1). It is within the realm of the systems staff to design an entire language at the assembly level, in which a special purpose processing module can then be written (3). The operating system has several modules which include altered logic.

OPERATING SYSTEM

The design of the operating system for executing on-line real-time experiments in cardiology and neurophysiology led to the selection and organization of the PB 440 computer. The need to interactively 1) collect data, 2) analyze it, and 3) continue the experiment based upon the results, led to the named-file data structure organization discussed below. In the analytic phase of the interactive closed-loop system, the ability to perform large and complicated analyses, as if each were a single operation, was essential. Some of the analyses required were developed in the BMD series of batch processing programs from UCLA Medical Computing Center. These were adapted into interactive modules for the system, providing such analyses as Pearson product-moment coefficient of correlation, cross correlation, and Efroymson stepwise multiple regression. At about the same time, the Culler-Fried system was independently under construction at UC-Santa Barbara. Both the NVAR and the Culler-Fried (level II) systems provide symbolic, implicitly indexed vector algebra*. This feature allows the user to do real-time programming in FORTRAN- like statements which operate upon the symbolically identified vectors.

The first version of the NVAR system, launched in early 1966, ran on two tape drives. Later, after the disk drive was added, version 2 moved off tape reels and onto a disk pack. A subsequent version 3 featured rebuilt, more efficient modules, error recovery I/O drivers, and a better framework for furthur expansion.

The current version of the system is organized in the following way. The disk and main memory (with the exception of a small common block) are treated as volatile storage. The system consists of named files on a tape that is bootstrap-loaded onto the disk via the system loader, TLLD. As the files (including TLLD itself) are loaded, the first two disk tracks are used to construct a master file directory (MFD), details for which are provided below. The MFD is copied by each module as an inverted array into main memory, starting at main memory's highest address; updates in that copy do not become viable outside the module until the MFD in memory is copied back onto the disk. Thus, if a module is not allowed to finish, the file structure is maintained as if the module had not started. The operating system is in five parts: 1) the system loader (TLLD), 2) the interactive user communication and linkage routine *TP, 3) the assembly language logic elements and I/O drivers (BIAX), 4) processing modules, and 5) the micro-subroutine and macro-libraries. System expansion is accomplished with processing modules designed specifically for that purpose. The sum of these files, along with the MFD, occupies the first 120 disk tracks of the 1000 available, the remaining tracks being available to the user. The last task done in TLLD is the loading of *TP -- the user communication module. *TP, like other processing modules, is self-starting. The flow for this module is shown in the left half of figure 4. *TP first types the prompt '>' on the on-line console, to indicate

---

*Christine Moss, personal communication.

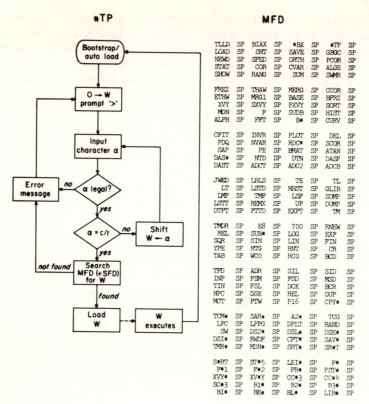

Figure 4. The flow in the operating system's user communication module *TP, and a list of the processing modules in the system.

that the system is waiting for a processing module name, and a 4-character, right-to-left shifting buffer (W) is cleared. The last 4 characters typed by the user prior to a carriage return (c/r) are assumed to specify a module name and the MFD is searched to locate that module. A current list of the names that appear in the MFD is shown in the right half of figure 4. If the name is not found, then the subfile directory (SFD) is searched for the name. If the name still is not found, then the message 'NO' is typed to indicate that the name is undefined. As will be seen later, the system creates an entry for a subfile directory in the MFD if the user has supplied programs and/or data, and the system must look for processing modules there. Use of a reserved symbol in a module evokes the message 'ILLEGAL'. When a module name has been correctly supplied, that module is found and loaded. Each module gathers in the information from the user that it needs to perform its function, issuing prompts to indicate what is to be supplied. When the module finishes, *TP is reloaded by the module to link to the next processing module through the user. Some modules link directly to another, but the last in a sequence links to *TP.

*TP can be reloaded at any time via a bootstrap initiated by the user, by pushing a button on the console. This is used to cold start the system when it is already resident on the disk, to clear a deadend interlock, or to interrupt a processing module during execution (cancel and restart).

Some of the modules interact with the user rather than returning directly to *TP. These rely upon the user to specify a function on each pass, including the operating system return function. The text editor module TE executes in this fashion. Other modules may require a user interaction which is more suited to their purposes. For example, the Digital Analog Simulator (DAS) uses its own assembly language to implement a compiler level language, in which the user writes the programs (3). DAS starts with the operating system data structure, embellishes that structure internally, and has its own mechanisms for the user to interact with running simulations.

Central to the user's interaction with the system is the file structure that the user will be manipulating. The master file directory (shown in the left half of figure 5) consists of the paired entries ('name', address) which define the existence of the file named, and where that file is located (starts) in the virtual memory. The addresses are shown in both their physical and virtual forms in figure 5. These paired references can point to processing modules, to a temporary working file, to a subfile directory, to null files, and indirectly to user files. The MFD is used in a software simulation of an associative memory in the following way. A memory fetch is requested by name. That name is looked up in the MFD and its address returned. A memory storage space request is entered by name and desired length. The associative memory simulator locates the first null file (blank name) large enough to satisfy the request, returns its address, and the supplied name replaces the blank

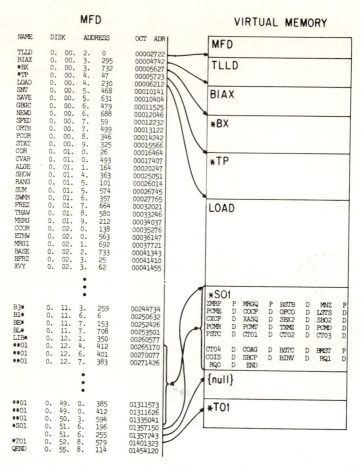

Figure 5. The organization of the virtual memory and the master file directoy which maps the named files.

name. The MFD is separated two places at this point to insert a null file name and starting address, which recovers the leftover, unused piece of the null file (the null file is usually larger than needed). File length is determined by subtraction of adjacent MFD addresses. Often, there is no null file large enough to accommodate the request. In such cases, the space used to satisfy the request is located at the end of the MFD, and becomes the highest file space occupied. The MFD-end entry (QEND) and its address are updated to point to the highest unused address in the virtual memory. Operations will occur which convert currently occupied file space to null space as a file is modified. Eventually, as the user progresses from one processing module through the next, a patchwork of null file spaces will develop, with holes scattered throughout the user virtual memory space. Nothing is done in the system to completely remove null space (e.g., by packing defined files), but contiguous empty files referenced in the MFD can be collected into one null file by executing the garbage collection module (GBGC). The user inserts GBGC into the processing stream now and then to keep the storage space tidy. GBGC collapses contiguous null files by retaining a null file pointer in the MFD discarding all immediately contiguous subsequent null file entries if they exist, and pushing the remaining MFD entries up to fill the gap.

While the user proceeds through a task stream, one of the most mobile files in the system is the temporary working file (*T01). User data files are not modified directly in the system, but are copied to the work area where their contents and dimensions can be changed. At the development point in the working file, when the user wishes to retain the working file's current contents, these contents are copied to a new, nonvolatile file under a new and unique, user- defined name. Each time a processing module makes a change in the *T01 file, a new file is created, the data are modified as they are copied to the new *T01, and the old *T01 becomes a null file.

The virtual memory is logically, though not physically, separated into system space and user space. System space is entirely accounted for in the MFD, but user space is handled through a subfile directory (SFD) and that directory's entry (*S01) in the MFD. The user can expect the system space to contain the system, but the user's space is erased by the user when a processing stream is finished. *S01 references a file that contains all user-defined file names and their locations. Such a directory -- *S01 -- is shown in the right half of figure 5. The name **01 is entered in the MFD for each user file, along with that file's address, to account for the space having been used, but all accesses to user-defined files use the SFD to locate them. File access is organized in this way so that there is a single directory (MFD) to account for the entire virtual memory, and a separate directory (*S##) to account for each user's collection of files. Though there is presently only one user at a time on the system (##=01) it was designed to be developed into a multi-user system using terminal number (##) to keep user files (*S##, **##, and *T##) separated. The SFD is structured like the MFD with paired entries, but these entries are not address-ordered. To make a subfile entry, the SFD is fetched using the MFD, and the MFD is searched as described above for the first null file space large enough to accommodate the entry. The name **01 is inserted over the null name, and the remainder of the null file is retained, as with a direct MFD entry. The address returned is paired with the user-defined file name, and is appended to the end of the SFD. This requires that a new *S01 file be created that is two words longer than the old, that the old be copied to the new, that the final two words be entered for the user-defined file, and that the old *S01 file pointer in the MFD be changed to a null file pointer. Removing a user defined file from the SFD is accomplished by removing its name and address from the SFD, copying the old SFD to a new SFD that is two words shorter, and changing the corresponding **01 and the old *S01 entries in the MFD to null file pointers. The second originates from user modification of an existing file, such as compiling a program (text) to produce an object module (binary). The third originates from files loaded from user file library -- a file-structured magnetic tape. The first two sources need no explanation in concept, but their mechanisms will be presented below (see TE). However, the user file library requires some explanation. Once a user has created a file from some input medium, and has perhaps modified it into one or more additional files, the user runs SAVE to add files to the end of a library tape. If the tape is empty, SNT is executed to format the tape for file additions. This format is as follows: the first record is a tape file directory (TFD) consisting of three-word entries {name, type, number or words} for each file on the tape, followed by a long null space for future TFD expansion, and then an end-of-file (EOF). After this EOF, the files themselves are written, in multiple blocks if required, followed by an end block and another EOF. Once having saved the user file on tape, it can be accessed another time by running LOAD, which transfers designated files from the tape into the virtual memory, creating the SFD as it executes. The file header (5 words) is loaded from each file not designated, as well as the abstracts from numeric data bases that were not loaded. This is done to ensure that the user does not assign a new file name that has already been used or a file which is stored on the user library tape.

Since all user space is expected to be available to the next user, it behooves the current user to save all files on the tape library that it is essential to keep and then, when the processing stream is finished, to release the user space by running NEWD. NEWD removes the SFD pointer *S01 from the MFD, as well as all **01 entries, thereby erasing access to the user's files.

User files can be user-defined, executable modules, or data, the data being text, symbolic program text, decimal matrices and miscellaneous data files, such as plot blocks. Text enters the system, or is modified, symbolic programs become executable modules, and numeric text is transformed into decimal matrices through the text editor (TE). TE is a text input and editor, which also links text to appropriate compilers and interpreters. The scope of the inner workings of TE is too large to discuss here, but it will be necessary to describe the decimal matrix generation portion in order to see where the data in subsequent discussions is generated. A matrix of column vectors starts as a text file, where the first line contains symbolic names for each column, the names being separated by at least one space. The second and following lines contain the row values, in BCD, for each column, separated by at least one space -- one line for each row in the matrix. The last line, as for all text files, is ',END'. Text file input is a function in TE that accepts text from any of several possible sources. Once the matrix text file is in the virtual memory, TE links it, as directed by the user, to the matrix interpreter (R). R then requests the user to supply the number of columns in the matrix and an estimate of matrix length. The data are then transformed into decimal matrix in *T01, and the dimensions m and n, and the vector names are stored in a region in main memory (COMMON-locations $2-77_8$). An example of TE card input and the R interpreter is shown in figure 7 at the second system prompt (>). All modules that process vectors expect to find the matrix of columns in *T01 and the dimensions and vector names in COMMON. It is customary for the user to make a frozen file of the TE-generated *T01 data by executing FREZ. FREZ requests the user to supply a file name and a short text abstract

describing the data, whereupon the data in *T01 and
COMMON along with the abstract are copied to a fixed file
under the supplied name. The conceptual organization for
the generated file is shown in figure 6. Once the file
is frozen, *T01 can be altered without losing the
original data. When several matrices are available for
processing, either through TE input or from the user's
library, any of these matrices can be manipulated by
moving the desired one to the working area. This is
accomplished by running THAW, which requests the user to
supply the file name, finds that file, moves the matrix
to *T01 can then be processed in the manner prescribed by
the user in a real-time function-directed stream. This
"seat of the pants" data processing method can become
quite innovative as the user chooses a processing path
based upon unforeseen results from previous operations on
the data.

The vector processing modules are of two varieties.
The first computes using vector elements without changing
any element or the dimensions of the matrix. STAT is
such a module. Its purpose is to compute the mean,
variance, and standard deviation over all the elements
for each vector specified. The user specifies vectors by
supplying their names and each module checks to see that
these names are defined for the matrix. Stepwise
Multiple Regression (SWMR) and Pearson Product Moment

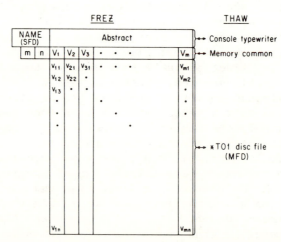

Figure 6. The format for a frozen matrix of column
vector numeric data bases and the relationship of the
matrix to the thawed copy which the user manipulates.

Correlation Coefficients (COR) are also modules of the first type. The discussion of figures 7 and 8 will show the modules in action in the context of a specific analysis. The second variety of module either makes changes in elements within vectors, appends new vectors, deletes vectors, lengthens or shortens the matrix, or a combination of these. A new *T01 is created in each of these cases. DEL is a type two module; it deletes the

```
>LOAD
HELP (Y,N)? N
CT12   D L  CC01  D L  CT02  D X  CT01  D X
PSTC   D X̄  PCME  D X̄  MRGQ  P L̄  ZMAP  P X̄
   X  P X̄  XY*C  D X̄
>TE
FUNCTION ?  I
DEVICE?  C
ERROR
±ERROR
±ERROR
±
ID CC02
FUNCTION ?  C

ID CC02
COMPILER ?  R
NO. VARIABLES = 16
N = 1000
MULTIPLE LINES (Y,N,H)?  N
N = 744
>FREZ
ID? PST2
REMARKS END WITH A ;
CCP AND TPK CARD 2 PRE C POST C MID S AND COR;
>TE
FUNCTION ?  C

ID CC01
COMPILER ?  R
NO. VARIABLES = 15
N = 4000
MULTIPLE LINES (Y,N,H)?  N
N = 3614
>FREZ
ID? PST1
REMARKS END WITH A ;
CCP AND TPK CARD 1 PRE C POST C MID S AND COR;
>ALGE
CD=CD1-MCD<CD1, 100>/100;
```
(A)

Figure 7. On-line console communication between the user (text underlined) and the system, showing (A) the input of data.

```
> BASE
CD=12+CD=21;
N = 1458  KEEP(Y,N)? Y                                    OOIX
                                                          OCD1
> GBGC                                                    OCD2
> MRGQ                                                    DPTH
ID?                                                       OPY1
PST2                                                      PYSD
SV1 SQ1                                                   PYFQ
SV2 SQ2                                                   OCF1
CCP AND TPK CARD 2 PRE C POST C MID S AND COR             CFSD
NVAR = 31                                                 CFTH
                                                          CFFQ
> ETHW                                                    OIF1
HELP (Y,N)? N                                             IFSD
*T, CT12;                                                 IFTH
> DEL              ------------                           IFFQ
NVAR = 31                                                 OSQ1
N = 1759 ___                                              OCH1
> FREZ                                                    CHSD
ID?  POCR                                                 CHTH
REMARKS END WITH A;                                       CHFQ
CARDS 1 AND 2 FOR POST C CFT,TPK, + CCP  N = 1759;        OIH1
> SAVR                                                    IHSD
NO                                                        IHTH
                                                          IHFQ
> SAVE                                                    OPY2
HELP (Y,N) ? N                                            OPY3D
CC02  D S PST2   D S PST1   D S POCR  D S                 OPY4D
                                                          OPY5D
> LT                                                      OPY6D
XY*C  D     X    P  ZMAP  P  MRGQ  P                      OPY7D
PCME  D   PSTC   D  CT01  P  CT02  D                      OPY8D
CC01  D   CT12   D  CC02  D  PST2  D
PST1  D   POCR   D  END

> NEWD
UNSAVED USER FILES CANT BE RECOVERED AFTER NEWD.
ABORT OR CONTINUE (A,C)?   C
>                          (B)
```

Figure 7. (*continued*) On-line console communication between the user (text underlined) and the system, showing (B) the preparation of a working matrix.

vectors specified from the matrix. ALGE is another type two module, which provides for vector algebra capabilities. The vector equations are specified in terms of the vector names, and indexing is implied as with all vector processing modules; that is, C = A + B is taken to mean $C_i = A_i + B_i$, for i = 1 to n. ALGE accepts algebraic and/or logic statements from the user, compiles them, and then executes the resulting program on the vectors indicated in the statements. These statements are FORTRAN-like, with relational operators, logic operators, and function subroutines. Error detection is done on a line-by-line basis as each line is input and compiled, with an errant line deleted; in order to proceed, the user must replace it with a correct line. [The first version of this module, called NVAR (the system's namesake), employed a left-to-right incremental compilation method, initially with +, -, *, /, and = as the only operators. Parentheses and function calls were soon added, and re-implementation (ALGE) has now reached

```
> LOAD
HELP (Y,N)? N
PCOR   D L PST1   D X PST2   D X CC02   D X
CT12   D X CC01   D X CT02   D X CT01   D X
PSTC   D X PCME   D X MRGQ   P X ZMAP   P L
    X  P L XY*C   D X
> ALGE
NO DATA THAWED
> THAW
ID: POCR
CARDS 1 AND 2 FOR POST C CFT,TPK, + CCP  N=1759

> ALGE
TCD=MOD(CD1,100)-MCD(CD1,10)/10
ACD=MOD(CD1,109;ERR.
ACD=MOD(CD1,10);
> FREZ
ID? POC2
REMARKS END WITH A;
SAME AS POCR BUT WITH TCD AND ACD ADDED;
> BASE
CF1 > 0;
N = 1692 KEEP(Y,N)? Y

> STAT
CF1

                   CF1

MEAN           1.79694E01
VARIANCE       0.94416E02
STAND DEV      0.97168E01

> MDN
CF1
                .25         MEDIAN       .75

OCF1          1.38999E01   1.67998E01   2.05000E01

> THAW
ID: POC2
SAME AS POCR BUT WITH TCD AND ACD ADDED

>ZMAP
LINEPRINTER ON; PUSH START
15 BINS OR MORE REQUIRE WIDE PAPER
HELP(Y,N)? N
ID: CELL COUNT BY CLASS    CELL COUNT BY CLASS

X: TCD=0,4,1
Y: ACD=0,2,1                  1 72  8 04  6 91   76    16    1759
COUNT(DIGIT, VECTOR)
INCL ZERO (Y,N) Y             *  *  *  *  *  *   *  *  *
PLOT(A, T, P)  A              52  2 51  2 67     15     7    592
PLOT ZERO (Y,N) N                                   *
BASE (Y,N)  N                 44    59  2 12     23     2    340
> BASE                                                 *
TCD=2*CF1 > 0*IF1 > 0*CH1 > 0*IH1 > 0;
N = 398 KEEP(Y,N)? Y          76  4 94  2 12     38     7    827
                         (A) *  *  *  *  *  *   *  *  *
```

Figure 8. On-line console communication between the user (text underlined) and the system, showing (A) an analysis of a population sample.

```
>STAT
CF1, IF1, CH1, IH1

                    CF1             IF1             CH1             IH1

         MEAN       1.91492E01      3.21144E01      3.90081E01      4.55025E01
         VARIANCE   3.95154E01      0.77331E02      1.21079E02      1.46040E02
         STAND DEV  6.28613         0.87938E01      1.10036E01      1.20846E01

>MDN
CF1, IF1, CH1, IH1

                    .25             MEDIAN          .75

         OCF1       1.46997E01      1.79497E01      2.20000E01

         OIF1       2.57998E01      3.06992E01      3.65996E01

         OCH1       3.10000E01      3.76992E01      4.50000E01

         OIH1       3.68984E01      4.30496E01      5.15000E01

>COR
CF1, IF1, CH1, IH1

                    CF1             IF1             CH1             IH1

         CF1        0.99999
         IF1        0.36571         0.99999
         CH1        0.23496         0.57137         1.00000
         IH1        0.14753         0.61101         0.75085         1.00000

>SWMR
HELP(Y,N)?   N
INDEP. VARS?   IH1
DEP. VAR?      CH1
PARAMS(Y,N)?   N

                    SUM SQUARES     DF
         TOTAL      0.48068E05      397
         RESIDUAL   2.09684E04      396
         REGRESSION 2.71002E04      1

         MULTIPLE RSQ=   0.56378             R= 0.75085
         STD. ERROR OF EST.= 7.27661
         CONSTANT = 7.89879

         VARIABLE   COEFFICIENT     ERROR             RSQ
         OIH1       0.68368         0.30220E-01       0.56378

>ALGE.
CHY=CH1*10
IHX=IH1*10
X=0
X=LAST(X)+2.5
Y=.684*X+78;
                                  (B)
```

Figure 8. (*continued*) On-line console communication between the user (text underlined) and the system, showing (B) an analysis of the subset.

a level of sophistication equal to current day compilers. The programs from ALGE can be stored as user-defined modules intead of executing them immediately. The user assigns a name to the compiled program; the program becomes a named file in the SFD and can be executed on a call-by-name basis, as with any other system or user processing module. However, such a program is only defined for matrices that contain the same vector names as the matrix for which the program was defined.

The system contains nearly 120 processing modules, over half of which operate upon vector matrices. This

```
>XVY
IHX,CHY
XX,Y
E
>ZMAP
LINEPRINTER ON; PUSH START
15 BINS OR MORE REQUIRE WIDE PAPER
HELP(Y,N)?  N
ID: M SET

 X: CF1=6,40,2
 Y: DPTH=0,2400,200
 COUNT(DIGIT, VECTOR)
 INCL ZERO (Y,N)  Y
 PLOT(A, T, P)    A
 PLOT ZERO (Y,N)  N
 BASE (Y,N)  N
>THAW
ID: POC2
SAME AS POCR BUT WITH TCD AND ACD ADDED

>BASE
TCD=1*CF1 > 0;
N = 785 KEEP(Y,N)?  Y

>STAT
CF1

                CF1

MEAN          1.67678E01
VARIANCE      2.40417E01
STAND DEV     4.90323

>MDN
CF1

              .25         MEDIAN       .75

OCF1       1.31997E01   1.56997E01   1.91992E01

>ZMAP
LINEPRINTER ON; PUSH START
15 BINS OR MORE REQUIRE WIDE PAPER
HELP(Y,N)?  N
ID: S A  SET

 X: CF1=6,40,2
 Y: DPTH=0,2400,200
 COUNT(DIGIT, VECTOR)
 INCL ZERO (Y,N)  Y
 PLOT (A, T, P)   A
 PLOT ZERO (Y,N)  N
 BASE (Y,N)  N
>NEWD
UNSAVED USER FILES CANT BE RECOVERED AFTER NEWD.
ABORT OR CONTINUE (A,C)?  C
>
```

(C)

Figure 8. (*continued*) On-line console communication between the user (text underlined) and the system, showing (C) an analysis of the subset and comparison to another subset.

vector representation of data is applicable in a wide variety of disciplines and is a major element in the system's usefulness, because it maintains a uniform data structure for the user. The system is expected to be resident when the user comes on-line, with the MFD structured for the system modules. The user then loads or creates user files, which causes the MFD to be expanded and a user SFD to be generated. Processing modules are then sequentially executed on a call-by-name basis as the user manipulates data through a real-time directed processing stream. When the user is finished, the files to be retained are saved on a user library tape, and the virtual memory is cleared of user files in preparation for the next user.

POPULATION SAMPLE DATA

The neuron population sample data are prepared in the matrix of column vectors as described above. The character of the data, however, requires some additional conventions, as shown in Table 1. Though the outer bound

Table 1. The set of matrices in which a population sample is prepared for processing and the vector names and descriptions for each variable.

MATRIX 1			MATRIX 3		
CD1	Class code		I11	CF 1st sp & i = 1.0	
CD2	Modality code		I1SD	s/d	
DPTH	Depth (in microns)		I21	CF	1 = 0.5
AD1	Antidromic 1st latency		I2SD	s/d	
ADSD	" (AD)	s/d	F11	CF 1st sp & f = 2.0	
ADFQ	"	1/ff	F1SD	s/d	
CF1	Contrafore 1st latency		F21	CF	f = 4.0
CFSD	" (CF)	s/d	F2SD	s/d	
CFTH	"	threshold	F31	CF	f = 6.25
CFFQ	"	1/ff	F3SD	s/d	
IF1	Ipsifore 1st latency		F41	CF	f = 10.0
IFSD	" (IF)	s/d	F4SD	s/d	
IFTH	"	threshold	SQN	Sequence number	
IFFQ	"	i/ff			
SQN	Sequence number				
MATRIX 2			MATRIX 4		
CH1	Contrahind 1st latency		CF2		
CHSD	" (CH)	s/d	IF2		
CHTH	"	threshold	CH2		
CHFQ	"	1/ff	IH2		
IH1	Ipsihind 1st latency		I12	2nd spike latency	
IHSD	" (IH)	s/d	I22	for all inputs	
IHTH	"	threshold	F12		
IHFQ	"	1/ff	F22		
AD2			F32		
AD3			F42		
AD4			SQN	Sequence number	
AD5					
AD6			MATRIX 5,6,7,...		
AD7					
AD8			CFi	Repeats of the form	
SQN	Sequence number		IFi	in Matrix 4 for the	
			.	3rd, 4th, 5th,...	
s/d = spikes per discharge			.	spike latencies	
f = frequency			.		
ff = frequency following					

on matrix dimensions is theoretically 55 by 32k, it would be impractical to deal with a matrix that large. If all the elements for the largest population sample on hand were entered into one matrix, it would be 250 by 4k -- a size equally impractical. As will be seen below, this would accommodate all the elements for a supra-responsive neuron, but substantially more than half of the matrix elements would be empty (zero) with some vectors being entirely empty, except perhaps for one or two elements. For the sake of better storage efficiency, the population sample data are kept in an ordered set of matrices -- each containing 11 to 16 vectors. This presents no problem in the analysis, because the user can collect the appropriate vectors for a sequence of calculations from the set of matrices into a single working matrix of tractable dimensions. In particular, matrices comprising the set are as follows: The first contains 15 columns, each containing data for each neuron sampled from the population. The last vector, as in all the matrices, is a sequence number identifying the specific neuron; the other 14 vectors contain coding elements, static measures, and the 'on focus,' initial response properties. The second matrix has 16 columns, 15 of which contain 'off-focus' initial response properties, along with other orthodromic responses. Since many neurons show no off-focus response, the length of the second matrix is often about 1/3 that of the first. The sequence number in each matrix is the key to row alignment when vectors from matrices of different lengths are side-by-side merged into a working matrix. The third matrix conatins data from high frequency (iterative rate) and low intensity stimulation. These first three matrices provide latency information on the first spike of each response (burst of spikes). The next several matrices contain second-through last-spike latency data, one matrix for each spike, resulting from all stimulus sites tested (in the sense that spikes in a single discharge are sequentially counted). Each of these matrices is shorter in length than the preceding, as fewer and fewer neuron responses show additional spikes. There are two additional matrices, the size of the first, which contain response classification and probability information; this information will not be considered here. The column vectors in the population samples have a unique, predetermined name to allow different population samples to be end-to-end compatible. There are null elements in all of the matrices where data are not available, or where the neuron was unresponsive to a particular test, but the quantity of these is considerably less than it would be with fewer, yet larger, matrices.

The neuron population sample matrices are input from punched cards through TE, as described earlier, and are converted to working files in *T01. These are copied via FREZ to frozen files, under user-defined names, and are then copied to user library by SAVE. Figure 7 shows this process, step by step, as the user prepares a matrix. The underlined part of the dialog indicates the user's part of the interaction. In Figure 7A, the user starts by LOADing the files from the library. LOAD asks if the user would like help with operating instructions. Many of the modules which require more than one response from the user provide a tutor option. The user in figure 7A answered 'NO' to the help question: the file names from the TFD were then listed one by one, and the user answered either L or X to load or skip each file. The user called TE, the input function 'I' was indicated, and the device 'C' was specified to read a deck of cards. Three card reader errors occurred which were user-corrected (↑). The text file was named CC02 by the user. The next function indicated in TE was a compile 'C', and the matrix interpreter 'R' was specified for 16 vectors and a maximum of 1000 single lines. The file contained 744 lines, and the decimal matrix in *T01 from a frozen file, via THAW. In figure 7B, the desired matrix was already in *T01, and a combination of ALGE and BASE had been used to select the population sample for the specific recording site shown in figure 1B (details for this procedure will be described below in the context of subset selection). The user executed GBGC, and then MRGQ was executed in order to row-to-row merge the frozen file PST2 to that in *T01, using the seqence number vectors to align rows, and filling in null elements for *T01 rows which had no mates in with identical names from the separate population sample CT12, taken at the same sampling site, to the vectors in *T01. DEL listed the vector names one by one, and the user answered either '(sp)', to keep the vector, or 'D', to discard it (this has been excised and placed to the side in figure 7B to save space). The user then froze the 26,385-element matrix in *T01, naming it POCR, and SAVEd the files on all the rows in *T01, in which the vector element TCD is exactly 2, and the vector elements CF1, IF1, CH1, user answering after each file name either 'S' for save or 'X' for skip.

POPULATION ANALYSIS

The user begins the analysis at the first level for the population sample data, prepared as described above, by calculating statistics for the sample as a whole. The data are initially on a user library tape, and the files containing the desired vectors are first LOADed, thereby establishing the user's SFD (top of figure 8A). The vectors involved in projected calculations have been assembled into one matrix via THAW, MRGQ, and DEL, as shown in figure 7B. The resulting file--POCR--was frozen with FREZ, so that it could be recovered later in the analysis. On completion of loading, the user THAWed the matrix POCR. ALGE was then used to decode the two least significant digits from the coding vector CD1. Note that a programming error occurred in the second equation, where a '9' was typed instead of a ')'. The equation was correctly replaced, the list was terminated (;), and was executed. The resulting matrix was frozen, and named POC2. BASE (described below) was then used to clear the CF1 vector of null entries before the statistical modules STAT and MDN (median) were executed. After re-establishing POC2 with THAW, the remainder of the level one processing consisted of counting neurons jointly by type (TCD and ACD) with ZMAP (also described below), as shown in the table insert at the bottom of figure 8A. The second level analysis consisted of first selecting a subset from the population sample, calculating response properties, and then inferring from the results the likely functional relationships this subset has with other subsets. Constructing a subset is accomplished with the module BASE, which selects rows from the matrix currently in *T01, and assembles them into a new *T01. BASE expects the user to type a logical expression which uses the boolean operators to combine relational statements, comparing named vector elements to each other, or to constants. Parentheses can be used to specify precedence, where necessary. For example, TCE = 2*CF1 > 0*IF1   0*CH1 > 0 (bottom of figure 8A) selected the tape library. The procedure in SAVE is the reverse of LOAD; the SFD is listed rather than the TFD with the and IH1 are all non-null. This subset of 398 neurons is the collection of wide-field neurons (m neurons) for which there is a known first-spike latency to all of the somatosensory stimulus sites tested in the experiment. Some of this subset's response properties are calculated using STAT, MDN, and COR, as shown in figure 8B. The relatively high degree of synchrony indicated between CH1 and IH1 in the correlation matrix is shown in the scattergram insert (top of figure 8C), which is a photograph of the cartesian display generated by the module XVY. IH1 is on the abscissa, and CH1 is on the ordinate. The straight line fit in the photo was calculated by running the stepwise multiple regression module, SWMR (figure 8B), on CH1 and IH1, and then using ALGE to calculate an x, y pair of vectors for the line. X and y were then displayed via XVY, and a second exposure made. XVY has a fixed 0-to-1024 range in x and y, so data must be scaled (in ALGE) to this range for display.

The lower correlation between CF1 and the other three latencies indicates a possible external influence from some other subset upon the responsiveness of the m neurons to contralateral forepaw (CF1) stimulation. There is a subset of neurons that is responsive to contralateral forepaw stimulation only (sa neurons). In order to compare the response properties of sa neurons with m neurons, another calculation is made for the m neurons before proceeding to the sa neurons: the joint density in time and depth is calculated with ZMAP (top of figure 8C). This module partitions two variables into bins along perpendicular axes, and counts the number of x, y pairs that fall in each bin. The result can be viewed as a 3-dimensional surface, contours routinely being used to represent it, much as in topographic maps of terrain. Such a joint density calculation is shown in figure 9A for the m subset, using CF1 as x and DPTH as y. ZMAP expects the user to specify the variables, counting-space dimensions, and the type of output desired when the result is printed.

The sa subset is collected (center of figure 8C) in much the same way as the m subset but in this case, TCD = 1*CF1 > 0 is sufficient in BASE to select the proper rows from the reTHAWed population sample. STAT and MDN are run to compare the results with those from the m subset

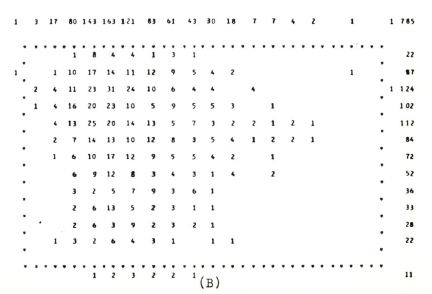

Figure 9. Line printer output from the joint density calculations done by ZMAP for the (A) and (B) neuron subsets.

CF1 properties (figure 8C). The mean and median latencies are smaller for the *sa* neurons than for the *m* neurons. Other properties show a distinct difference between these two subsets, as well (analysis not shown in figure 8). The joint CF1, DPTH density produced by ZMAP (figure 9B) shows the earlier position in time for the *sa* neurons, and also shows that the *sa* neurons are concentrated at a shallower depth in the cortex than the *m* neurons. The user then cleared the SFD with NEWD to finish the run. Further analysis of the *m* neurons (not shown) shows that they concentrate in two regions -- a shallower and later group in layer III and another, somewhat earlier and deeper group in layer V. The *sa* neurons, however, concentrate earlier and more superficially than both *m* groups. These results, along with others (calculations not shown), have led to the circuit diagram shown in figure 10, which depicts the

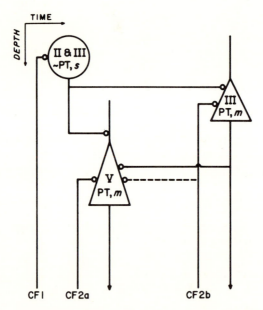

Figure 10. Circuit diagram for three elements of the functionally definable subsets in postcruciate cerebral cortex (of the cat). From Towe et al., Experimental Neurology 20:510 (1968).

organization of functional elements in the tissue, with their monosynaptic and multisynaptic inputs from the periphery. This model has been discussed elsewhere (5), and is much like the models deduced by others at various locations in the spinal cord, midbrain, and cerebellum. Continued experiments and analysis are done in an attempt to test this model, and refine it, to account for further experimental observations in the mammalian nervous system.

CONCLUSION

The ability to manipulate these data in a real-time setting has accelerated the developments in the population sampling technique to a level where it is conceptually manageable. For example, one cloudy afternoon in April, 1967, the state of the model and some recent experimental observations led to the conclusion that one of the population sites must contain a heretofore unrecognized subset that is only scarcely present in the other sites. This subset would have response properties somewhat like the sa neurons and would comprise 25% of the sample. The difference between the postulated sb neurons and the sa neurons was that the sb subset would respond to stimulation of both forelimbs and upper trunk, but would be unresponsive to stimulation of the hindlimbs or hindquarters. Twenty minutes later, this subset was identified (these neurons had previously been classified as m neurons), and it was shown to possess the response properties predicted. More recently, examination of the relationship between sa neurons and m neurons has led to the formulation of the model in terms of coadunate behavior (5). The 'off-focus' response for the subset is used to predict where (in time, and with respect to the responding population) an unmodulated 'on-focus' response should occur; the ratio of observed to expected response latencies is then calculated. If the ratio is less than one, then the sa influence on that neuron is facilitatory. On the other hand, if the ratio if greater than one, then the net influence is inhibitory.

The understanding which has been gained over the last several years, and the development of the neuron population analysis technique, would not have progressed nearly as rapidly without the NVAR system. The features that make NVAR useful are the data structuring done by the system rather than by the user, the ability to program in real time, and the ease with which rather sophisticated analyses can be acccomplished.

DISCUSSION

MCCORMICK: You said you could select rows. It's not a piece of code but one formal statement?

HARDING: Yes. But it uses relational operators and vector names, so it can be quite long. Nesting with parentheses is allowed. You can get as simple or complicated as you like.

REFERENCES

1. Kehl, T. H., and Moss, C. Systems programming on-line. Comp. Biomed. Res. 1:550-555 (1968).
2. Kehl, T. H. Interactive real-time computation. Comp. Biomed. Res. 1:590-604 (1968).
3. Kehl, T. H., Moss, C. B., and Keck, T. S. Interactive continuous system simulation. Comp. Biomed. Res. 7:71-82 (1974).
4. Towe, A. L. Neuronal population analysis in the cerebral cortex. *Proceedings of the Symposium on Information Processing in Sight Sensory Systems*, ed. P. W. Nye, pp. 143-156, California Institute of Technology, Pasadena, (1965).
5. Towe, A. L., Whitehorn, D. and Nyquist, J. K. Differential activity among wide-field neurons of the cat postcruciate cerebral cortex. Exp. Neurol. 20:497-521 (1968).
6. Towe, A. L., and Harding, G. W. Extracellular microelectrode sampling bias. Exp. Neurol. 29:366-381 (1970).

APPENDIX

The processing modules available on the system along with a brief description of their functions are listed from a text file on the system library tape.

**** **** PROGRAMS IN DISK LIBRARY **** ****
SEPTEMBER 1973

0. TEXT INPUT-OUTPUT

TE		TEXT EDITOR: INPUT (CARDS, TYPEWRITER, PTAPE, MAG TAPE)
		ANY TYPE OF TEXT IS MADE INTO DISK FILES
	TE	TEXT EDITOR: ALTER (USES SCOPE DISPLAY)
		TEXTS ON DISK CAN BE ALTERED MAKES NEW FILES
TL		TEXT LIST(TYPEWRITER, LINE PRINTER, PAPER TAPE)
		OUTPUT OF ANY TYPE OF TEXT ON DISK
	SAVE	SAVE DISK FILES ON USER'S LIBRARY MAG TAPE
		FOR TEXT, COMPILED PROGRAMS, AND DATA FILES
	LOAD	LOAD FILES FROM MAG TAPE TO DISK
		FOR USER LIBRARY TAPES GENERATED BY SAVE

1. GENERATING NEW NVAR DATA BASES (VECTOR ARRAYS)

   1A. IN WORKING AREA

SUDB	SET UP DATA BASE
	GIVES IX FROM 1 TO N FOR FURTHER CALCULATION
TE	TEXT EDITOR: COMPILER R
	CONVERTS STORED NUMERIC TEXTS TO NVAR DATA BASES
THAW	THAW DATA BASE FROM DISK
	COPIES AN EXISTING BASE TO WORKING AREA
ETHW	EXTENDED THAW: EXTENDS WORKING BASE WITH MORE ROWS FROM DISK
	KEEPS COMMON VARIABLES MAKING LONGER COLUMNS
MERG	MERGE DATA ON DISK WITH DATA IN WORKING AREA
	ADDS NEW VAR'S (COLUMNS) TO DATA SET
MRG1	MERGE DATA ON DISK WITH DATA IN WORKING AREA
	ADDS SELECTED NEW VAR'S TO WORKING DATA
JWED	RAW DATA MAG TAPE SCOPE DISPLAY AND EDIT
	SELECTS TAPE RECORDS, CONVERTS THEM TO NVAR BASES

   1B. STORED (FROZEN) ON DISK

LOAD	LOAD USER MAG TAPE (SYSTEM FORMAT)
	DATA BASES SAVED ON TAPE ARE LOADED ON DISK
FREZ	FREEZE DATA FROM WORKING AREA
	COPIES DATA ONTO DISK WITH IDENTIFIER AND LABEL
BFRZ	SELECTIVE FREEZE DATA FROM WORKING AREA
	ONLY FREEZES ROWS SATISFYING GIVEN LOGICAL RELATION
DTN	DAS TO NVAR
	RECASTS WORKING DAS FILES AS FROZEN NVAR BASES

2. OPERATIONS ON NVAR DATA BASES IN WORKING STORAGE

    2A. FORTRAN-LIKE ALGEBRAIC AND LOGICAL OPERATIONS

      NVAR   NEW VARIABLES
                SIMPLE ALGEBRA DIRECTLY FROM TYPEWRITER
      ALGE   EXTENDED ALGEBRA WITH CONDITIONAL EXECUTION
                ALGEBRA FROM TYPEWRITER WITH LOGICAL EXPRESSIONS
  F   MINI FORTRAN (A COMPILER OPERATING ON STORED TEXT)
                ALGEBRA, LOGIC, AND CONDITIONAL JUMPS (CAN LOOP)
  BASE   REDUCES SIZE OF DATA SET BY DROPPING ROWS
                SAVES ROWS SATISFYING INPUT LOGICAL EXPRESSION

  2B. STATISTICAL OPERATIONS (DATA UNCHANGED UNLESS NOTED)

    SORT   SORTS ROWS BY CHOSEN VARIABLE
            ROWS REORDERED SO VARIABLE INCREASES THROUGHOUT
    HIST   HISTOGRAM OF INPUT VARIABLE
            DISPLAYED ON SCOPE
    NHST   NEW HISTOGRAM
            NEW DATA BASE FORMED WHICH IS A HISTOGRAM
    RANG   RANGE OF VARIABLES
            TYPES MAX AND MIN OF INPUT VARIABLES
    MDN   MEDIAN
            TYPES MEDIAN AND QUARTILES OF CHOSEN VARIABLES
    STAT   STATISTICS
            TYPES MEAN, VARIANCE, AND ST DEV OF VARIABLES
    ORTH   ORTHOGONALIZATION
            ORTH, SWMR, AND CFIT ARE QUITE SIMILAR
    SWMR   STEPWISE MULTIPLE REGRESSION
            TYPES OUT LINEAR MEAN SQUARE REGRESSION
    CVAR   COVARIANCES

    COR   CORRELATION COEFFICIENTS OF INPUT VARIABLES
              TYPES OUT
    PCOR   PARTIAL AND MULTIPLE CORRELATION COEFFICIENTS OF
            INPUT VARIABLES ARE TYPED OUT
    CCOR   CROSS-CORRELATION OR AUTO-CORRELATION
            USES UNBIASED ESTIMATOR FORMULA (NO TAPERING)
    SCOR   CROSS-CORRELATION OR AUTO-CORRELATION
            USES BIASED ESTIMATOR WITH SMALLER MEAN SQUARE ERROR
    CFIT   CURVE FIT - FINDS LINEAR COEFF'S TO APPROX INPUT VAR
            BY OTHER INPUT VAR'S (LST SQRS) SEE CURV
    CURV   CALCULATE FITTED CURVE AS NEW VAR IN DATA BASE
            USES COEFFICIENTS SAVED BY CFIT
    FFT   FAST FOURIER TRANSFORM - FORWARD AND INVERSE
            REAL + IMAG COEFF'S + PWR SECTRUM ADDED TO DATA BASE
    INVR   INVERT MATRIX
            TYPES OUT
    BMAT   BOOLEAN MATRIX
            TYPES SINGLE AND JOINT OCCURRANCES OF INPUT INTEGER

2C. OUTPUT AND DISPLAY OF WORKING DATA

- SHOW   NUMERIC OUTPUT OF SELECTED VARIABLES (ENTIRE COLUMNS)
  TYPEWRITER, LINE PRINTER, OR PAPER TAPE
- PXVY   SCOPE DISPLAY OF ONE VARIABLE AGAINST ANOTHER
  TYPE P TO SAVE AS A FILE FOR CALCOMP PLOTTER
- SXVY   LIKE PXVY
- XVY   LIKE PXVY EXCEPT DISPLAYS X VALUES IN
  NON-DECREASING ORDER (SORTED)
- ALPH   ALPHANUMERIC SCOPE DISPLAY OF TYPED IN CHARACTERS
  FOR LETTERING TO ACCOMPANY SCOPE DISPLAY PHOTOS
- PLOT   CALCOMP PLOTTING PACKAGE: TEXT, AXES, LINES, SYMBOLS
  USES FILES SAVED BY PXVY AND TEXT EDITOR TEXTS

2D. OTHER MANIPULATIONS OF DATA BASE

- DEL   DELETE SELECTED VARIABLES (COLUMNS) FROM WORKING DATA
  REDUCES SIZE OF DATA, LISTS ALL VARIABLE NAMES
- PDQ   PUSH DOWN QUEUE: PUSHES SELECTED VARIABLES DOWN INPUT
  NUMBER OF ROWS, REDUCES SIZE OF DATA SET
- SUM   SERIAL SUMMATION: REPLACES SELECTED VARIABLES BY THEIR
  SERIAL SUM, LIKE AN INTEGRAL

3. OTHER USER PROGRAMMING SYSTEMS OR LANGUAGES

3A. BASIC

- TE   TEXT EDITOR: INPUT (CARDS, TYPEWRITER, P TAPE, ETC)
  FOR INPUTTING BASIC SOURCE PROGRAMS (TEXT)
- TE   TEXT EDITOR: COMPILER B (COMPILES BASIC SOURCE PROGRAMS)
  COMPILED PROGRAM SAVED AS FILE ON DISK

3B. DIGITAL ANALOG SIMULATOR

- TE   TEXT EDITOR: INPUT (CARDS, TYPEWRITER, PTAPE, ETC)
  FOR INPUTTING DAS SOURCE PROGRAMS (TEXT)
- TE   TEXT EDITOR: COMPILER D (COMPILES DAS SOURCE PROGRAMS)
  COMPILED PROGRAM SAVED AS FILE ON DISK
- DASF   FREEZE DAS FILE FROM WORKING AREA TO DISK
  COPIES DAS DATA FILE TO DISK WITH IDENTIFIER
- DAST   THAW DAS FILE FROM DISK TO WORKING AREA
  MAKES FILE AVAILABLE TO DAS PROGRAMS
- NTD   NVAR TO DAS
  RECASTS FROZEN NVAR FILE AS WORKING DAS FILE
- DTN   DAS TO NVAR
  RECASTS WORKING DAS FILE AS FROZEN NVAR FILE
- PE   PUNCH ,END AND LEADER ON PAPER TAPE
  TO TERMINATE DAS PUNCHED TAPES IN TEXT EDITOR FORMAT

3C. TEACHING MACHINE

    TM   TEACHING MACHINE INPUT ROUTINE

    TMDR TEACHING MACHINE DATA RETRIEVAL

4. MAGNETIC TAPE INPUT/OUTPUT

  4A. USER LIBRARY TAPES

    SNT   START NEW TAPE: INITIALIZES A NEW TAPE FOR SAVE
            CAUTION: WIPES OUT ALL FILES ON TAPE
    SAVE SAVE DISK FILES ON USER LIBRARY TAPE
            FOR TEXT, COMPILED PROGRAMS, AND DATA FILES
    LOAD LOAD FILES FROM MAG TAPE TO DISK
            LOADS FILES WRITTEN BY SAVE
    LT   LIST TAPE: CHECKS THAT ALL TAPE RECORDS ARE READABLE
            LISTS IDENTIFIERS OF FILES. USE AFTER SAVE AS
            DOUBLE CHECK
    LSTT LIST TAPE FILE IDENTIFIERS AT FRONT OF TAPE
            TYPES ID'S WITHOUT CHECKING RECORDS

  4B. DISK SYSTEM LIBRARY TAPE (TLLD TAPE)

    TLLD LOADS DISK OPERATING SYSTEM FROM TLLD TAPE
            ALTERNATIVELY USE MAG TAPE BOOTSTRAP AT CONSOLE

  4C. USER RAW DATA TAPES

    ADCT ANALOG TO DIGITAL INPUT TO TAPE
            CONTINUOUS INPUT, RECORDS GAPPED AT EXTERNAL TRIGGER
    ADC2 ANALOG TO DIGITAL INPUT TO TAPE
            LIKE ADCT BUT FASTER SAMPLING
    ADCB ANALOG TO DIGITAL INPUT TO TAPE
            SAMPLES FOR SPECIFIED INTERVAL AFTER TRIGGER
    LBLS LABELS: TYPES LABELS AND COUNTS DOWN RECORDS
            FOR POSITIONING RAW DATA TAPES FOR EDIT
    JWED GENERAL RAW DATA MAG TAPE SCOPE DISPLAY AND EDIT
            SELECTS TAPE RECORDS, CONVERTS THEM TO NVAR BASES
    RNEW SPECIALIZED RAW DATA EDITOR FOR DR. SCHER

    E8   SPECIALIZED 8 CHANNEL RAW DATA EDIT FOR DR. SCHER

    DUMP OCTAL WORD BY WORD DUMP OF ANY TYPE OF TAPE
            PRINTS OUT ON LINE PRINTER

5. DISK FILE MANAGEMENT FOR USERS

    LOAD LOAD FILES FROM MAG TAPE LIBRARY TO DISK
            LOADS FILES WRITTEN BY SAVE
    SAVE SAVE DISK FILES ON USER LIBRARY MAG TAPE
            FOR TEXT, COMPILED PROGRAMS, AND DATA FILES
    LSTD LIST DISK FILE DIRECTORIES
            TYPES IDENTIFIERS OF FILES CURRENTLY ON DISK

SFED	SUB FILE EDIT: DISCARDS UNWANTED FILES AND IDENTIFIERS	
	REDUCES NUMBER OF USER FILES (SUB FILE AREA)	
NEWD	NEW DISK: WIPES OUT ALL USER FILES ON DISK (SUB FILE AREA) USED WHEN A NEW USER TAKES OVER COMPUTER	
REMX	REMOVE X'ED IDENTIFIERS FROM SUB FILE DIRECTORY	
	WEEDS OUT NOT-LOADED ID'S: MAKES ROOM FOR NEW ONES	
GBGC	GARBAGE COLLECTION: SIMPLIFIES FILE DIRECTORY AFTER COMPLEX DISK OPERATIONS. MAY HELP REDUCE DISK ERRORS	
TLLD	RELOADS DISK OPERATING SYSTEM FROM TLLD TAPE	
	USUALLY MAG TAPE BOOTSTRAP IS USED FOR THIS	

6. PROGRAMS, ASSEMBLERS, AND DIAGNOSTICS FOR PROGRAMMERS

SAP	.SYSTEM ALTERATION PROGRAM	
	TO MAKE BINARY LEVEL ALTERATIONS OF OPERATING SYSTEM	
GLIB	GENERATE LIBRARY EXTERNAL FILE	
	LIBRARY BLOCK FOR SAR ASSEMBLER	
DTPT	DISK TO PAPER TAPE: PUNCHES COMPILED BINARY PROGRAMS	
	TAPE IS INPUT VIA PTTD	
PTTD	PAPER TAPE TO DISK: INPUTS COMPILED BINARY PROGRAMS	
	FOR TAPE FROM DTPT	
UP	UTILITY PROGRAM TO TRANSFER NAMES BETWEEN SUB AND MAIN FILE	
	FOR ADDING NEW ASSEMBLED PROGRAMS TO OPERATING SYSTEM	
SDMP	SYSTEM DUMP: WRITES MAG TAPE OF MAIN FILE PROGRAMS	
	FOR GENERATING NEW TLLD TAPE OF OPERATING SYSTEM	
700	RAYTHEON 700 SERIES COMPUTER ASSEMBLER	
REL	RELOCATABLE LOADER: GIVES CONSOLE DISPLAY/ENTER FNCT'S TO TYPEWRITER. ACCESSES CORE AND REGISTERS	
BIAX	RELOADS BIAX (FAST MEMORY) FROM DISK	
LMF	LIST MAIN FILE ON LINE PRINTER	
	GIVES EXTENSIVE MAP OF MAIN FILE	
LSF	LISTS SUB FILE ON LINE PRINTER	
	GIVES EXTENSIVE MAP OF DISK SUB FILE	
EXDT	EXAMINE DISK TRACK	
	BRINGS CHOSEN TRACK TO CORE, END	

chapter 23

AN ON-LINE MEAN-SQUARE-ERROR ANALYSIS TECHNIQUE USING WHITE NOISE INPUTS

D. P. O'LEARY, C. WALL, III AND L. TRAINI

Department of Otolaryngology
University of Pittsburgh School of Medicine
Pittsburgh, Pennsylvania

INTRODUCTION

The application of white noise inputs for the analysis of neurobiological system is useful for obtaining small-signal linear characteristics on-line and also either nonlinear characterizations or low-order linear parameters from a subsequent off-line analysis (Marmarelis and Naka, 1973a, b, 1974; O'Leary and Honrubia, 1975). These inputs are particularly valuable in that they provide information-rich stimuli which test the full dynamic range of a system, and through the use of either cross-correlation or Fourier Transform techniques, yield system characterizations in a relatively short time.

One advantage in using a pseudorandom binary sequence as a white noise probe for this purpose is that the sequence can be repeated and the ensemble response from several successive periods can be averaged, which

---

Supported by grants from the National Institute of Neurological Diseases and Stroke number NS-09692, University of Pittsburgh School of Medicine and the Eye and Ear Hospital, Pittsburgh, Pennsylvania.

tends to minimize certain nonsystematic errors (Briggs, et al., 1965; Davies, 1970).

It is necessary for the experimenter to determine when sufficient data have been accumulated in order to provide an accurate estimate of the system's parameters. In many cases, a subsequent off-line analysis reveals that either too little data was recorded to describe the system's parameters with sufficient accuracy, or that more than enough data was recorded which resulted in a waste of valuable experimental time. It appeared useful therefore, to develop a method that would provide an objective on-line decision criterion for determining when sufficient data has been obtained for an accurate characterization of a system.

A classical criterion for the determination of the goodness-of-fit of a system characterization is the mean-square-error (MSE) of the system's output relative to that predicted from the system characterization.

As shown by Norbert Wiener (1958), the MSE computed from the predicted and experimental responses to white noise inputs provided a criterion that was in a sense optimal. Although Wiener's development was based on the use of Gaussian white noise inputs, the MSE criterion appeared to be useful also with newer forms of digital white noise, such as those based on pseudorandom binary test perturbations. The use of these inputs offers certain computational advantages for on-line characterizations of linear systems, and they are amenable also to both frequency domain descriptions and nonlinear characterizations as described in recent applications in both engineering and sensory systems (Briggs, et al., 1965; O'Leary, et al., 1974, 1975).

Two important advantages of the use of pseudorandom binary sequences (PRBS) as white noise inputs for neurobiological systems are the reproducibility of the white noise properties over sucessive periods of the pseudorandom input and also the on-line computational capabilities for the determination of small-signal linear characterizations. O'Leary and Honrubia (1975) described both the theoretical and practical aspects of the use of PRBS inputs to characterize neuronal system with spike train or analog voltage outputs, and they illustrated the use of this technique by examples from a system's identification of afferent responses from the guitarfish (*Rhinobatos productus*) semicircular canal. In this report, we describe a complementary extension of this analysis in the form of on-line computations of the mean-square-error following each period of the stimulus, as a criterion for the goodness-of-fit of the resulting system characteristics.

THEORY

The technique is composed of two separate on-line computations which have theoretical bases in the dual areas of system identification and linear prediction (Eykhoff, 1974). A system is considered "identified" when an analysis of the input-output characteristics produces a model which summarizes the dynamic response characteristics of the system over the bandwidth of interest. A linearized system can be characterized as an equivalent filter in the time domain by its unit-impulse response (UIR), which can be computed by cross-correlating the system's response to a white noise input (Lee, 1960). For the particular form of white noise with only two amplitude states that are switched by a deterministic digital algorithm, called a pseudorandom binary sequence (PRBS), the cross-correlation computation reduces to merely gating, adding and subtracting the system's output which is delayed in time relative to the input (Davies, 1970). In the identification of a sensory system with a spike train output, the cross-correlation of the spike train responses with a PRBS input with state durations Δt and amplitudes $\pm a$ can be scaled appropriately to result in the UIR $h(\tau)$ by the equation

$$h(\tau) = \frac{\Delta y_\tau}{a(N + 1)\Delta t} \quad (1)$$

where $\Delta y_\tau = y_+ - y_-$ (2)

and $y_+ = \sum_+ y_i$ (3)

represents the number of spike counts that occurred only during $+a$ states of the input for a specified relative shift τ between input and output, and y_- is defined similarly for negative input states (O'Leary and Honrubia, 1975).

In practice, the computation of the UIR $h(\tau)$ identifies the system in the sense that the outputs $y(t)$ resulting from all other inputs can be predicted from the UIR by the convolution equation

$$y(t) = \frac{1}{T}\int_{-\infty}^{\infty} h(\tau)x(t-\tau)d\tau \qquad (4)$$

where $x(t-\tau)$ is the input which is delayed in time relative to the present. The UIR thus characterizes or identifies the system by the manner in which past inputs are filtered in order to result in the present output. It can be considered a model of the system represented as an equivalent linear filter, although this form of model is "nonparametric" in the sense that it is represented by a large number of discrete points plotted as an amplitude vs. time curve. In practice, however, the UIR is useful as a preliminary model of the system, and other forms of parametric models can be obtained from the UIR in a subsequent off-line analysis.

The second stage in the application of this technique is the computation of the predicted output to the binary white noise input, which allows a comparison of the prediction with the system's experimental output. A conventional measure of the relative goodness-of-fit is the error obtained by subtracting the prediction $y_p(t)$ from the experimental output $y_e(t)$. This error is then squared and the average value is computed over the time period of interest. This conventional mean-square-error is expressed as

$$MSE = \frac{1}{T}\int_0^T (y_e(t) - y_p(t))^2 dt \qquad (5)$$

This error calculation can be normalized by subtracting the mean $\bar{y}_e$ to transform y_e to a zero mean process, dividing by the mean-square value of the experimental data and multiplying this result by 100%.

$$\%MSE = \frac{\int_0^T (y_e(t) - y_p(t))^2 dt}{\int_0^T (y_e(t) - \bar{y}_e)^2 dt} (100\%) \qquad (6)$$

It is appropriate to normalize with this parameter since the UIR $h(\tau)$ is used to predict the dynamic output and not the mean value. Thus, the percent MSE compares the mean-square value of the error with the variance of the experimental response.

As a summary of the different stages in the on-line calculation, the PRBS stimulus is repeated periodically to obtain a cumulative experimental response. A cumulative cross-correlation is then calculated and used to derive an impulse response $h(\tau)$ by Eq. (1). The UIR $h(\tau)$ is then convolved with the PRBS using Eq. (4), and the result is scaled by multiplying it times the number of trials to obtain a cumulative predicted response $y_t(t)$. The percent mean-square-error is then computed from Eq. (6), and used to compare the cumulative predicted response with the cumulative actual response. This error estimate is plotted as a histogram which is updated after each successive PRBS period.

When the percent MSE stabilizes about a plateau, the use of further periods in the experiment will result in only minor improvements in the overall signal-to-noise of the UIR. The stabilization of the percent MSE can then serve as an on-line decision criterion for the application of sufficient numbers of periods of the input. Although this technique decreases errors which are due to random fluctuations in the response of the biological system, it is of course subject to errors resulting from time dependent or nonstationary aspects of the system.

This technique has been applied to a study of the input-output characteristics of the semicircular canal, and examples from this application will be described to illustrate its various aspects.

METHODS

A PDP-11/40 computer program was written for the on-line error analysis described in the THEORY section. The program performed four basic functions: (1) It generated a pseudorandom binary sequence (PRBS) stimulus as a white noise input for receptors of the vestibular system. This stimulus controlled the direction of rotational acceleration of a rate-of-turn table, used for stimulation of the semicircular canal, via a digital interface between the table and the computer system (Contraves/Goerz model 404D). The resulting motion of the table was a pseudorandom rotational acceleration with a direction that was controlled by the PRBS algorithm. The implementation of this algorithm as a white noise stimulus for sensory systems is described in O'Leary and Honrubia (1975). (2) The times of occurrence of nerve impulses from single afferent fibers innervating the semicircular canal which occurred in response to the white noise stimulus were digitized as 32-bit values with 10 microsecond resolution. The spike times were stored as a disk file under the RT-11 foreground/background operating system. (3) The duration of each individual PRBS state determined the high and low frequency cutoffs of the white noise bandwidth according to the equation

$$\frac{1}{N\Delta t} < f < \frac{1}{3\Delta t} \qquad (7)$$

approximately, where the frequency is in Hertz, and N is the total number of PRBS states in a single period (Davies, 1970). A program interrupt was generated at the completion of each PRBS state, and this interrupt led to the sequence of programmed events described as a flow chart in figure 1, which includes both the selection of a

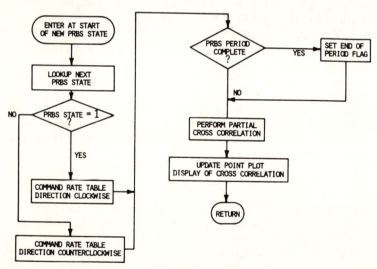

Figure 1. Flow chart for processing during one state of the pseudorandom binary sequence (PRBS). The value of the state (-1 or +1) determines the direction of table rotation, clockwise (CW) or counter clockwise (CCW). A fractional part of the input-output cross-correlation is made during each PRBS state, so that the correlation is complete at the end of the entire PRBS period.

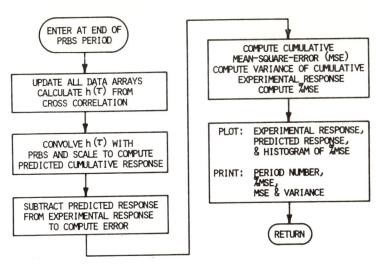

Figure 2. Flow chart for processing after the completion of an entire pseudorandom binary sequence (PRBS) period. The impulse response, $h(\tau)$ is calculated from the cross-correlation, and is then convolved with the PRBS to obtain a predicted cell output. The predicted and actual responses are compared for the mean-square-error (%MSE). All the quantities are then displayed on a Tektronix 4010 display terminal.

new PRBS state and also the necessary updating required for an on-line cross-correlation of the PRBS input with the spike train output. (4) At the completion of each PRBS period, the percent mean-square-error (%MSE) was calculated using the program described as a flow chart in figure 2. Because all of the cross-correlation, convolution and percent MSE computations were performed using assembly language, the results were available for display in the first few seconds following the end of each PRBS period. The display included the percent MSE plotted as a histogram vs. the number of periods, which was used to determine when this value had stabilized to sufficiently low value to terminate the protocol on a given afferent cell. Examples of the displays as plotted on Tektronix 4010 and Digital Equipment Corporation VR14 graphics terminals, are described in the RESULTS sections.

This technique was applied for a systems identification analysis of the semicircular canal of elasmobranch fish (O'Leary, et al., 1974). Isolated labyrinths from skates or rays were prepared by pithing the animal, removing the chondocranium and exposing the nerve innervating the horizontal semicircular canal. Single unit afferent responses were recorded using forceps electrodes after mounting the preparation at the center of rotation of the rate-of-turn table. Ten to twenty periods of a 255 state PRBS stimulus with Δt = 92 msec for a frequency bandwidth from .04 to 4 Hz was used for each afferent in the analysis.

RESULTS

Example of Single Afferent Responses to a PRBS Input

In figure 3a is shown the response of a typical spontaneously active semicircular canal afferent neuron to a portion of a PRBS input, shown as the table velocity in figure 3b. The first-order afferents innervating this receptor exhibit excitatory responses for a rotational acceleration of the head in one direction, and inhibitory

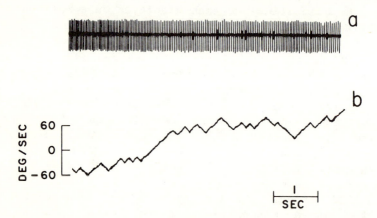

Figure 3. The output of a semicircular canal afferent unit is shown in a response to a portion of a pseudorandom binary sequence (PRBS) of rotational acceleration shown as the rotating table's velocity in b. The table accelerated at a constant magnitude of 100 deg/sec , and the direction was switched under command of the PRBS algorithm with an individual state duration of 92 msec.

responses for rotational accelerations in the opposite direction. The modulation of the interspike interval durations in figure 3a is therefore reflected in the recent history of the switching pattern of the PRBS input. In that sense, the response pattern was determined by the filter properties of the system in the sense that past inputs were weighted, or filtered, to result in the present output as expressed by the convolution equation 4. These weighting coefficients can be expressed as the amplitude of points along the contour of the impulse response, obtained by cross-correlating the white-noise PRBS input with the spike train output as described in the THEORY section. However, because of band-limiting in the PRBS command signal and also the rotational table's own response properties, it is necessary to apply correction factors to the impulse response contour in order to compensate for these nonideal experimental parameters. O'Leary and Honrubia (1975) showed that the correction factors required for impulse responses computed from spike trains could be applied to individual points of the impulse response in an off-line analysis after a preliminary estimation of the equivalent filter's time constants.

Examples of On-Line Computations

Figure 4 illustrates the computed results which are displayed at the completion of each PRBS period. Discrete point estimates of the impulse response $h(\tau)$, as displayed on the VR14 graphics terminal, are shown in 4a for a typical semicircular canal afferent cell. The particular form of the point contour provides the experimenter with an estimate of the dynamic characteristics such as bandwidth, decay time constants and degree of overshoot for each afferent. Figures 4b-e are displayed on the Tektronix 4010 storage terminal, and can be retained during on-line use via the attached hard copy unit. The histogram in 4c represents the modulated firing rate of the cell as averaged over 17 periods of the white-noise input. The predicted firing rate, shown in 4d, was obtained by linear convolution (Eq. 4) of the white noise input with the impulse response in figure 4a. The error estimate, obtained by subtracting the predicted response in 4d from the actual response in 4c, was the

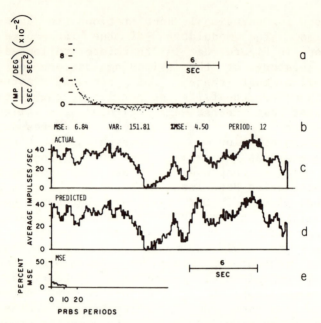

Figure 4. Results of on-line processing for linear cell.
a. On-line cumulative cross-correlation (normally displayed on VR 14).
b. Display of current mean-square-error (MSE), variance (VAR), percent mean-square-error (%MSE) PRBS period number, see text for further explanation.
c. Actual cumulative cell response in spikes for each of the 255 "bins" or time intervals of the PRBS.
d. Predicted cumulative cell response obtained by convolving the impulse response, $h(\tau)$ with the PRBS.
e. Histogram of %MSE vs. number of PRBS periods.

basis for numerical variance and percent mean-square-error (%MSE) estimates displayed in 4b. The histogram of %MSE in 4e illustrates how the error is reduced by period averaging. The experimenter uses this plot to determine when the error has reached an acceptably low value. For the data shown in figure 4, the stabilization of the %MSE to a plateau of 4.50% signalled the end of the white noise protocol for this cell.

Figure 5 illustrates an identical display for a semicircular canal afferent cell that exhibited a significant nonlinearity resulting from complete inhibition of firing during a portion of the input stimulus. The impulse response in 5a included quasi-periodic oscillations resulting from this form of nonlinearity. Despite the oscillations, the calculated impulse response still estimates the linear small signal characteristics of this cell. In addition, the examples shown in figures 4 and 5 illustrate the systems and error parameters which we have found valuable as on-line estimates for the analysis of vestibular receptor systems.

It should be emphasized that these results are an on-line preliminary view of the system characteristics. Since all essential input-output information is stored as "raw" data on a disk file, more complete analyses such as quantitative parameter estimation can be computed off-line.

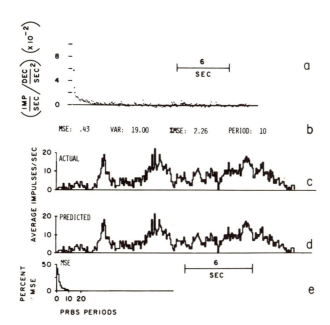

Figure 5. On-line processing for cell having a nonlinearity.
a. Cumulative cross-correlation.
b. Display of mean-square-error, variance, percent mean-square-error and period.
c. Cumulative cell response.
d. Predicted cell response.
e. Histogram of percent mean-square-error vs. period.

CONCLUSION

The on-line analysis technique for a neurobiological system described in this report is valuable because it provides the experimenter with a preliminary estimation of the dynamic characteristics of the system. This is due in large part to the advantages of the PRBS white noise stimulus, which, in addition to the information-rich properties of white noise, affords the advantage of cumulatively averaging the response output during the cross-correlation computation to obtain the impulse response. Moreover, because period averaging requires stationary data it is possible for the experimenter to determine empirically whether the output data is sufficiently stationary to result in reliable averaged response characteristics. This preliminary view of the system's characteristics can be used to further refine the experimental protocol with the same preparation. Specifically, the bandwidth of the PRBS stimulus could be changed to obtain a better match with that of the system undergoing analysis. This is particularly advantageous for a sensory system such as the semicircular canal in which afferents innervating different parts of the receptor have been found to exhibit systematic variations in their bandwidth characteristics (O'Leary, et al., 1974). Moreover, because an updated estimate of the percent mean-square-error is available following each period of the stimulus, it is possible to optimize use of the available experimental time by terminating the protocol when the error is reduced to an acceptable level.

DISCUSSION

D. O. WALTER: Have you extended this study to the nonlinear terms?

O'LEARY: No, we haven't. Unlike Gaussian white noise, the higher order Wiener kernels are not orthogonal. However, there is a lot of work being done on the higher-order kernels and also the Volterra functionals for this type of input, because it's so convenient to use.

RHODE: Have any comparisons with the intact animal been done?

O'LEARY: That's in progress.

REFERENCES

1. Briggs, P. A. N., Hammond, P. H., Hughes, M. T. G. and Plumb, G. O. Correlation analysis of process dynamics using pseudo-random binary test perturbations. Proc. Inst. Mech. Eng. 179:37 (1965).
2. Davies, W. D. T. *System Identification for Self-Adaptive Control*, Wiley-Interscience, New York (1970).
3. Eykhoff, P. *System Identification*. John Wiley Sons, Inc., New York (1974).
4. Lee, Y. W. *Statistical Theory of Communication*. John Wiley Sons, Inc., New York (1960).
5. Marmarelis, P. Z., and Naka, K. Nonlinear analysis and synthesis of receptive-field responses in the catfish retina. I. Horizontal cell ganglion cell chain. J. Neurophysiol. 36:619 (1973a).
7. Marmarelis, P. Z., and Naka, K. Experimental analysis of a neural system: two modeling approaches. Kybernetik 15:11 (1974).
8. O'Leary, D. P., Dunn, R., and Honrubia, V. Functional and anatomical correlation of afferent responses from the isolated semicircular canal. Nature (Lond.) 251:225 (1974).
9. O'Leary, D. P., and Honrubia, V. On-line identification of sensory systems using pseudorandom binary noise perturbations. Biophys. J. In Press.
10. Wiener, N. *Nonlinear Problems in Random Theory*. M.I.T. Press, Cambridge Mass (1958).

chapter 24

THE TABLE AS A BASIS FOR DATA ANALYSIS

B. J. RANSIL

Clinical Research Center
Beth Israel Hospital
Boston, Massachusetts

BACKGROUND

Most biomedical research data readily lends itself to tabulation, in as many dimensions as there are independent variables. The most general mathematical representation of such organization is the n-dimensional array of which the row and column-labeled table, the vector, the matrix and the group are special cases, distinguished from one another by their respective definitions and algebras.

To the clinical investigator studying multivariate functional dependencies, the table is a natural, congenial and conceptually simple form of data organization just as the plot is often the simplest and most useful way to visualize it. It is therefore not surprising that, as computer technology and expertise grew of age, the link between table storage and data processing was readily and rapidly forged to provide a convenient and powerful tool for data analysis.

One of the earliest programs for numerical and statistical analysis based on stored tables is OMNITAB (1), which is perhaps the first successful attempt by a well-defined user group (physical scientists at the National Bureau of Standards) to construct a do-it-yourself data analysis program that was "not restricted to those who have mastered programming, even the 'simple' kind," and which permitted the user to communicate with the computer via English sentences with tolerable loss in efficiency. Developing independently, more recently, and reinforcing the same philosophy of "user-orientedness" is PROPHET (2-9) which combines table-based data processing, a simple English command language with interactive graphics and a simplicity of on-site terminal opertion to produce an information handling system that truly design difference between the two systems is that the PROPHET system treats the table as a permanent file structure; OMNITAB does not.

MAKING TABLES IN PROPHET

Table making, entry and processing in PROPHET is straightforwardly simple. Triggered by the simple English command, MAKE TABLE tablename, and guided by clearly worded cues in conversational mode, the investigator rapidly enters his research data by typing it on the terminal keyboard, observing its registration on the CRT display screen. When all the data is entered, table making is terminated by typing the command EXIT which clears the screen and displays the completed table for proof-reading and correction. Correction is readily accomplished by a simple typed command that clears the offending cell (a cell is a specific row-column item) and enters the correct value. Thereafter anlaysis is accomplished by addressing respective column/ row numbers or an entire table, depending on the operation being performed.

Once made, a table is permanently stored on disk in the user's space. It may at any time be altered in any way--addition or deletion of any row(s) or column(s), data alteration in any cell or entire columns, row or column rearrangement, breakup into sub-tables--by simple English commands. By the ARCHIVE and DEARCHIVE commands, tables may be removed from the on-line disk, stored on an off-line disk-pack and restored to the user's space when desired.

Analysis from the table is accomplished with the use of a small but powerful repertoire of table-based commands and procedures. These include:

```
FIT LINE TO COL No vs COL No OF Tablename
FIT POLYNOMIAL TO COL No vs COL No OF Tablename
MAKE SAMPLE Name FROM COL No OF Tablename
MAKE GRAPH Name FROM Tablename AS COL No vs COL No
```

Both the FIT LINE and FIT POLYNOMIAL commands generate a brief dialogue which asks if the investigator wants goodness of fit measures and the like. In the case of the polynomial one may specify the lowest degree polynomial coefficients and goodness of fit measures.

The MAKE SAMPLE command takes the indicated column and computes central measures, median, mode, moments, skewness and kurtosis, all of which may be displayed quickly by the TYPE or DISPLAY commands. The central measures, median and mode provide the usual information about the distribution, which is assumed to be normal. Extension to Poisson and binomial distributions is trivial. The moments, skewness and kurtosis are useful measures for determining whether a distribution departs significantly from the normal.

Another powerful column-based tool is the DERIVE COLUMN option that is built into the MAKE TABLE dialologue. Initially, when making tables and doing column entries for each column, one is offered the following options:

DERIVED ENTER NEXT EXIT

Typing ENTER permits one to enter the data (number or text) row by row. NEXT shifts to the next column. EXIT, as mentioned above, stops table-making and displays the table.

The option DERIVED however responds with the prompt:

ENTER DERIVATION

at which point one may enter a functional derivation using as arguments the data from already existing columns. For example, the derivation

SIN(CELL(1)) + COS(CELL(6))

computes the indicated function from the data of columns 1 and 6 row by row, and enters it in a new column in the appropriate, corresponding rows.

This is an invaluable tool to the clinical investigator who often desires a variety of physiological quantities such as clearances, cardiac outputs or efficiencies on large amounts of data. Once computed and tabulated the derived quantities may be analyzed in the same way as non-derived data.

Data stored in tables may be quickly transferred to another user's work area by the command:

MAKE PUBLIC TABLE Name(1) FROM Name(2)

This places the table Name(1) into a special disk area under the new name of Name(2). When any other system user wants to use this table he merely accesses it by a short command for storage into his own work area.

Plotting from a PROPHET table is quickly accomplished with the above MAKE GRAPH command or by typing the alternative command:

MAKE GRAPH Name

which generates cues in a conversational mode allowing the investigator to specify graph title, axes and curve label, symbols and whether the points should be connected. There is a quick, simple way to alter data or any detail in the format to produce graphs that conform exactly to the investigator's requirements, within the space constraints imposed by the screen size of the CRT display tube.

A large variety of statistical and mathematical operations can be performed on whole tables or specified sections of tables by using PROPHET procedures which are programs written in PL/PROPHET or FORTRAN that are easily accessed and utilized by typing the commands:

RESTORE AND COMPILE Procedure Name
CALL Procedure Name (Table Name, other arguments where called for)

The desired analysis is performed on the specified data and the results automatically displayed on the screen and printed.

The table-based operations currently available on PROPHET are the following:

3-Dimensional Plotting
Multiple Exponential Fitting
Central Measures of Rows and Columns
Multiple Linear Regression
Variance Analysis
Multiple Range Testing

Chi-Square
Roots of Polynomials
Testing for Normal Distributions
Sorting
Michaelis-Menton Kinetics
Dose-Response Analysis
Curve Fitting to Arbitrary Functional Forms (MLAB)

In all these maneuvers the table is the primary mode of organization of input and output that is perceived by the investigator. As he works, his analytic train of thought is never interrupted and he can produce answers to his questions in concise, logical, sequential tabular form almost as quickly as he is able to formulate them.

TABLE TEXT

The foregoing description of table-based analysis assumed a numerical data base. While the capabilities for handling textual data in the table format at the command language level are currently considerably less than those for handling numerical data, the discrepancy arises more from developmental choices, given available funding and user priorities, rather than design constraints. Nevertheless there are several very useful capabilities. At the outset of the MAKE TABLE dialogue the investigator is asked:

ARE THERE ROW NAMES?

By answering Y, the investigator may construct tables with patient names, numbers, dates or any other type of identifier in the initial or "zeroth" column. Upon completion of this step, he then may continue the column-making process by multiple-choice cues, in the course of which he is given the following options for type of data to be entered into the column under construction:

NUMBER TEXT CHEMTEXT DESCRIPTOR TABLE GRAPH

NUMBER is self-explanatory. TEXT can be any alphanumeric word, phrase or expression. CHEMTEXT refers to chemical formulae. DESCRIPTOR may be any kind of symbol (e.g. +, -, * etc.) or letter assigned to characterize or qualify some property in a previous column (e.g. Headache in column 2, with *** in column 3 to denote severity). TABLE allows one to make tables of tables introducing a multidimensional structure. GRAPH allows one to make tables of graphs, an efficient way to store families of curves.

By judicious and imaginative use of all the available categories, but especially number, text and descriptor, the investigator may construct a versatile permanent clinical subject file in table format that is a composite of numerical and textual data for sort, retrieval and processing applications. The formal text manipulation repertoire available to him at command language level is currently limited to the very efficient TABLESORT procedure (which does alphanumeric sorts on a major column and any number of minor columns) and the capability for making and editing texts via the Text Editor. However, these capabilities, combined with command lists and procedure writing, can be used to produce remarkably sophisticated data sort-retrieval processing systems for mixed data bases stored as tables.

PERMANENT DATA BANKS

Because tables are a congenial way of storing data for many applications, they readily lend themselves to use as data banks. The design requirements for data banks include permanent storage, easy accessibility, a capability for indefinite extension by row and column; a simple means for altering the values of single items and entire columns, rows or blocks of data; the ability to delete rows and columns; the ability for multidimensional organization and for accessing by row, column or block. The table file structure, as implemented in PROPHET, provides all of these capabilities, most of them at command language level; the remainder may be achieved by writing specific procedures to meet storage and data processing specifications. Data banks may be easily achieved on off-line disk space and restored ad lib as needed. They may be quickly scanned by the SCAN command, or printed in numbered sequential pages by the PRINT command. They may be readily transferred to other users' workspaces by a few simple commands as described above.

THE TABLE AS TOOL

While the usefulness of accessing and operating upon columnar (or row) data with statistical, curve-fitting and graphical procedures is apparent, there are several other capabilities of tables, as implemented in PROPHET, that make them a powerful analytical tool.

The first of these was described above-- the ability to derive the most complicated analytical function from previously entered columnar data. From this capability it is a simple matter to construct histograms of samples and superimpose upon them the appropriate distribution (Gaussian, binomial, Poisson, Lorentzian) and perform the appropriate distribution testing procedure. Physiological parameters are readily computed, tabulated and graphed or correlated. Non-linear multiple regressions can be easily performed by deriving columns of the desired power and then invoking the multiple linear regression procedure. Interpolation is a trivial operation once an appropriate function has been determined.

Another useful application is linearization. From a graph of the data, one may decide upon the most probable functional fit. In a derived column the assumed function is computed from the arguments and then plotted against the independent variable. A significant straight line fit is considered strong evidence for a true functional fit. The derivation may be adjusted until the desired degree of linearity is obtained.

A third time-saving application is the solution of transcendental equations which can arise in multicompartmental analysis and complex transport problems. In respective columns, each side of the equation is computed for reasonable estimates of the parameters (obtained from boundary conditions or other knowledge of the experimental system) and appropriate ranges of the independent variables. One column is then plotted against the other and the parameters varied (simply by adding more data to the appropriate columns or clearing columns and entering new data) until the functions intersect each other on the graphs. The value of the parameters at each point of intersection is a solution. Intuition can be most helpful in obtaining fast convergence to these solutions.

A fourth and final application-- which by no means

exhausts the topic-- is the use of non-parametric statistics. Many of these depend on an ordering or ranking process of either numerical or symbolic (i.e. descriptor) data which is easily accomplished by TABLESORT. A quick scan and reference to appropriate tables achieves the desired result. The Kolmogorov-Smirnov test for normally distributed data is easily performed, as is the Signed Rank test, with a few commands and inspection of the printed tabular results.

In all these applications, the computer is used to do what it does best-- organize, tabulate, sort, count and display; while the trained human eye and judgement are used for what they do best-- quickly distinguish the desired effect, determine its significance and make branching decisions.

CONCLUSION

The table is a natural organizational and analytical tool capable of almost unlimited utilization and adaptation. In combination with high level interpretive language it is a powerful and versatile tool for many biomedical research activities. Its use is recommended whenever the data base comprises discrete collections of numbers, descriptors and text requiring reduction and analysis.

DISCUSSION

D.O. WALTER: Could you enter table data row-wise instead of column- wise?

RANSIL: Not initially, but once a table's made you can add new rows.

D.O. WALTER: How much does it cost to use the PROPHET system?

RANSIL: Subscription costs $5000 per year. This puts it in your lab and gives you full support, and allows you 1250 CPU minutes. Thereafter, additional time costs $1.00 per CPU minute.

MCINTYRE: Have you had any experience with APL? This looks much the same, with lots of added prompts and interaction. Of course the data structures are identical.

RANSIL: It could be. The computer group has a vast array of experience from systems all over the country.

MCCORMICK: Why do you say this isn't applicable to time-series analysis?

RANSIL: I meant real-time analysis. PROPHET wasn't designed for neuroscience type projects, but my feeling is that if a neuroscience investigator got together with BB&N on specific applications, the majority of them could be done with PROPHET.

KEHL: Can you window through a table?

RANSIL: Not by high-level command. You could do it with a written procedure; BB&N could write you one.

BROWN: Given the use of Tektronix terminals, could we set it up so my computer could stream data into yours?

RANSIL: PROPHET's design includes that feature. I can send messages or make tables public.

BROWN: The reason I ask is that I frequently find I can go from large volumes of raw neurophysiological data to reams of statistical descriptors, but then I have to hand-enter them into data sets for system packages like SPSS for analysis. If I had access to an appropriate network package, I could probably generate these tables using PROPHET, making it a good interface between laboratory systems and large analytical packages.

RANSIL: This is intended as an end-analytical tool, but that application makes sense.

KEHL: We have found it useful to be able to move data around between systems. How do you do that? How do you, say, pick up your data and get it into a simulator? Are there hooks so you can go to FORTRAN?

RANSIL: Yes, you can go back and forth between FORTRAN and PROPHET with your data.

REFERENCES

1. OMNITAB: *A Computer Program for Statistical and Numerical Analyses*. National Bureau of Standards Handbook 101, (1966).
2. PROPHET is a chemical/biological information handling system sponsored by the CBIH Program of the Division of Research Resources, NIH, designed and developed by the Medical Computer Systems Group of Bolt Beranek and Newman, Inc., Cambridge, Mass. and operated by First Data Corporation, Waltham, Mass.
3. Castleman, P. A., Russell, C. H., Webb, F. N., Hollister, C. A., Siegel, J. R., Zdonik, S. R., and

Fram, D. M. Implementation of the PROPHET system. Proc. Natl. Comput. Conf. 43:457-468, (1974).
4. Hollister, C. A. *Users' Manual for the PROPHET system.* Bolt Beranek and Newman, Inc., periodically updated.
5. Johnson, C. L. Applications of PROPHET system in molecular pharmacology -- structure-activity relationships in monoamine oxidase inhibitors. Proc. Natl. Comput. Conf. 43:473-476, (1974).
6. *Public Procedures Notebook.* Technical Document No. 47. Bolt Beranek and Newman, Inc.
7. Ransil, B. J. Applications of PROPHET in human clinical investigation. Proc. Natl. Comput. Conf. 43:477-483, (1974).
8. Weeks, C. M., Cody, V. C., Pokrywiecki, S., Rohrer, D. C. and Doux, W. L. Applications of the PROPHET system incorrelating crystallographic structural data with biological information. Proc. Natl. Comput. Conf. 43:469-472, (1974).
9. Ransil, B. J. Use of table file structures in a clinical research center. Fed. Proc. 33:2384-2387, (1974).

chapter 25

A NUMERICAL APPROACH TO DIFFERENTIAL EQUATION MODELS IN NEUROBIOLOGY

H. C. HOWLAND
Department of Physiology and Biophysics
Cornell University
Ithaca, New York

Most physiological events play themselves out in time, and hence are appropriately described in differential equation models. Such models are not only descriptive, but also have predictive and didactic value. Their creation may be divided into three phases: 1) recognition of the broad outlines of the rate processes involved, 2) detailed writing of the model, 3) generation of model "solution(s)". Each of these phases has its own appropriate notations for the development of differential equation models and their numerical solutions. A flow diagram notation is presented here and its use in model formulation is discussed. Some remarks on the suitability of various notational conventions for the detailed description of a model are made and, finally, a simple FORTRAN program for numerical solution of differential equation models is presented.

INTRODUCTION

I think anyone who teaches physiology at the university level today must wrestle with the problem of teaching their students something about the behavior of differential equation systems. The problem is one of Herculean proportions. In all likelihood neither the students nor often the teacher has had a formal course in differential equations. Ordinarily such a lack on either side would be sufficient condition for dropping the topic entirely. Yet the subject must be taught. Increasingly researchers are formulating their experimental results in terms of differential equation models. And increasingly, a gulf is widening between the research literature of physiology and the undergraduate and graduate training in that subject.

Nor can we physiologists or neurobiologists leave the problem to others to solve. It might have been thought at one time that physiology would become the step-child of engineering. However, though we may obtain welcome help from engineers and engineering curricula it does not appear that the engineering schools of today are going to provide for the physiologists of the future. Those of us who have taught physiology to engineering students recognize the skills those students bring to our courses, but we see also the lack of expertise that only experiments upon animals in the laboratory can engender. Moreover, if it is true, as I believe, that biology is a branch of analytical engineering, it is also true that it is its only branch. The traditions and motivations for analysis, wherever analysis may lead -- these must come from the physiologist. This is simply to say that the teaching of differential equation systems to physiologists is a physiologist's and not an engineer's problem.

I have not here considered the possibility that it might be accepted as the job of the mathematicians. Doubtless this may be a parochialism founded on the limited university experience I have garnered. However, I take it as an ominous sign that in many places in this country engineering faculties have been forced to construct their own departments of applied mathematics and to teach their engineers the mathematics needed for their curricula. I don't think it is realistic to expect that the mathematical needs of physiologists are going to be met by the mathematics departments of our universities, at least for a very long time to come.

I have belabored the educational plight of the physiologist for two reasons: Firstly, one should not lightly make proposals for the organization of teaching of subjects which are not well within one's professional expertise -- and I am a neurobiologist, not a mathematician. And secondly, though I recognise that we must concentrate on fulfilling practical research needs in this computer language project, yet the solutions we arrive at must, if they are to be permanent, be teachable solutions.

My own attempts to teach my students about differential equation systems have centered around computational approaches involving both analog and digital computers. The analog computer is peculiarly suitable for physiology students -- it has only slightly more knobs than our polygraphs and, like a polygraph, it draws curves in real time. It is thus an easy task for a teacher to interest his physiology students in an analog computer.

Digital computing is entering the undergraduate curriculum in ever increasing ways. Virtually all students who go on in research will have recourse to the computer in solving statistical problems. Over the past five years the number of students in our upper class physiology course who have experience with digital computing has grown from a small percentage to a majority, and will probably continue to rise. Thus the digital computer also affords a practical method for teaching students about differential equation systems.

These two methods of computing have given rise in the literature to what I will call graphical and algorithmic notations. By graphical notations I mean the systems flow diagrams whose lines represent variables, and whose boxes indicate functional relations between the input and output variables. By algorithmic notations I mean to include those logical flow diagrams and logical step notations, regardless of their graphical content.

In attempting to teach physiology students about differential equation systems I have developed what I would like to call "hardware free", precise notations which are appropriate to the formulation of a specific differential equation system in terms of the generation of specific analog or digital computer solutions. Before discussing these notations I would say a few words about the importance of freeing them from any hardware restrictions.

Because the physiology student knows little of dashpots and springs, or resistors and capacitors, many of the physical analogies which our engineering colleagues use as inductive devices to teach their students about the behavior of physical systems are simply lost upon the understanding of physiology students. Hence notations involving these items of hardware are also inappropriate for the description of physiological systems.

Fortunately we are also not burdened, nor do I have any desire to create (as some ecologists have done), any substitute analogies for physiologists. Nor do I think it wise to employ the hardware of computers for this purpose. Thus in the notations that follow I have attempted to embody in them only mathematical concepts. The ultimate goal of these notations is to aid in the precise mathematical formulation of the physiological system.

In pursuit of this goal, the mathematical description of physiological systems, I long believed that all flow diagrams should be mathematically precise. Experience and frustration have modified that view, and I now am the first to admit that we need the ability to be graphically vague, at least in the developmental phases of understanding of a physiological system. We need to be able to draw flow diagrams where some function boxes are left undefined, or where only the general tendency of the ouput with regard to the input is specified. I am not saying we should publish such diagrams. We shouldn't, except in very rare cases where we want to carefully circumscribe our ignorance, but we should not be afraid to write such diagrams. For this purpose I confess that an empty box, perhaps with a negation sign at its input to indicate that the output was a generally monotonically decreasing function of the input, has served my thinking well.

For those parts of the systems diagram which one can model precisely I have adopted the notation of figure 1. Its resemblance to the notation of analog computing on which it is based is strong. However, note the following differences between it and analog computer diagrams:

1) The elements do not invert. Inversion is indicated by a dot at the input.

2) There are no voltages (unless the modeled variable is a voltage), no potentiometers, nor any scaled parameters.

3) I have adopted some simple conventions for multiplication, division, flip-flops, etc.

4) The complete equation system can be written from inspection of the diagram.

What are the advantages of displaying an equation system in this way? I think there are several.

1) While the precision of the equation system is retained the entire system-- together with whatever supplementary anatomical and physiological information is necessary -- may be displayed and appreciated at a glance. (Naturally this applies only to small systems of say less than 10 or 20 elements.)

2) It is possible to build up an appreciation of blocks of such a diagram and to recognize low pass, high pass, and second order units, and so to infer rapidly some aspects of the behavior of the system.

3) Naturally the system may be easily programmed for solution on an analog computer from the block diagram.

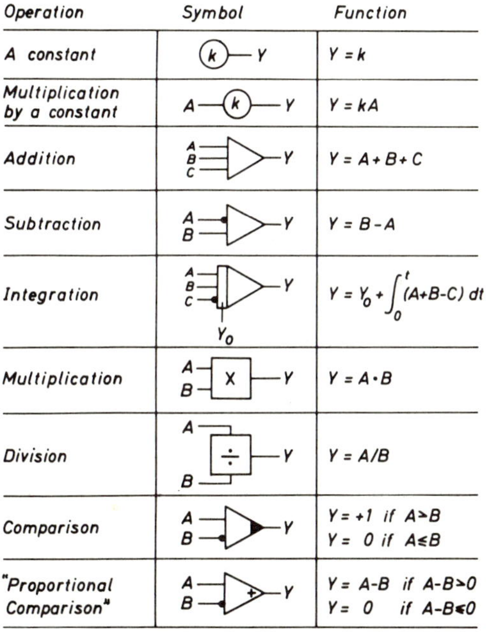

Operation	Symbol	Function
A constant	(k)—Y	$Y = k$
Multiplication by a constant	A—(k)—Y	$Y = kA$
Addition	A,B,C ▷—Y	$Y = A+B+C$
Subtraction	A,B ▷—Y	$Y = B-A$
Integration	A,B,C ▷—Y, Y_0	$Y = Y_0 + \int_0^t (A+B-C)\,dt$
Multiplication	A,B —[X]—Y	$Y = A \cdot B$
Division	A,B —[÷]—Y	$Y = A/B$
Comparison	A,B ▷—Y	$Y = +1$ if $A > B$; $Y = 0$ if $A \leq B$
"Proportional Comparison"	A,B ▷—Y	$Y = A-B$ if $A-B > 0$; $Y = 0$ if $A-B \leq 0$

Figure 1. Mathematical flow diagram notation. The major difference between this notation and that of conventional analog computing is the representation of inversion, here a dot at the input. Note that the denominator input in a division operation is denoted by placing it *below* the numerator input.

It is perhaps worth while pointing out one difference between this notation and that which employs transfer functions for its linear portions. This notation allows the specification of initial conditions which may, in some cases, be important for the modeling of particular physiological situations.

I would also point out that while the lumped system transfer function is an admirable tool for investigating the behavior of linear systems, the flow diagram notation here is perhaps better adapted for explicitly mapping the mathematical model on the underlying physiological reality.

It is probably not profitable here to discuss the merits of particular notational conventions that I have adopted. In principle I would be happy with any precise and concise hardware-free notation that lent itself to easy computer implementation.

An application of this notation is given in figure 2 wherein a model for the vestibular stabilization system of a fish is presented (Howland, 1971). Those of you accustomed to looking at analog computer diagrams will probably easily recognize the second order mechanics of the semicircular canals and the fish itself and the high pass or pseudo-derivative response to the otolithic organs.

I have found in my teaching that it is easy to relate these analog computing-like flow diagrams to the transfer function diagrams employing the Laplace operator. Indeed, it is relatively easy to teach students to write the transfer functions for given equation systems and to derive the steady state sinusoidal responses by substitution of $j\omega$ for the Laplace operator, s, and to find the impulse and step responses of the system from table look-up. These techniques are discussed extensively in Milsum (1966), and I shall not pursue them further here.

ALGORITHMIC NOTATIONS

I have emphasized that the goal of modeling is the precise description of the system to be modeled in terms of a set of equations-- usually differential equations where the derivatives are taken with respect to time.

The physiologist may well be interested in the general behavior of his system, but in practice he may be

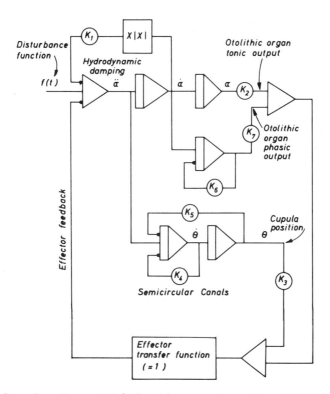

Figure 2. Angular stabilization system of goldfish about pitch or roll axis to illustrate the notation of Fig. 1. The second order dynamics of the semicircular canals are readily recognizable. The function, $y = x^2$ for positive x and $y = -x^2$ for negative x, is represented by writing it as x times its absolute value in a function box. For further explanation of the control system see Howland (1971).

satisfied with specific solutions of his equation system with specified inputs and initial conditions, and a specific parameter set.

While analog computers are excellent devices for building intuition and for teaching students about the behavior of simple systems in general, unless one has made a very large investment, analog computers are not suitable for the solution of practical equation systems encountered in research. For such systems, digital computers present a far better, if somewhat slower, solution. Furthermore, their handling of the documentation problems that accompany systems simulation is far superior to that attainable with an analog computer.

Now it is a fortunate fact that the assignment statements of a wide variety of digital computing languages are essentially the same, perhaps with a few differences in verbs (such as SET or LET) or an absence of any verb at all, as in the FORTRAN assignment statement.

Thus if one adopts some very reasonable conventions such as the inputs of all integrators (i.e. derivatives) will be designated as D(1), D(2)... D(n) and the outputs of all integrators, i.e. Integrated variables will correspondingly be labeled Y(1), Y(2)...Y(n), and all other parameters K(1), K(2)...K(n)-- if one does this it is very easy to tell a student how to write his equations system in such a form that a digital computer could produce specific solutions of the equation system for given values of parameters and initial conditions.

To take a textbook example, Simon (1972, pp 73-75) discusses diffusion between two compartments. The equations to be solved are:

$$V_1 \frac{dC_1(t)}{dt} = K(C_2 - C_1) \quad (1)$$

and

$$V_2 \frac{dC_2(t)}{dt} = K(C_1 - C_2) \quad (2)$$

letting K(1) = V_2 and K(3) = V_1 and D(1) = $dC_1(t)/dt$

and D(2) = $dC_2(t)/dt$ these equations become

from Eq (1) and

$$D(1) = K(1)^*(Y(2) - Y(1))/(3) \quad (3)$$

$$D(2) = -D(1)^*K(3)/K(2) \quad (4)$$

Equations (3) and (4) are now in a form which can be plugged into a computer differential equation solving routine together with the necessary initial conditions, values of constants, step size and length of run. Such routines are easy to write and implement from standard books on the subject. I have included such a routine as an appendix to this chapter: It is written in FORTRAN suitable for running under WATFIV compilers or, (since it uses no subroutines) on any of a number of smaller computing systems. Elsewhere (Howland, 1972) I have published a similar routine in FOCAL. These programs are based on a second order Runge-Kutta algorithm as given in Conte (1965). The output of the FORTRAN routine, together with a computation of the closed form solution for the diffusion problem is given in figure 3.

I do not pretend that such routines are without pitfalls. But it is also true that, with any reasonable care, they usually work. It is certainly true that the average university student can obtain solutions to systems which far exceed the capabilities of the largest analog computer he is likely to find in the university. Moreover, these simple differential equations solving routines are far less expensive to run than the fancier simulation languages such as CSMP.

Now I want to make it very clear that I do not think such numerical methods should replace a formal treatment of differential equation systems. That would surely be a mistake. However such numerical treatments have the to the student and researcher, that they are simple to teach and master, and that they afford a way of checking results obtained by formal methods.

The importance of this last point should not be underestimated. Unless there are obvious errors in a mathematical work we tend to take it for granted that it is right, particularly if it appears in print. However, as anyone knows who works in quantitative biology, errors creep in. Thus, even when a person has full mastery of a range of formal mathematical techniques, numerical solutions can provide a valuable confirmation of the correctness of a solution.

A GENERAL ALGORITHMIC NOTATION

I would like to conclude this chapter by suggesting a solution, doubtless rashly, to a very vexing problem posed by the need for an easily understood but nonetheless powerful algorithmic notation suitable for the description of computing algorithms which will be implemented on a digital computer in FORTRAN or one of the FORTRAN-like languages, e.g. BASIC or FOCAL.

```
PROGRAM DESOLV OUTPUT

  TIME     Y(1)    Y(2)...
  0.000   0.200   0.000   0.200
  2.500   0.177   0.005   0.177
  5.000   0.157   0.009   0.157
  7.500   0.140   0.012   0.140
 10.000   0.125   0.015   0.125
 12.500   0.112   0.018   0.112                    *
 15.000   0.101   0.020   0.101                   *
 17.500   0.092   0.022   0.092                  *
 20.000   0.084   0.023   0.084                 *
 22.500   0.077   0.025   0.077                *
 25.000   0.071   0.026   0.071               *
 27.500   0.065   0.027   0.065              *
 30.000   0.061   0.028   0.061             *
 32.500   0.057   0.029   0.057            *
 35.000   0.054   0.029   0.054           *
 37.500   0.051   0.030   0.051           *
 40.000   0.048   0.030   0.048          *
 42.500   0.046   0.031   0.046         *
 45.000   0.045   0.031   0.045         *
 47.500   0.043   0.031   0.043        *
 50.000   0.042   0.032   0.042        *
 52.500   0.040   0.032   0.040       *
 55.000   0.039   0.032   0.039       *
 57.500   0.039   0.032   0.039       *
 60.000   0.038   0.032   0.038      *
 62.500   0.037   0.033   0.037      *
 65.000   0.037   0.033   0.037      *
 67.500   0.036   0.033   0.036      *
 70.000   0.036   0.033   0.036      *
 72.500   0.035   0.033   0.035     *
 75.000   0.035   0.033   0.035     *
 77.500   0.035   0.033   0.035     *
 80.000   0.035   0.033   0.035     *
 82.500   0.035   0.033   0.035     *
 85.000   0.034   0.033   0.034     *
 87.500   0.034   0.033   0.034     *
 90.000   0.034   0.033   0.034     *
 92.500   0.034   0.033   0.034     *
 95.000   0.034   0.033   0.034     *
 97.500   0.034   0.033   0.034     *
100.000   0.034   0.033   0.034     *

TIME STEP =       0.5000
NO. OF INTERVALS =    200
```

Figure 3. Output of the simultaneous differential equation solving program DESOLV set up to solve the two compartment case discussed in the text. The concentrations of the two compartments as computed numerically are given in the second and third columns. The fourth column gives the formal solution of the concentration of the first compartment. Note that, to the accuracy given, it is identical to the numerical solution. The graph reflects the concentration in the first compartment. The program which produced this output is given in the appendix.

Many would argue, and cogently, that such a language is the computer language itself. Indeed, Conte (1968) wrote the algorithms of his book on numerical methods in FORTRAN, format statements and all.

Others resort to logical flow diagrams, while still others describe their algorithms in idiomatic English (e.g. Simon, 1972).

I find all these solutions unsatisfactory. FORTRAN, I think, is too long-winded and full of unnecesary conventions. Logical flow diagrams take up an enormous amount of space and require a draftsman to look pleasing to the eye. Idiomatic English is also long-winded and often ambiguous in very subtle ways.

My solution to the problem involves a modification of a FORTRAN- like language stripped of its unnecessary conventions, and altered to include some useful features of other languages. Figure 4 gives an example of a program translated into this language which I will call Algorithmic FOCAL.

In this language it will be seen that I have substituted new symbols for the input (a question mark) and output (an equals sign followed by a colon). I have adopted the FORTRAN logical IF and assignment statements, the former with a slight modification, and retained the FOCAL iteration statement but written it in such a way that the sequence is just the same as in FORTRAN and BASIC. I have dropped statement numbers where they are not needed and retained the FOCAL subroutine convention.

Now, I am not so much interested in this particular algorithmic notation as the general precepts which it illustrates and which I think should be incorporated in any algorithmic notation one might suggest for general adoption.

These are:

1) The notation should be easily translated into FORTRAN or a FORTRAN-like computing language.

2) The notation should be divorced from all hardware or input and output formatting considerations.

3) The notation should be concise, but not contain special symbols.

4) The notation should have input, output, assignment, logical branching, subroutine, function and iteration statements.

```
        1.1 C APPROXIMATION FOR 10**X
            C REF. LYUSTERNIK ET AL. 1965
            ? X
            IF ((X.LT.0).OR.(X.GT.1)) 1.9
            A(0) = .99942
            A(1) = .57548
            A(2) = .17031
            A(3) = .03246
            FOR J = 0,3; DO 4
        4.1 SUM = SUM + A(J)*X**J
        1.2 Y = (SUM**2)**2
            =: "10**X = ", Y
            HALT
        1.9 =: "INVALID ARGUMENT"
```

Figure 4. Algorithmic FOCAL. The program accepts a variable and computes a polynomial approximation of 10^x using only exponentiation to the powers 0, 1, 2, and 3. Note the following differences from FOCAL: It is assumed that all variables are set to zero unless otherwise defined. Statement numbers have been omitted where they are not needed as labels or to delimit sections. Symbols for input (?) and output (=:) have been changed and the assignment verb, SET, has been eliminated. The logical expressions of FORTRAN have been incorporated into a simple two-way branch. The iterative FOR statement and the subroutine DO statements have been retained, but the iterated statements are written directly under the FOR statement to improve legibility for BASIC and FORTRAN users. The FORTRAN notation for exponentiation (**) has been used to avoid special symbols.

I think the only serious extant contender for an algorithmic language is ALGOL. Through usage, or more precisely, lack of usage it has become apparent that whatever virtues ALGOL may have as an algorithmic language, they have been far outweighed by practical considerations. I think one important reason why ALGOL is not used as an algorithmic language is given in my first precept. The international scientific computing language is FORTRAN, not ALGOL. It is also important that an algorithmic language be easily translated into a computing language, in order that the algorithm be checked.

My second precept, freedom from hardware and formatting considerations, has been a recurrent theme in this chapter, and reflects my fundamental belief in the importance of the mathematical structures themselves.

The virtues of conciseness, I hope, are self evident, and exclusion of special symbols is a practical measure. The fourth precept simply gives what I regard as a minimal statement set commensurate with a powerful language.

Again, I would be happy with any algorithmic language which approximately fulfilled these precepts, and I would like to close by suggesting that the construction of such a language, upon whose features a number of persons could find agreement, might not be an unworthy goal of these workshops.

DISCUSSION

RANSIL: I just want to comment that ALGOL was originally formulated as an international algorithmic language.
HOWLAND: I know, and isn't it sad what happened, really? Because people don't use ALGOL. What I want is a language I can easily get into, like a major scientific computing language or one of its variants. I perceive that as FORTRAN. Like it or not, that's what scientists use, the world over. BASIC and FOCAL are so FORTRAN - like that they're just minor dialects.
MCINTYRE: There is a nice ALGOL available now for the PDP-8E.

REFERENCES

1. Conte, S. D. *Elementary Numerical Analysis, An Algorithmic Approach.* McGraw Hill, New York, New York (1965).
2. Howland, H. C. The role of the semicircular canals in the angular orientation of fish. Annals of the New York Academy of Sciences 188:202-216 (1971).
3. Howland, H. C. Digital and Analog Computing in

General Animal Physiology. Proceedings of the 1972 Conference on Computers in the Undergraduate Curricula. Southern Regional Educational Board, Atlanta, Georgia (1972).
4. Lyusternik, L. A., Chervonenkis, O. A., and Yanpol'skii, A. R. *Handbook for Computing Elementary Functions*, Pergamon Press, Oxford, England (1964).
5. Milsum, J. H. *Biological Control Systems Analysis*. McGraw Hill, New York, New York (1966).
6. Simon, W. *Mathematical Techniques for Physiology and Medicine*. Academic Press, New York, New York (1972).

APPENDIX

```
*WATFIV
C PROGRAM DESOLV
C BOT= MINIMUM VALUE OF PRINTER- PLOTTER RANGE
C D() = DERIVATIVE VECTOR
C FINT = MAGNITUDE REPRESENTED BY ONE SPACE OF PRINTER PLOTTER
C H = WIDTH OF TIME STEP
C I = INDEX VARIABLE
C IP = COMPUTED TIME STEPS FOR PLOTTED POINT
C IPN = TOTAL NUMBER OF COMPUTED INTERVALS
C IPRNT = LOGICAL DEVICE NO. OF PRINTER
C IPT = TOTAL NUMBER OF PRINTED OR PLOTTED POINTS
C ISW = LOGICAL SWITCH VARIABLE
C IT = INDEX VARIABLE
C JP = PRINTER PLOTTER VARIABLE
C K() = CONSTANT VECTOR
C KS() = SYMBOL VECTOR FOR PRINTER PLOTTER
C L = INDEX VARIABLE
C LINE() = PRINTER PLOTTER OUTPUT VECTOR
C ND = NUMBER OF INTEGRATORS
C OL() = INTERMEDIATE VALUE VECTOR IN INTEGRATION
C R() = INTERMEDIATE VALUE VECTOR IN INTEGRATION
C TM = REAL TIME DURING PROGRAM RUN
C TQ = QUITTING TIME
C TS = STARTING TIME (USUALLY 0)
C TOP = MAXIMUM VALUE OF PRINTER PLOTTER RANGE
C X = PRINTER PLOTTER VARIABLE
C Y = INTEGRATOR VECTOR
C DIMENSION SECTION
1        INTEGER LINE(80),KS(2)
2        REAL K(40)
3        DIMENSION Y(20),OL(20),D(20),R(20)
4        DATA KS/' ','*'/
5        LINE(1) = KS(1)
6        X = 0.
7        BOT = 1.
8        FINT = 1.
C******
C USER DETERMINED SECTION 1. INITIALIZATION
C 1A RUN TIME PARAMETERS
C
9        ND = 2
10       TS= 0.0
11       TQ = 100.
12       IPT = 40
13       IP = 5
```

Figure A1

```
      C 1B PLACE PRINTER PLOTTER STATEMENTS HERE
14          BOT=0.
15          TOP =.25
16          FINT = (TOP-BOT)/40.
      C 1C PLACE LOGICAL DEVICE NO. OF PRINTER OUTPUT IN FOLLOWING IPRNT STA
17          IPRNT = 6
      C 1 D DEFINE ANY SUBSCRIPTED K CONSTANTS HERE
18          K(1) = .5
19          K(2) = 50.
20          K(3) = 10.
      C
      C END OF SECTION 1
```

Figure A1 (*continued*)

```
      C******
21          WRITE(IPRNT,101)
22      101 FORMAT(//,' PROGRAM DESOLV OUTPUT',//)

23          WRITE(IPRNT,209)
24      209 FORMAT('    TIME    Y(1)    Y(2)...')
25          H = (TQ-TS)/(FLOAT(IP)*FLOAT(IPT))
26          TM = TS
27          DO 10 J=1,ND
28          Y(J)=0.
29          OL(J)=0.
30          D(J) = 0.
31          R(J)=0.
32       10 CONTINUE
      C******
      C USER DETERMINED SECTION 2. OPTIONAL INTEGRATOR INITIALIZATION
      C PLACE ANY INITIAL NON-ZERO VALUES OF INTEGRATORS HERE
      C
33          Y(1)=.2
      C REF SI-72 P 55
34          AL = K(1)*(K(2)+K(3))/(K(2)*K(3))
35          C10 = Y(1)
36          C20 = Y(2)
37          CT = 0.2
      C END OF SECTION 2
      C
      C******
      C WRITE INITIAL VALUES
38          ISW=1
39          X=0.
40          GO TO 333
41      334 CONTINUE
42          ISW =2
43          IT=1
44      400 CONTINUE
45          DO 31 I=1,IP
46          TM=H*FLOAT(((IT-1)*IP)+I)+TS
47          DO 41 L=1,ND
48          OL(L)=Y(L)
49       41 R(L)=0.
50          DO 42 L=1,2
      C******
      C USER DETERMINED SECTION 3 EQUATION SYSTEM
      C PLACE DIFFERENTIAL EQUATION SYSTEM HERE
      C
      C VOLUMES ARE K(2) = V2 AND K(3) = V1
51          D(1) = K(1)*(Y(2)-Y(1))/K(3)
52          D(2) = -D(1)*K(3)/K(2)
53          CT = C10 +(1.-EXP(-AL*TM))*(C20-C10)*K(2)/(K(3)+K(2))
      C END OF SECTION 3
      C
```

Figure A2

```
      C******
54          DO 191 M=1,ND
55      191 R(M)=H*D(M)+R(M)
56          IF(L-1)192,192,193
57      192 DO 194 M=1,ND
58      194 Y(M)=Y(M)+R(M)
59      193 CONTINUE
60       42 CONTINUE
61          DO 43 L=1,ND
62          Y(L)=OL(L)+R(L)/2.
63       43 CCNTINUE
64       31 CONTINUE
65      333 CONTINUE


      C******
      C
      C USER DETERMINED SECTION 4. OUTPUT OUTPUT
      C PLACE VALUE TO PLOT ON RIGHT SIDE OF = SGN IN NEXT STATEMENT
66          X = Y(1)
      C
      C END OF SECTION 4
      C******
67          JP = (X-BOT)/FINT
68          IF(IABS(JP-40)-39)300,300,303
      C REJECT POINTS OUT OF RANGE
69      300 DO 302 J=1,80
70      302 LINE(J)=KS(1)
71          LINE(JP)=KS(2)
72      303 CCNTINUE
      C******
      C USER DETERMINED SECTION 5. OUTPUT VARIABLES AND FORMAT
      C CHANGE THE FOLLWING FORMAT TO CHANGE FORMAT OF INTEGRATOR OUTPUT
      C
73      100 FORMAT(1X,4F7.3,80A1)
      C ADD DESIRED VARIABLES AFTER Y(1) LIST
74          WRITE(IPRNT,100) TM,(Y(I),I =1,ND),CT,(LINE(J), J=1,JP)
      C END OF SECTICN 5
      C
      C******
75          GO TO (334,335) ,ISW
76      335 CONTINUE
77          IT=IT+1
78          IF(IT-IPT)400,400,30
79       30 CONTINUE
80          IPN=IPT*IP
81          WRITE(IPRNT,210) H,IPN
82      210 FORMAT(///,' TIME STEP = ',F10.4/,' NO. OF INTERVALS = ',I6)
83          STOP
84          END

      *DATA
```

Figure A3

chapter 26

COMPUTER SYSTEM ARCHITECTURE AT THE UCLA BRAIN-COMPUTER INTERFACE LABORATORY

JACQUES J. VIDAL AND R. H. OLCH
Computer Science Department
University of California
Los Angeles, California

INTRODUCTION

Computer technology plays a growing role in neuroscience laboratories, taking up tasks of ever increasing complexity. Common requirements include multi-channel data acquisition with high sampling rates, complex experiment control sequencing, on-line data management and analysis, and effective graphic representation of results. It is often desirable for these processes to be available simultaneously, to operate virtually in real-time, and to be controlled interactively by the experimenter on an on-line basis. Facilities of this type have been the goal, either stated or implied, of the planners of a number of computing facilities for neuroscience research (e.g. Estrin, Abraham, Sanders, Betyar, Baskin, Macy); yet to our knowledge, few of them, if any, have been fully operational in the sense of meeting the above mentioned challenges.

The problems encountered are rather typical. Tradeoffs must be made between centralized versus distributed (local) computing power, data storage, and process control hardware, between software and the

---

This research was supported in part by NFS grant DCR 75-02612.

development of special hardware and so on. Reliance upon a dedicated or shared staff of programming and hardware specialists provides more power at high costs, while keeping flexibility and intellectual manageability in the experimental environment and conserving resources remain all important. Making the correct decisions in these matters may mean the difference between a successful project as compared to a nightmare of constantly increasing delays, confusion and expenses.

While there is no simple methodology for solving these problems, the approach taken in developing the architecture of the UCLA Brain-Computer Interface Laboratory system provided some experience and appears retrospectively as a reasonable, if sometimes idiosynchratic, set of compromises in the search for the perfect system.

THE BCI LABORATORY

As the name indicates, the system was to have in the most literal sense, the capability to act as a direct I/O interface between computer and brain processes by interpreting the neuroelectric responses in direct context, i.e. within the real time constraints of the experiment or application. While the realization of this concept in its grandest form has still a few challenges in reserve for the future, this system is now capable of performing EEG pattern recognition in real-time over a vocabulary of pattern primitives sufficient for a practical demonstation of such a direct communication process. This closed loop approach is schematically illustrated in figure 1. Experiments performed in this manner can provide information that is not attainable in the usual passive mode of data recording and off-line processing.

The BCI laboratory focuses almost exclusively on EEG evoked responses obtained with scalp electrodes from human subjects. Yet the problems it faces are probably not very different from those found with other neuroscience applications, many of which share the characteristic of dealing with epoch-oriented data, produced in considerable volume and requiring, at least in part, statistical treatment.

Aside from considerable historic influences, there are a number of justifications for development of the complex system of interconnected computers that has been developed for this purpose. A typical problem we **have**

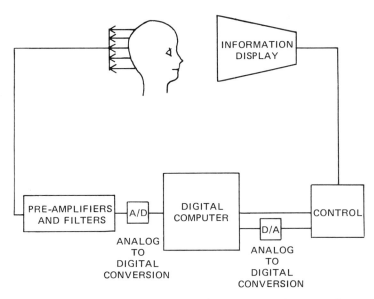

Figure 1. Closed loop approach to experimentation.

tackled has been the classification of single epoch evoked responses to multiple stimuli as represented by scalp potential measurements over a number of electrodes or channels. Among the many discrete parameters which must be specified for each experiment of this type are the electrode configuration, stimulus types, sampling times and rate, event conditions, recording formats, subject identification, on-line analysis conditions, and so on. As the experiment progresses, the operator must observe on-going changes in data, process control conditions and real-time analysis results, preferably in a highly condensed, quickly understandable form. For each such experiment, hundreds of epochs must be stored and individually tagged with information on stimulus type, analysis decision, autonomic subject state, and many other experiment control parameters. Epochs must be analyzed and classified within a few seconds after stimulus presentation if feedback to the subject is required.

In the same example, the computer system must periodically evaluate its classification performance and the subject's learning rate, recalculating classification criteria using the few minutes of subject time avilable between groups of epochs (called epoch strings). Due to subject state variability, and the nature of the operant control paradigm, there is no way to perform such experiments if the usual hours or days elapse between experiment preparation and the return of the results. Yet to outrun these usual constraints leads to some complexity (some of which is historic rather than technical in origin as pointed out earlier). The system has been in constant evolution. The current and apparently relatively stable state of this development process is shown in figure 2. This multicomputer architecture consists of five CPU's interconnected by high-speed, parallel data links, supported by a large complement of peripherals and interactive software systems.

Three qualities were sought in the system, namely some degree of fault-tolerance, a distribution of tasks between processors that uses their best features and a flexible, hands-on interface with the experimenter. Thus, redundancy in data transfer paths, peripherals, data and program storage media, and software has been sought. Considering that the most valuable resource is the subject's and the experimenter's time, alternate ways are provided for continuing a task in the face of software or hardware failure, at some loss of speed of versatility. For example, IMLAC normally loads programs by transfering from the 930 drum via the 930-920 and 920-IMLAC links. Should the 930 link or drum fail, the same transfer can be effected directly from tape to 920 and on to IMLAC. Furthermore, 920 or IMLAC itself could acquire the same data from the 360/91. The software protocols between these machines enable them to detect system failures and compensate for them by one of these dynamic reconfigurations in a way transparent to the user except for the additional time lags.

The hands-on advantages of a dedicated system of this stature are clear. The system allows the experimenter maximum freedom to exeriment with the process control and analysis environments, thus molding the system to his requirements, and a rapid turnaround in trial and error tests. Fast response is obtained in high level heuristic interactions and process control; fast and convenient implementation and modification of major system software or hardware additions is readily possible.

An obvious disadvantage of this approach is the increased amount of technical support and the higher

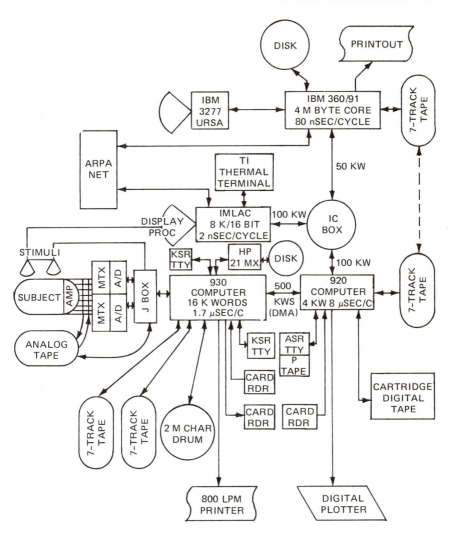

Figure 2. Multicomputer architecture.

initial equipment costs required over those of a system tied to an autonomous time-shared machine. However, the increase in experiment potential has more than compensated for these, perhaps in part because of the considerable fringe benefits associated with operating from within a university Computer Science Department. In fact, a flexible combination of in-house and shared resources has been found to be most desirable; the BCI lab can also act as a highly effective remote batch input source for 360/91 data processing. With the help of the interactive MONITOR software, the system can actually drive the 91 as an oversized peripheral processor in real-time protocols of on-line, closed-loop data collection and analysis. Finally, the connections from the 91 or IMLAC to the ARPA Network provide access to shared data management and analysis facilities unavailable locally.

The third directing principle of the multicomputer architecture was the idea that each machine should be dedicated to those tasks for which it is best suited. A task-segmented multiprocessor system is inherently more reliable and easier to debug than a multiprocessing software system inside of a larger machine. It also provides the capability for real-time parallel processing, parallel software development, hardware testing, and data storage. Table 1 gives a breakdown of principal tasks at which the five main CPU's in the BCI system are earning their keep.

Efficient operation in this multiprocessor environment requires high-speed data transmission between CPUs and efficient line protocols to coordinate their activities. Figure 3 shows the main data transfer links in the system. All links are 24-bit parallel and fully asynchronous. Interrupts are provided between CPUs to initiate transmission protocols which typically require a few words to identify the requested operation, followed by another interrupt and the data, with a final interrupt and status code sent back to the initiating CPU to verify successful completion of the task or error conditions.

The two main communication hardware sub-systems are the Intercomputer-Communication Box (ICB) and the 930 Station Access Control (SAC). The ICB acts as a two-way multiplexer for IMLAC and 920, connecting these machines together or either one of them to the 360. Communications with the 360/91 is via a 2701 parallel data adapter on a multiplexer channel. This setup does not provide for interrupts to originate from the 360 channel control program to an external system. Therefore, the 360 must be polled by the external device to initiate data transmission. This polling function is accomplished as the 920 or IMLAC interrupts the 360/91, which then takes one command word from the ICB via a channel program previously set up by the BCI 360 MONITOR software system. The MONITOR then interprets the command

SYSTEM ARCHITECTURE 417

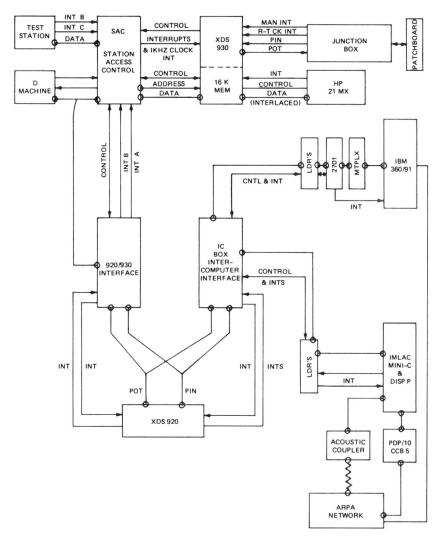

Figure 3. Main data transfer links in the multicomputer system.

word and sets up the channel for a read or write of the appropriate data.

In contrast, the SAC system provides the 920 with a 24-bit parallel direct memory access path to the 930 core. The transmission protocol is initiated by either machine when it sends an interrupt to the other. At that time, the 920 sets its end of the interface into either a read or write mode and waits for its one-word buffer to fill or empty. At the same time, the 930 sets a word count and starting address for both read and write into fixed core locations and with a single instruction, sets these words into counters within the SAC. The SAC then autonomously provides addresses to 930 memory as it buffers words to and from the 920. 930 memory operations are then interlaced between SAC and CPU requests on a priority-demand basis. That is, SAC demands for memory cycles have higher priority than CPU demands but are made only when the 920 is ready for the next word transmission. The 930 is approximately five times faster than the 920 and only 5 percent of available memory cycles are lost from 930 CPU operations under the interlace. Thus, real-time data collection and analysis algorithms are almost unaffected by SAC transmissions.

The following sections will be devoted to additional details on tasks performed by each machine in the multicomputer net and to the experiment control procedure which binds them.

THE XDS-930 EXPERIMENT TASK SCHEDULER

In satisfying the requirements of experiment process control for real-time data acquisition and manipulation, several tradeoffs must be considered. Among these is the problem of resource distribution among concurrently active tasks. These tasks might include A/D conversion, data buffering and archival storage, interactions with experimenter controls and the subject environment, data pattern recognition and so on.

The BCI Scheduler is a mix of real-time, priority selected multi-tasking, and of simple sequenced task execution. Indeed, data sampling throughout experiments occurs in bursts (epochs) and, therefore, the data collection periods may be thought of as a time-bound activity interspersed with non-critical intervals available for data manipulation. Decisions whether to store (acquire) or reject data, however, must be made in real time by implementing a circular buffer or a tracking pattern filter. A special peak-detection filter is employed in the Scheduler to automatically delete artifact-corrupted data epochs. As shown in figure 4, artifact detection proceeds throughout pre-epoch delay sampling, the adjustable warning and stimulus delays and on to the end of the data collection period. Artifacts

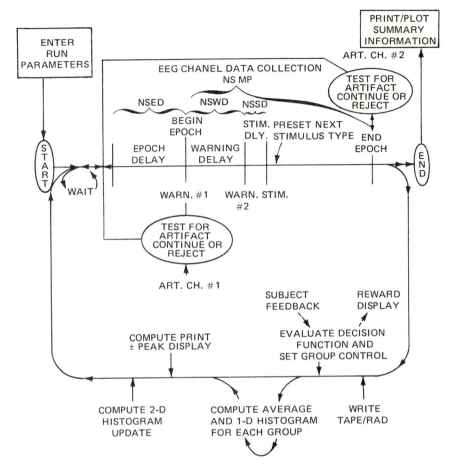

Figure 4. Artifact detection.

detected during the epoch delay restart the epoch dalay or in a separate option, initiate the stimulus and data collection for experiments where artifact-bound epochs are sought. Artifact peaks detected during the epoch will restart the collection sequence following the end of the epoch. Two artifact channels are provided.

Data epochs which have survived artifact detection are stored on an archive tape and a drum file for transfer to subsequent analysis programs. Typically a classification function is applied to the epoch data and the most probable group is identified, and amplitude averaging, peak shape and distribution analyses may also be carried out by the Scheduler on the current epoch before the next is taken. All these including data collection, artifact rejection and analysis are completed for all channels within three seconds between epochs. The Scheduler also controls the subject environment either directly from the 930 using the Junction Box process control interface shown in figure 5 or by sending commands through the 920 to IMLAC (under the 920-IOPS program) for background display generation.

The off-line subsequent analysis of the EEG data collected in this manner requires the use of an effective data management system. All BCI data are transferred between collection and analysis programs in the standard BCI Format (Appendix 1) which specifies each experiment or epoch string as one physical file. Each epoch within a file consists of fixed-length records, the first of which is a contact file. This epoch descriptor contains all essential identifying information about the particular experimental and computer environment which existed at the time the epoch was taken. By carrying along with the data information such as experiment and subject name, the date and time, and a record of past computer processing, the data may then be incorporated into an epoch-oriented file structure. This structure may be accessed by in-house application programs or transferred outside (to the 360/91 and beyond) for analysis. Through higher level heuristic data manipulations the results of many experiments may be rapidly sorted by contact file key words to provide correlations between subjects, stimulus-type subsets, intra-subject variabilities, and variations in experiment conditions such as sampling rate or subject feedback.

It is clear that this concept of epoch-oriented data is applicable not only to evoked response EEG but to the organization of a wide variety of information generated in neurophysiological experiments. On-going spike trains may be viewed as long epochs arbitrarily sequenced by time periods or changes in the preparations. For example, large quantities of data accumulated from studies of cochler potentials have been handled in a similar manner (Ronken and Eldredge) using Fourier transforms and other transformations.

THE 920 IOPS AND IMLAC PDS-1

To effectively operate the CPUs of a multicomputer system in parallel on interrelated tasks, it is necessary

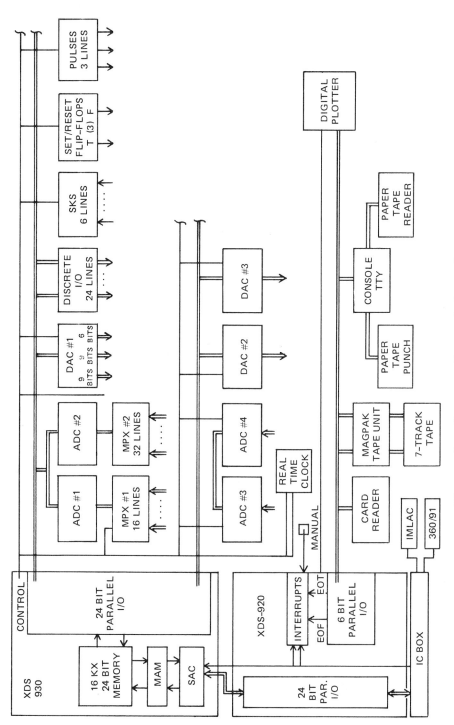

Figure 5. Process Control interface.

to provide an efficient intercomputer communication system. The XDS-920 Input/Output Processing System (IOPS) and the IMLAC Resident constitute the primary resident portions of such a distributed system at BCI (figure 6). The other processors, i.e. the 930, HP21MX, and 360/91 support similar elements which are called into foreground operation by the priority interrupts which initiate all transmissions.

These initial interrupts alert the IOPS in the receiving CPU to accept a series of command words that include the data source and destination, the intervening transformation, a file name, word count and the like. If the requesting CPU is also to be the data input or output device, the CPU handling the request sends a similar protocol back to the requestor which respecifies the nature of the data to be transferred. The requestor then sends or receives the specified data and finally receives a return code from the IOPS indicating any abnormalities in the transaction. This process of protocol software

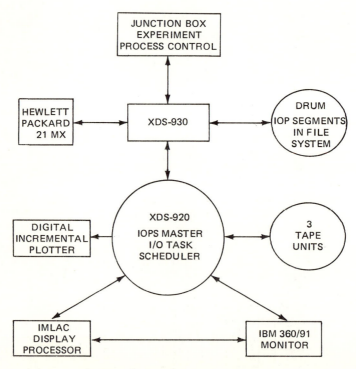

Figure 6. IOPS and IMLAC.

handshaking allows a CPU to make a request, return to background work, and later, at the return request of the IOPS, buffer in or out the appropriate data, later still checking that the transfer had been successful. Thus, processing may continue in the initiating CPU as its I/O request is transparently cascaded through other processors toward the target file or I/O device. In particular, the 360/91 often imposes unpredictable delays as polling interrupts often wait several seconds for servicing because of time-sharing constraints. Logical unit assignments entered into each processor's IOPS provide on-line and dynamic real-time modification of the IOPS environment to reach the various data storage mediums whose physical characteristics need not be assumed in the original request. Input/Output Processors, segments of code, each of which perform a particular input or output function are either resident or external. While most operations require only core-resident IOPs, an IOPS will load and execute an external IOP on requestor demand, transparently contacting other CPUs if the load module resides on a remote input device.

Besides intelligent buffering of data between CPUs, there are several other missions that increase the complexity of the communication system. In the system, either tape on the 920, drum on the 930 or disc on the HP21MX can provide bulk storage to IMLAC which has no peripherals of its own. The file management system can provide IMLAC assemblers, applications software, and display lists and reciprocally hard copies of any IMLAC display can be produced on the 920 controlled incremental plotter. Higher level plotting software being developed will support any number of user defined fonts for alphanumeric strings as well as macros to simplify labeling and scaling of graphic plots. The system also relieves programmers from the considerable difficulty and duplication of effort which would be incurred if only hardware link protocols were specified for each CPU. In the present design the full power of the distributed IOPS will be invoked through simple subroutine calls in FORTRAN on the 930 and by similar calls or direct statement in the graphic compiler language developed for the IMLAC.

This language, called GRAL, was also developed as part of the BCI research effort at UCLA. It was designed around the concept of an intelligent satellite graphic

Table 1. Principal tasks of five main CPUs.

XDS-930

- interactive experiment set-up
- experiment process control scheduling
- discrete stimulus generation
- tape generator for archive data
- transfer data written on drum
- real-time data analysis and artifact recognition/rejection
- on-line pattern recognition for discriminant function generation
- exercises discriminant function for pattern classification

XDS-920

- I/O Processing System for data exchange between 920, IMLAC and 360
- on and off-line digit plotter control
- maintains IMLAC software library and display files
- transparent, virtual data buffering and IOP loading/execution

IMLAC-PDS-1

- display generation for on-line visual stimulus to subject
- interactive experiment status display to operator
- on and off-line terminal for data analysis and report generation

IBM 360/91

- data and program storage on on-line disc
- large-scale off-line numerical computation, methodology development
- URSA time-shared services
- BCI MONITOR System services
- ARPA NET through TSO for data file transfer between Network hosts

HP 21MX

- intelligent disc handler for rapid access file system
- BCI interface to cassette and 8-level paper tape

system involving two general purpose CPUs. The main CPU (here the 360/91) is remotely accessed by a minicomputer (IMLAC) local to the graphic station which provides the intelligence to the terminal display processor. This configuration allows a division of labor between the main CPU and the graphic support minicomputer, freeing the main CPU from more or less trivial or repetitive graphic operations. With exchanges between CPUs limited to general commands with slower turnaround, the speed requirement for the communication link is also considerably relaxed.

The graphic language and cross-compiler were designed for extended stand-alone, real-time graphic animation using to fullest advantage the IMLAC-PDS-1 architecture which consists of an independent 16-bit minicomputer and display processor interlaced into a single 8k memory. The language (Gonzalez and Vidal) includes many of the features of mathematically oriented languages such as vectors, floating point computation and mathematical functions. Graphic variables are used to store graphic values and are manipulated by graphic operators that handle line drawings (including invisible and dotted lines). The use of string variables allows the manipulation of alphanumeric information in the same fashion as graphic variables. The present implementation of the compiler is written in PL/1 and runs on the IBM 360/91 producing IMLAC assembler source code. This source is run through an IMLAC assembler in the 360/91 with the object finally sent to IMLAC through the intercommunication link (IC Box in figure 2).

GRAL directly supports IMLAC IOPS communications which are processed by a small, resident IOPS interface. The integration of GRAL into the system considerably reduces the time required to implement new experimental displays or the graphic representations of results. Thus, GRAL effectively facilitates important machine-man communication in the laboratory.

THE IBM 360/91 MONITOR SYSTEM

In addition to the processors available in-house, the laboratory makes extensive use of the broad range of facilities available at the UCLA Campus Computing Network (CCN). This system, based on an IBM 360/91 offers an exceptionally large, 4M-byte, core memory and very high speed arithmetic processing capabilities, along with an array of compilers, assemblers, file management and timesharing systems, documentation text editors, and ARPA Net access.

In order to effectively utilize this computing power within the ambitious, real-time constraints discussed earlier, it was necessary to provide a direct data communication path between the laboratory computers and the 360/91 and to develop the attendant 360 software development that would allow the monitoring of all transactions. This last item proved to be a monumental one which took over two years to develop. It is referred to as the MONITOR system, and it runs on the IBM system in a priviledged position within the operating system (OS/MVT) environment to optimize response time. Certain installation routines provided by CCN have made the MONITOR somewhat installation dependent. The system's principal hardware components are shown in figure 7. Although only one user is shown in this diagram, the MONITOR system can simultaneously support several users, each tied into the 360 via his own hardware interface, and each utilizing a different set of subtasks of the MONITOR. Typically users would communicate to the 360 using an IBM 3277 terminal. The MONITOR subtasks may include complex data analysis programs, such as elements of the UCLA Biomedical Data Computing System (Dixon) or the user's own application software. A typical program, when running under the MONITOR, can acquire data from the laboratory, process it and return graphic results or control decisions in as close an on-line, real-time manner as possible within the time sharng constraints borne by the central facility. The cost for such on-line service is relatively low since even though the MONITOR emerges as a fairly massive piece of software, it does not accrue computing costs until it is awakened by an interrupt from the laboratory. Figure 8 indicates the subtask attachments process and subprocessor control of MONITOR system status.

The Line Handler is the MONITOR's logical eqivalent and handshaking partner of the 920's IOPS. The Line Handler is treated by the MONITOR as just another subtask despite its central function for data transfer. The Line Handler controls all I/O functions between the 360 and the laboratory computers using the data link. Figure 9 shows some of the system options handled by this routine. These include on-line modification of experiment control parameters and access to any sequential or partitioned disk data set. A procedure is available which allows the

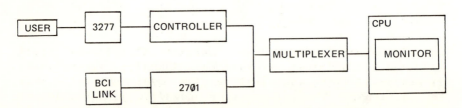

Figure 7. MONITOR system hardware.

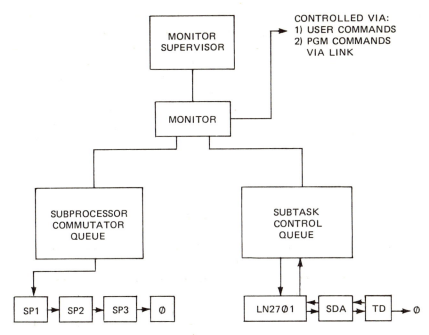

Figure 8. Subtask attachment process and subprocessor control of MONITOR system status.

laboratory computers to automatically create new, named files on demand which can then access existing data management routines at CCN for remote storage and retrieval.

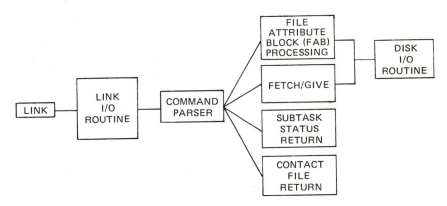

Figure 9. Line Handler system options.

It must be noted that the BCI MONITOR system, in contrast to previous efforts in shared-resource computing systems for the neurosciences, is a non-disruptivce insertion into an existing facility with a broad scope of users; that is, the MONITOR is simply a job run under a much larger time-shared system. Of course it means that the time required to service lab interrupts to the MONITOR is on par with the wait time experienced by other CCN console users on the UCLA timesharing system (URSA). However, these time delays, which in off-hours are often negligible, have no significant effect on the laboratory operations which are designed to permit variable delays between critical data taking periods. A second major difference between the MONITOR and systems such as SLIP (Estrin) and ACME (Sanders) rests with the confinement of specialized operations in the laboratory, so that MONITOR access puts only standard demands on the remote facility. While the later installations concentrated most (though not necessarily all) analog conversion equipment at the central facility, the MONITOR's interaction with data is purely digital, a configuration that makes sense in view of the declining costs of analog conversion and control hardware. Thus, the data is packaged in a standard form, sent out for processing and results are received after a variable delay. Of course, cabling costs to the central CPU are also reduced from that of the parallel coaxial lines needed for the analog data.

Each user of the MONITOR is allowed full control on-line over all MONITOR functions associated with his tasks as if he had his own copy of the system; other users are unaffected should one of them crash a subtask.

BCI EXPERIMENT SYSTEM CONTROL

The preceding sections have shown the BCI laboratory computer architecture to be composed of a flexible mixture of independent CPUs interconnected by a high-speed communication network. But perhaps the most important element of such a system is the means by which its many components are coordinated into an effective pool during live experiments.

The primary problems encountered in this coordination effort are finding an appropriate division of work between machines and separate programs, supplying those programs with useful parameterizations of their control variables, standardizing formats of data transfer between programs, and finally sequencing parameter changes and program calls in various ways including some that will be contingent upon previous computational results. The chosen approach centered on the centralization of all higher-level control within the XDS-930 FORTRAN link system. Under RAD MONARCH, the 930 operating system, any FORTRAN program, with optional assembly subroutines, may be written on the RAD as a numbered link or overlay. Whenever a sequence is chosen, constituted by a number of these links, simple MONARCH JCL begins their execution by loading FORTRAN RUN-TIME and the first requested link. RUN-TIME contains a push-down stack of link numbers which can be manipulated by any of the links. For instance, a simple "CALL NEXT LINK" overlays the current link with the one specified in the top of the stack, and pops the stack. Since a link is stored in the RAD as the core image of a complete load module, overlays occur very rapidly (it requires approximately 22 milliseconds to randomly access, load and begin execution of an 8k instruction link). Since the overlay process does not affect COMMON, links may communicate through such regions. Given this feature and the ability for a link to completely alter the contents of the stack, very complex link execution sequences may be assigned with a minimum of programming effort.

This linking capability also became the key to the experiment control system by using links which interpret Experiment Control Language (ECL) statements. Typical ECL statements associate symbolic names with link numbers, set initial program run parameters and call for the recursive execution of sequences of links. Decisions made on the basis of data or results of link executions may change run parameters and alter link execution sequences. Thus, a high level of heuristic programming may be specified in a few ECL statements to organize the automatic execution of a number of live EEG experiments, interleaved with processing.

A valuable advantage of this modular system is that instead of having a group of software modules as subroutines of a single experiment control program, functional elements may be designed as separate programs. Each such program becomes a link which may be independently purged from the system, recompiled/assembled, and replaced in RAD storage without disturbing

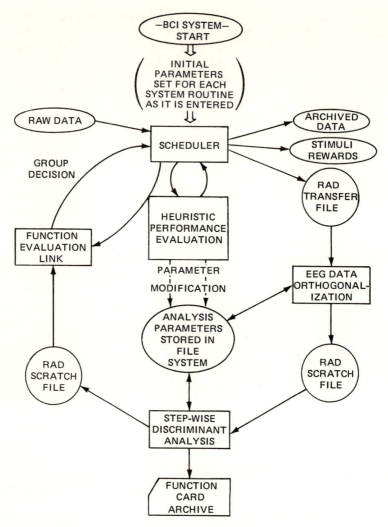

Figure 10. An ECL linking sequence.

other elements. A typical link sequence would include functional elements responsible for data collection scheduling, stepwise discriminant analysis, graphic plotting, data reformatting, and so on. An ECL program sequence of this type is illustrated in figure 10. Raw multichannel data is being taken by the scheduler, orthogonalized by channel and passed to the Stepwise Discriminant Analysis Program which computes a discriminant function for the classification of data epochs by stimulus type. This function is then passed on to an evaluation link which is exercised on the next pass of the scheduler on every incoming epoch. In this and subsequent passes, the subject can be fed-back the results of decisions based on the EEG acquired a few seconds before.

This immediate feedback is of course the key to a whole class of experiments involving operant conditioning. The avilability of results from the processing of the data while the experimental session is in progress also has important implications. The general paradigm, made possible with this integrated experiment control and data processing, is illustrated in figure 11. It has proven to be an extremely powerful strategy in

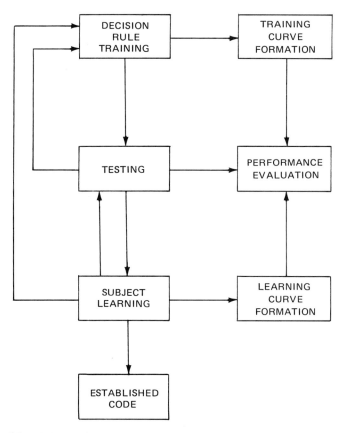

Figure 11. Experiment control and data processing paradigm used in experiments at BCI Laboratory.

current experiments at the Brain-Computer Interface laboratory, aimed at discovering codes for sensation and perception in EEG evoked responses. The codes that are identified (i.e., the components and corresponding time windows in the evoked responses) are validated in a rigorous (and previously unattainable) manner by the ability of an automaton (the discriminant algorithm) to successfully perform real-time classification of the data. Thus baseline measurements are taken at the start of each experimental session (the training phase) to establish the discriminant function. The latter is then tested against new data by performing a real-time classification of each evoked response. As classification results are available for each epoch within a fraction of a second, information can be fed back to the subject (learning phase) and learning hypotheses can be tested (for instance, regarding modification of evoked response components or simply avoidance of interfering EEG activity).

In this strategy a complete independence of measurements is maintained between subjects and for each subject between experimental sessions. The invariance of code activity no longer needs to be assumed beyond the time span of a relatively short experimental session. Rather, long term invariance, if present, will be unplanned and resulted from external factors. Behind it, however, lies a considerable design effort that addressed real-time problems that will have to be faced in advanced laboratory systems regardless of any updating of hardware technology. The complexity of these problems and of their solution (hardware and software) has continuously slowed down progress and defeated many optimistic experiment schedules. In the last year, however, the full potential of the system is beginning to unfold and momentum is accruing rapidly. Experiments that would have taken months in a conventional environment (if feasible at all) were completed in days. Planning is now discovered from the useable data and can then be separated from the shifts due to biological cycles or other causes.

CONCLUSION

The system described in this chapter has evolved through three years of trials and errors. Clearly a fraction of the present complexity and heterogeneity was underway for a more modern implementation. In particular, a miniaturized instrument for the testing/training of EEG discrimination, using microprocessors, is being evaluated for possible application to prosthetics. It is hoped that some of the experience acquired with this effort will be valuable to other projects and will help in the establishment of the much hoped for standard for a neurosciene computer language and data format.

DISCUSSION

BROWN: I'm sure everyone asks you this, but to what extent could this be used for practical man-machine communication?

VIDAL: I once thought this was far in the future, requiring unforseeable technical advances. But I'm beginning to change my view. We've found that in fact it's possible to do something that works practically. We can actually run a maze in fairly good time. We display a maze on our IMLAC graphics, and the subject is forced to make quaternary decisions at intersections. He can do that. I don't think he could do better with a light pen. Of course, he might make less mistakes with the light pen, but he's about as fast. And this setup is far from optimized. So I begin to think that it's not completely crazy.

MOISE: With regard to your artifact detector, what proportion of the epochs do you eliminate?

VIDAL: That depends on threshold settings. In general, with a good subject, you shoot for about 10%. The subject has to be trained, however. Naturally, with the feedback loop, they learn not to make eye movements at the wrong times. They learn that it's perfectly all right to blink at some times and not others.

REFERENCES

1. Abraham, F. D., Betyar, L., and Johnston, R., An on-line multiprocessing interactive computer system for neurophysiological investigations. Proc. Spring Joint Comp. Conf. (1968).

2. Baskin, E. B. et al. A modular computer sharing system. C. A. C. M. 12, no. 10, (1969).
3. Betyar, L., A user-oriented time-shared on-line system. C. A. C. M. 10, no. 7, (1967).
4. Dixon, W. F., *BMD Computer Programs*, Los Angeles Health Sciences Computing Facility, Department of Preventive Medicine and Public Health, School of Medicine, University of California, (1976).
5. Estrin, T. et al, Facilities in a brain research institute for acquisition, processing and digital computation of neurophysiological data. In *Proceedings of the Conference on Data Acquisition and Processing in Biology and Medicine* (1962).
6. Gonzalez, C. and Vidal, J. J., GRAL--a graphic compiler language for intelligent terminals. In *Proceedings of the Conference on Computer Graphics, Pattern Recognition and Data Structure* (1975).
7. Macy, J. Jr., White, M. P., Design and development of a multiple-laboratory computer complex for biomedical research. In *Computer Technology in Neuroscience*. P. B. Brown, ed. Washington: Hemisphere Publishing and John Wiley & Sons (1976).
8. Ronken, D. A. and Eldredge, D. H. Retrieval of data by attribute using parameter-flagged data storage. In *Computer Technology in Neuroscience*. P. B. Brown, ed. Washington: Hemisphere Publishing and John Wiley & Sons (1976).
9. Sanders, W. J. et al. An advanced computer system for medical research (ACME project-Stanford). In *Fall Joint Computer Conference* (1967).
10. Vidal, J. J. New developments in EEG signal processing. In *Proceedings of the Third Symposium on Nonlinear Estimation* (1972).
11. Vidal, J. J. Toward direct brain-computer communication. Ann. Rev. Biophys. Bioeng. 2 (1973).
12. Vidal, J. J., Neurocybernetics and man-machine communication. In *1975 International Conference on Cybernetics and Society*. San Francisco, California, (1975).

APPENDIX I

BCI EEG DATA FORMAT

General Comments:

   Following is a tentative description of the format and layout of the EEG epoch-oriented data for all processing transactions involving the BRAIN COMPUTER INTERFACE (BCI) lab. All current and future BCI applications or systems programs will follow this format or an update version thereof. Also, all data from external sources must be pre-converted if use of any part of the BCI system is intended. Note that the format described below is compatible with standard IBM/360 word and byte sizes, which can be read or written by any high level language such as FORTRAN IV or PL/1, or even by machine dependent languages such as the as 360 Assembler Language. Thus the format is intended to be ultimately compatible with any application on 360/370 systems.

General Data Format:

   BCI EEG data is grouped into logical and physical units called 'epochs'. Each epoch corresponds to data collected during a discrete time period, while a specific task is being performed by the subject, or while the subject is being exposed to a given set of environmental conditions. For example, an epoch might consist of 8 channels of EEG data collected during a 1/2 second interval following a red strobe flash over a green background. There are usually many such epochs per experiment, and groups of epochs often have some common characteristic. They are grouped together into 'epoch strings', with n epochs in each string, (numbered 1 through n). Each epoch is broken down into two parts:

   (1) the epoch header
   (2) the epoch data.

The epoch header consists of useful and necessary information about the epoch, such as the subject's identification code, the task performed, the length of the epoch, how many channels of data are present, the sampling rate of the collection, and so on. The epoch data follows directly, containing the actual experimental data for that particular epoch.
   The logical length for the 'records' that compose each epoch may vary, depending on the data, from 40 to 80 bytes. 40 bytes is the minimum due to the size and form of the epoch header (see below). 80 bytes was chosen as the maximum because it is standard card image size, and because it appears most unlikely that more than 80 bytes will ever be needed (see below).

Epoch Format:

(1) Epoch Header

The epoch header (tentatively) is composed of 15 records. Each record contains information about the epoch data to follow. Because it contains the type of information presumably required with most existing or future BCI applications software, it was decided that the header should be a modified version of the input control cards to the UCLA Biomed program BMDX92 (auto- and cross-spectral analysis). These records are in 360 character mode, and as mentioned are no less than 40 characters in length. (Only 40 characters are used. If the logical length of the epochs is greater than 40, then the header records are padded to the right with blanks, as usual). Thus all unspecified columns below are blank.

Card 1:
    col 1-6:   'PROBLM'
    col 7-12:   6 byte run identification field
    col 13-16:   number of samples (FORTRAN 'I4' fixed point)
    col 17-18:   number of channels (FORTRAN 'I2' format)
    col 19-21:   number of frequency bands resolved ('I3')
    col 22-25:   sampling rate ('I4')
    col 26-40:   (reserved - blank - X92 compatible)

Card 2:
    col 1-6:   'PREFLT'
    col 7-40:   (reserved - blank - X92 compatible)

Card 3:
    col 1-6:   'TITLES'
    col 8-12:   'EPCT:'
    col 14-16:   epoch no. (location in string ('I3')
    col 19-23:   'EXPR:'
    col 25-27:   epoch string identification ('A3'), task id
    col 30-34:   'SUBJ:'
    col 36-38:   subject id - initials ('A3')

Card 4:
    col 1-6:   'TITLES'
    col 8-12:   'DATE:'
    col 13-16:   date of experiment/epoch - mmdd ('2I2')
    col 19-23:   'STMT:'
    col 24-27:   stimulus type (1-nstm) for this epoch ('I4')
    col 30-34:   'EPXS:'
    col 35-38:   number of epochs in each string ('I4')

Card 5:
    col 1-6:   'TITLES'
    col 8-12:   'NSMP:'
    col 13-16:   number of samples in the epoch
    col 19-23:   'NSED:'
    col 24-27:   number of samples in epoch delay ('I4')
                  (used for artifact detection prior to epoch data)
    col 30-34:   'NSWD:'

col 35-38: number of samples between beginning of data collection and warning stimulus ('I4')

Card 6:
 col 1-6:   'TITLES'
 col 8-12:   'NSSD:'
 col 13-16:   number of samples between warning and main stimulus ('I4')
 col 19-23:   'SINT:'
 col 24-27:   scan interval (millisec) or 1/rate ('I4')
 col 30-34:   'NACH:'
 col 35-38:   the number of channels

Card 7:
 col 1-6:   'TITLES'
 col 8-12:   'DERL:'
 col 13-16:   decision rule used for grouping of similar stimuli ('I4')
 col 19-23:   'ARPD:'
 col 24-27:   artifact detection threshold (levels/2048) ('I4')
 col 30-34:   'ERPD:'
 col 35-38:   evoked response peak detection threshold ('I4') (levels/2048 for peak detection)

Card 8:
 col 1-6:   'TITLES'
 col 8-12:   'NSTM:'
 col 13-16:   number of stimulus types ('I4')
 col 19-23:   'SEED:'
 col 24-27:   odd integer seed for any applications random number generator (zero if not applicable) ('I4')
 col 30-34:   'GRPS:'
 col 35-38:   no. groups or possible classifications ('I4')

Card 9:
 col 1-6:   'TITLES'
 col 8-12:   'CLOK:'
 col 13-16:   time the epoch was taken (hhmmss - '3I2') may be zero if not needed
 col 17-40:   reserved for future use

Card 10:
 col 1-6:   'TITLES'
 col 8-12:   'PCWD:'
 col 13-16:   I/O control word ('I4')
 col 17-40:   reserved for future use

Card 11:
 col 1-6:   'TITLES'
 col 8-12:   'EKGR:'
 col 13-16:   EKG rate ('i4')
 col 17-40:   reserved for future use

Card 12:
 col 1-6:   'TITLES'
 col 8-12:   'GPCW:'
 col 13-16:   group control word (varies from 0 to grps) ('I4')
 col 17-40:   reserved for future use

Card 13:
   col 1-38:   'NAMCHN C01 C02 C03 C04 C05 C06 C07 C08'

Card 14:
   col 1-38:   'NAMCHN C09 C10 C11 C12 C13 C14 C15 C16'

Card 15:
                      Fortran type variable format card which can
                      be used to read the data to follow, of the
                      form '(mmAnn)' or '(mmFnn.pp)', where 'mm'
                      is actually the number of channels, 'nn'
                      the width of each value field, and 'pp'
                      the number of decimal digits assumed (if any).

(2) Epoch Data

From the above description of the variable format it can be seen that each record contains one and only one scan. Each scan (record) contains one EEG sample value for all of the channels. Thus, if there are 8 channels involved, and 200 samples per epoch, there would be the usual 15 epoch header records, and 200 epoch data records, each of which contains 8 values (eg: 8F5.0).

A sample epoch follows:

```
PROBLMEXAMPL0200 8 64 128
PREFLT
TITLES EPCT:0002   EXPR:TRA2   SUBJ:FLC1
TITLES DATE:AP09   STMT:1438   EPXS:0007
TITLES NSMP:0200   NSED:0100   NSWD:0100
TITLES NSSD:0010   SINT:0008   NACH:0008
TITLES DERL:0000   ARPD:0600   ERPD:0050
TITLES NSTM:0002   SEED:0000   GRPS:0000
TITLES CLOK:1622221
TITLES PCWD:0008
TITLES EKGR:0200
TITLES GPCW:0008
NAMCHN C01 C02 C03 C04 C05 C06 C07 C08
NAMCHN C09 C10 C11 C12 C13 C14 C15 C16
(8F5.0)
+0106+0142-0118+0095-0057-0087-0038-0082
+0122+0159-0050+0138-0138-0178-0125-0083
+0221+0099-0130+0078+0100-0122-0229-0003
             .
             .
             .
+0104+0207+0017+0012-0048-0081-0225-0019
```

COMMENTS

   This epoch is therefore 15+200=215 records long. There are 0200 samples in each epoch for each of the 8 channels of data: the sampling rate is 128 samples/second. This particular epoch is epoch number 002 of a 7-epoch string called 'TRA2' for subject 'FLC1'. The epoch was taken on April 9 at 4:22:21. The data following the header is structured into 200 records containing 8 5-column values in FORTRAN 'F' floating point format. Each record forms one scan.

chapter 27

RETRIEVAL OF DATA BY ATTRIBUTE USING PARAMETER-FLAGGED DATA STORAGE

D. A. RONKEN AND D. H. ELDREDGE
Central Institute for the Deaf
St. Louis, Missouri

INTRODUCTION

Table-lookup procedures are often used to specify an arbitrary function. Figure 1 illustrates this familiar technique for the case where successive values of the function $F(x)$ were calculated at equally-spaced values of the argument Δx units apart and stored as a table. Future references to the function can then be obtained by using the integer argument as a pointer to the location containing the value of the function $F(N \cdot \Delta x)$. A somewhat different application of such lookup procedures uses a table that is filled with experimental observations, not calculated values. If the entries in the table correspond to observations of a dependent variable and the position in the table corresponds to values of an independent variable, then the table defines an experimentally observed functional relation. We have found that table-lookup methods may be used to extract complicated functional relations from large data tables that contain several independent and dependent variables.

We illustrate this data retrieval method by describing its application to a particular set of

---

This work was supported in full by a U. S. Public Health Service, Department of Health, Eduction and Welfare research grant NS 03856 from the National Institute of Neurological and Communicative Disorders and Stroke.

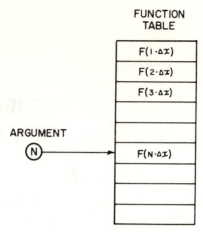

Figure 1. Simple table-lookup procedure where the table pointer is given by the value of a variable.

physiological data that has an underlying structure amenable to table-lookup methods. Other applications of the technique would of course proceed somewhat differently, but our illustration should serve to outline the general approach.

DESCRIPTION OF A PARAMETER-FLAGGED DATA SET.

With gross electrodes in the cochlea we can record continuous physiological potentials (cochlear microphonics) in response to sounds. Computer processing of these responses is made possible by periodic sampling with an analog-to-digital converter. In our case, the instantaneous value of the analog response is quantized to 8 bits and 256 such samples are used to represent a single wave form. Several instances of these sampled wave forms are added on-line to enhance the desired signal in relation to incoherent background noise, and the resulting digitized wave form is saved as a 256-point table. As many as four physiological responses are obtained in response to a single sound and saved as a group, along with the experimental parameters in effect at the time of recording. The parameters and their associated responses form what we call a waveform unit, as shown in figure 2. Included in the parameter fields are waveform identifiers and the settings of several

PARAMETER-FLAGGED DATA STORAGE 441

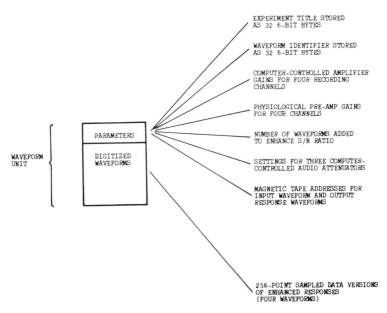

Figure 2. Detailed composition of a particular waveform unit.

laboratory instruments (amplifiers, attenuators, etc.) used during the recording process. Many of these parameters are needed to produce properly adjusted mesurements from unrefined recordings. Others are used principally for data retrieval, while still others only provide housekeeping services.

Several thousand such waveform units may be produced sequentially during the course of a single experiment. They are stored in temporal order on magnetic tape, as diagrammed in the left-hand portion of figure 3. When the tape records are later transferred off-line to a disc, the waveform unit corresponding to any parametric condition may be quickly found by examining only the parameter fields. Because of the crucial role that the parameter encoding plays in this retrieval process, we call this general scheme parameter-flagged data storage.

The physiological system being studied in this instance is frequency selective, so it is useful to work with the data in the frequency domain, rather than the time domain. Parameters associated with the response waveforms remain linked to the data after conversion to the frequency domain by a Fast Fourier Transform program

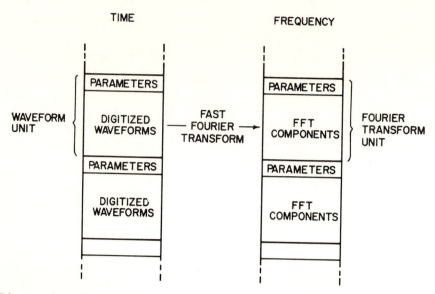

Figure 3. Format for disk storage of waveform units and Fast Fourier Transform units.

to form a Fourier transform unit, as indicated in figure 3. The data are still in tabular form, with each table entry giving the value for a single component of the Fast Fourier Transform. The location in the table corresponds to the discrete frequencies of the transform, all of which are integral multiples of a common fundamental frequency. These Fourier transform units are therefore tables that give amplitude and phase measurements for sinusoidal components which we may identify with the positive integers.

Level-dependent nonlinearities are observed in the response measurements, and it is this aspect that introduces much of the complexity and structure into the data. The existence of nonlinearities means that frequency components are present in the electrical responses that are not present in the sound stimulus. The nature of the nonlinearity is such that these extra frequency components (distortion products) occur at easily specified frequency components. Thus, if the frequency components corresponding to integers L and H are present in the sound, there may be frequency components in the electrical response $2L$, $3L$, $2H$, $H-L$, $2L-H$, $2H+L$, and so forth. We illustrate the data retrieval process by using a data set obtained from an ensemble of such two-tone sounds.

DATA RETRIEVAL BY SIMPLE TABLE-LOOKUP USING A PARAMETER VALUE AS ARGUMENT.

Figure 4 illustrates one instance of an off-line search through a parameter-flagged data set in order to inspect a possible experimental relation. This simple search corresponds to the straightforward table-lookup procedure of figure 1. In this case, we wish to retrieve the frequency component in the response that has the same frequency as the low frequency tone present in the stimulus. We want to examine only those stimulus conditions for which the low frequency was at component 10. Fourier transform units are processed serially, examining the value of the parameter L from each unit.

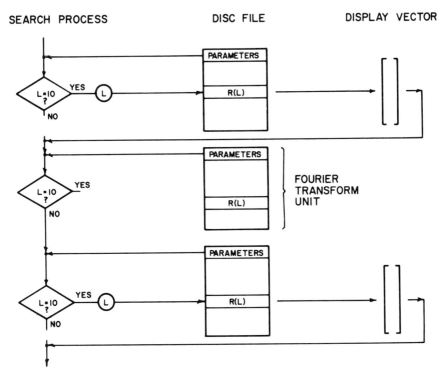

Figure 4. Example of a data retrieval search process using the simple table-lookup procedure.

If the parameter L is equal to 10, the table-lookup occurs which retrieves the desired response and places it into the display vector for graphical presentation. There is no need for the data set to conform to any particular serial order, since the parameter value serves as a flag for the presence of the desired datum. Actual searches typically involve more elaborate branch operations than shown in figure 3 because most often the conjunction of several parameter values is required for a retrieval condition to be satisfied. Even such simply-defined retrieval operations can be very helpful in filtering out experimental detail that may not be relevant to a given experimental question.

RETRIEVAL OF DATA WITH SPECIFIED ATTRIBUTES BY TABLE-LOOKUPS BASED ON A FUNCTION OF PARAMETER VALUES.

An extended form of the table-lookup scheme is shown in figure 5. In this case, two variables are needed to access the table. The actual pointer to the table entry is determined by the value of the function G, which has J and K as arguments. We have used this technique to retrieve data that do not reside at any fixed position within a table, but instead are specified only as satisfying a certain parametically-defined relation. A particular instance is illustrated in figure 6.

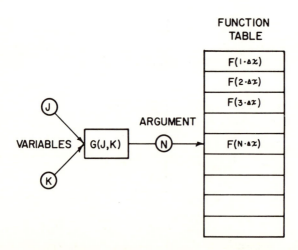

Figure 5. Extended form of a table-lookup procedure.

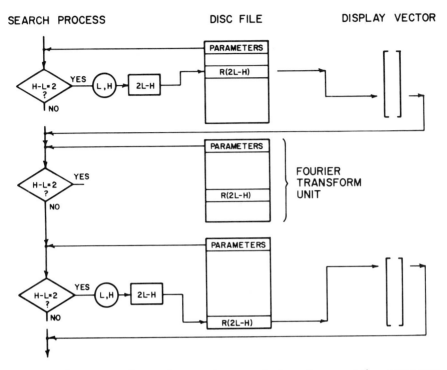

Figure 6. Example of a data-retrieval search process where the table pointer is a function of two variables.

In this example, we wish to examine that portion of the entire data set for which the high (H) and low (L) frequency components in the stimulus were separated by a difference of two Fourier components. This partitioning is readily accomplished by branching on the stimulus requirement that $H-L=2$. When the stimulus conditions have been met, the retrieval procedure fetches a particular response component that is the result of a nonlinear effect in the physiological system. The property or attribute of this particular nonlinear response is that its frequency component is given by the expression $2L-H$. From figure 6, it is apparent that because the pointer for the table is now a function of two parameter values that vary from unit to unit, successive retrieval operations will in general go to different positions within the table. The resulting display vector will contain only responses with the specified attribute (their frequency is $2L-H$ times the fundamental frequency) that were observed for the defined stimulus condition ($H-L=2$). Since the parameter assignments encode many properties of the stimuli and the values of the Fourier components represent many aspects of the responses, it is evident that many detailed relations within the data set could be extracted by this form of retrieval.

SUGGESTIONS FOR IMPLEMENTATION.

In order to use the parameter-flagged retrieval scheme outlined here, the data must be structured in the form of tables. This is not a severe constraint, for large classes of neurobiological observations fall quite naturally into this category. Any continuous potential will be in this format if digital signal processing is employed. Another large class includes all the data represented by counting occurrences of discrete events, such as the all-or-none response of a neuron.

It has been our experience that retrieval operations based on experimental relations allow the experimenter to concentrate on scientific findings without the distraction of procedural minutiae. The increase in intellectual perspective it has provided is similar to the change encountered in moving from machine language programming to FORTRAN coding. For this reason, we believe that systems for processing neurobiological data may often benefit from a parameter-flagged storage scheme.

DISCUSSION

BROWN: Why didn't you use white noise analysis?
RONKEN: We weren't sufficiently familiar with the technique when we started the study, or we might have used it.

chapter 28

A SAMPLING ALGORITHM FOR BANDWIDTH LIMITATION OF ACTION POTENTIAL TRAINS

A. S. FRENCH

Physiology Department
University of Alberta
Edmonton, Alberta, Canada

INTRODUCTION

The analysis of input-output behaviour in neurophysiological systems is often approached by linear systems theory. Such analysis involves measuring the impulse response, or its Fourier transform the frequency response, between the input and output signals of the system (D'Azzo and Houpis, 1966). Practical estimation of the system behaviour is more easily accomplished in the frequency domain because correlation functions are transformed into point-by-point multiplications and because movement between the time and frequency domains is made efficient by the fast Fourier transform (Cooley and Tukey, 1965). Unfortunately many neurophysiological systems are substantially nonlinear. However, methods are now being developed for generalized nonlinear analysis of biological systems, involving the Wiener functional expansion (Wiener, 1958), and in this area it has also been demonstrated that measurement in the frequency domain is advantageous (French and Butz, 1973).

The fast Fourier transform is a special case of the discrete Fourier transform (Bergland, 1969) and is thus subject to all of its limitations. The two most

---

Support for this work was provided by the Canadian Medical Research Council.

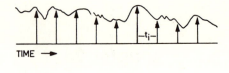

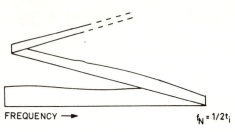

Figure 1. The effect in the frequency domain of regular sampling in the time domain. Above: a signal is sampled regularly which contains frequency components too high to be accurately characterized by the samples. Below: the components above the Nyquist frequency fold into the lower frequencies producing aliasing of components at incorrect frequencies.

important properties of the discrete transform which must be taken into account when using it for practical measurements are the finite time window, which affects the accuracy of each frequency amplitude, and the regular sampling rate, which causes aliasing or folding of frequency components above the Nyquist frequency $f_N = 1/2t_i$ where t_i is the interval between regular samples. Figure 1 illustrates the problem. Fluctuations in the signal between sampled points are not accurately characterized, but they do not merely escape from the measurement. Instead they appear at incorrect frequencies via a folding process which places all detected signal power between DC and the Nyquist frequency (Bendat and Piersol, 1966). It is therefore essential that signals to be processed via the fast Fourier transform be filtered in such a way as to remove all components at frequencies above the Nyquist frequency. This requirement may alternatively be stated in the form that the sampling rate must be chosen sufficiently fast as to ensure that no components are aliased. Unfortunately fast sampling produces a great deal of data, which may be difficult to deal with, and the high frequency resolution produced by the fast sampling may not be needed for the characterization. The problem becomes acute when we deal with neurophysiological signals containing action potentials since action potentials are relatively rapid fluctuations in voltage and therefore contain high frequency components. It is often assumed that the actual form of the action potential is not significant in carrying information but rather acton potentials are often treated as point events in time or Dirac delta functions in time. Although such a representation removes the frequency components associated with the time profile of the action potentials it does not limit the frequency bandwidth of the signal, in fact it extends the bandwidth to infinity. Figure 2 illustrates the time and frequency domain representations of some examples of

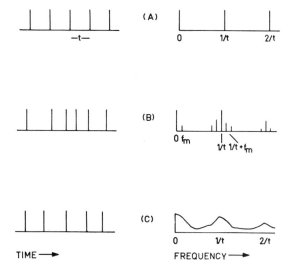

Figure 2. The time and frequency domain characteristics of three types of point processes in time. (A) a regular spike train with interspike interval t has a spectrum containing equal amplitude components at all multiples of the reciprocal of the interspike interval. (B) a sinusoidal modulated regular spike train has additional components at the modulating frequency, f_m, and at many sum and difference frequencies between the carrier frequency and the modulating frequency. (C) A randomly modulated regular spike train has power over the spectrum of the modulating process and spreading of the peaks due to the carrier rate.

point process signals. Figure 2 (a) shows a regular series of Dirac delta functions separated by times t. The spectrum of such a signal is a regular series of delta functions in frequency at intervals t. The other examples are of a sinusoidally modulated train of events (Bayly, 1968) and of a randomly modulated train of events (Stein, French and Holden, 1972). In all of these cases and in any signal consisting of delta functions there will be infinite bandwidth.

It is clear that if frequency domain analysis is to be used with action potential signals derived from them, it is necessary that some method be available for limiting their bandwidth before the fast Fourier transform is used. The method must optimally be capable of performing such bandwidth limitation without introducing bias or distortion into the measured signal and must also have efficiency approaching that of the fast Fourier transform if the speed of the frequency domain measurement is to be maintained. I shall briefly review some of the methods which have been used in attempts to overcome this problem and shall then describe in some detail an algorithm which appears to be the most satisfactory approach to the problem so far devised.

EARLIER METHODS OF BANDWIDTH LIMITATION

The most direct method is simply to pass the signal through an electronic filter with a powerful frequency cutoff characteristic. The higher frequency components are then removed and the signal may be treated as a continuous voltage in the normal way. The objections to this technique are practical. Since the characteristics of the filter are now introduced into the observed system it is essential to remove them from the final estimates of system behaviour. The most obvious way to do this is to pass both input and output through identical filters, thus removing their effects on the cross spectra, but it is quite difficult to build filters which are exactly identical, especially when they have very sharp characteristics. Any difference between the filters will distort the measurement. Since the whole signal is to be passed through the filter it will contain fluctuations which the experimenter may wish to remove if he considers that only the times of occurrence of the action potentials are significant. Probably the best way of applying this technique is to use a threshold device to produce a short, identical pulse for each action potential. The pulses are then passed through a sharp filter and the resultant signal treated as a continuous band-limited voltage. The input signal must then be passed through an identical filter (Koles, 1970).

Once the series of action potentials has been converted by a threshold device to a point process in time, they may be processed in a number of ways. Watanabe (1969) used a counting technique, in which the number of spikes occurring in each sample interval was used as an estimate of the process. However, this procedure introduces distortion in both gain and phase portions of the estimated frequency response function, as has been pointed out by Koles (1970).

Another more commonly used measurement is the average instantaneous frequency of a spike train, which is defined as the reciprocal of the interval between each spike and its predecessor in the series. A method of sampling such a function has been described by Shapley (1969). The main problem with this technique is that it assumes that a new function, the average instantaneous frequency, is the information carrying characteristic of the action potential signal. Although there is a close relationship between this function and the original signal for regular spike trains with small fluctuations from the mean rate, the relationship becomes complex beyond these boundaries. Terzuolo and Bayly (1968) have pointed out that the method introduces spurious troughs into the frequency response function at multiples of the mean carrier frequency.

Discussion of some other methods of obtaining frequency domain measurements from point processes are given by Lewis (1970), and French and Holden (1971a).

DIGITAL CONVOLUTION OF DELTA FUNCTIONS

The ideal filtering mechanism would remove all frequency components above the Nyquist frequency and leave those below the Nyquist frequency unaltered. Figure 3 shows the time and frequency domain forms of such a filter. The frequency envelope has a rectangular form, as specified above, and the time domain envelope is given by the function: $Sin(2\pi t f_N)/(2\pi t f_N)$ where t is

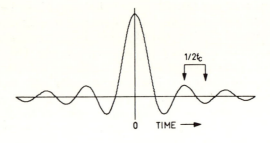

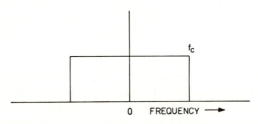

Figure 3. The time and frequency domain characteristics of a perfect sampling filter. The rectangular profile in the frequency domain treats all components below the cutoff frequency, f_c, identically. Above f_c the function is zero valued. The impulse response in the time domain is the function $\sin(2\ tf_c)/(2\ tf_c)$. This function has unity value at time zero and extends to plus and minus infinity in time.

time and f_N is the Nyquist frequency. This function, which is sometimes called the Sinc function, has unity value at zero time and extends to both plus and minus infinity in time. If such a filter were to be built it would have to predict infinitely ahead in time all of the signals which were ever to pass through it. It is therefore physically unrealizable. However, once signals have been sampled, digitalized and stored in computer memories time becomes more flexible since it has been transformed into a spatial index. The time domain envelope of figure 3 is the impulse response of the filter and so the signal to be filtered has to be convolved with it, but the signal which we are dealing with is a series of delta functions and so their convolution with the impulse response duplicates it with a shift in time. In other words we must replace each event or action potential with the envelope of figure 3 centered at the time of occurrence of the event. Clearly we can't extend the function to infinity, but it dies down quite rapidly and it may be possible to truncate it without great loss of accuracy. Having substituted each event with the Sinc function we would then have to sample it at regular intervals to obtain the time series for the fast Fourier transform. Since we have chosen the Nyquist frequency, f_N, the sampling interval is fixed at $1/2f_N$. This separation in time may be converted into distance along the Sinc function by substituting for t in the expression: $2\pi t f_N$ to give: $2\pi f_N/2f_N$, or π. This is the crucial relationship, since it means that the required samples can be obtained from a single Sinc calculation because:

$$\sin(x) = |\sin(x + \pi)| = |\sin(x + 2\pi)|$$
$$= \cdots |\sin(x + n\pi)|$$

where n is an integer. The samples can be obtained by computing the appropriate Sinc value and then dividing successively by incrementing intervals.

Figure 4 illustrates the practical situation with an event occurring at time T and a sampling interval of t_s. Since the numerator in the Sinc function remains constant

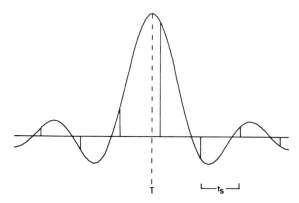

Figure 4. Diagramatic illustration of the convolving and sampling process. A spike has occurred at time T and is being sampled at regular intervals, t_s. After obtaining the initial Sinc value the succeeding values decay hyperbolically with alternating signs.

the samples are a hyperbolically decaying series with alternating sign. Figure 5 illustrates a basic flowchart for the process with the positive and negative time indices processed separately, which is usually the easiest way of programming the algorithm. The Sinc function may be obtained either from the series relationship:

$$\sin(x) \frac{x < \pi}{2} = x - x^3/3! + x^5/5! - x^7/7! \cdots$$

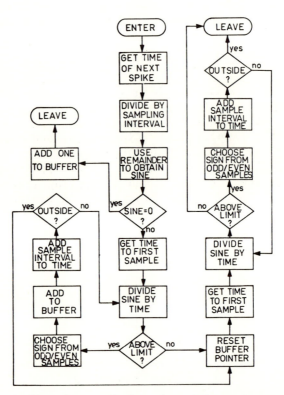

Figure 5. A flowchart of the algorithm as programmed for the PDP-8 and PDP-11 computers. Dividing the time of occurrence by the sample interval gives the first sample point index and the argument of the sinc function. Divisions are then performed in positive time until the end of the sample is encountered or the function falls below a predetermined limit. The process is then repeated in negative time.

or by the use of a stored digital quarter Sinc wave. A special case occurs if the event and a sample point exactly coincide. In this case the function has unity value at the time of occurrence and zero value at all other sample points.

Termination of the dividing process can be caused either by reaching the end of the sample epoch or by the absolute value of the computed function falling below some predetermined level. The former case is merely the effect of the rectangular data window which is bound to be applied to the data, and whose effects are well known. The effects of truncation of the Sinc function are more difficult to assess accurately. Since it is essentially a rectangular time window it would be expected to cause some spreading of the Fourier coefficients, but the hyperbolic decay of the function means that the effect should be substantially reduced. We currently use the criterion that when the function falls below 2^{-9} it will be truncated. This means that the function has to fall to about 0.02% of its initial value before it is truncated and an error of this level is probably similar in magnitude to many of the other errors in measurement which are difficult to avoid, such as analog to digital conversion and threshold device accuracy. This truncation criterion means that in the worst case the function will have to be computed for about 150 values before it is considered insignificant.

The algorithm has been programmed in both PDP-8 and PDP-11 machine languages and has been used successfully over the past four years. It has always been programmed in fixed point arithmetic with a floating scale to preserve maximum accuracy. The data may be converted to floating point format at the end of all of the convolutions. If the action potentials are occurring at a relatively low rate, about 100 per second on average, then it is possible to perform the convolutions on-line as the events are detected. In practice this condition cannot be guaranteed and it is best to use a foreground storage routine with a background convolution routine processing the data as fast as it is capable of doing so. Such a program will appear to be on-line for reasonably slow events, although it will take extra time at the end of the sampling process if events have arrived rapidly (French and Holden, 1971b). This procedure is often inconvenient and it is more desirable to store all of the data in digital fomat and then process it at leisure. In

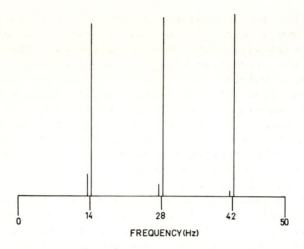

Figure 6. An example of filtering by the algorithm. A regular spike train of 14 events per second was sampled at a sample interval of 10 msecs. with a resultant Nyquist frequency of 50 Hz. Note the absence of any folded components at 44 Hz, (56 Hz), 30 Hz, (70 Hz), etc.

this case it is possible to store the events in terms of their times of occurrence and to process them by the convolution process afterwards (French, 1973).

Figure 6 illustrates a practical demonstration of the algorithm. A function generator was used to produce a regular train of impulses at a rate of 14 per second. The impulses were then sampled by use of the algorithm and then transformed into the frequency domain by use of the fast Fourier transform. Finally the power spectrum was calculated and is illustrated in Figure 6. As may be observed, there are the expected peaks at 14, 28 and 42 Hz. The next peak would be expected to occur at 56 Hz, and it would be of equal amplitude to the other peaks. Since the Nyquist frequency in this case was 50 Hz the first aliased peak would be expected to occur at 44 Hz. As can be seen, there is no detectable power at this, or any of the other expected aliased frequencies.

DISCUSSION

VIDAL: What is the basis for choosing the original bandwidth?

FRENCH: Purely from experimental considerations. I usually use white noise stimulation and use a sampling rate just higher than the bandwidth. Then the computer works out the Sinc function automatically from the sampling rate. This way I can sample at a rate which is independent of the actual firing rate.

CAPOWSKI: I thought the goal was to reduce the bandwidth of the incoming spikes, i.e., to have a lower necessary sampling rate.

FRENCH: Exactly. The slower you sample, the wider the Sinc function becomes, and the lower the bandwidth.

REFERENCES

1. Bayly, E. J. Spectral analysis of pulse frequency modulation in the nervous system, I.E.E.E. Trans. Biomed. Engng. BME-15:257-265 (1968).
2. Bendat, J. S., and Piersol, A. G. *Measurement and Analysis of Random Data*, New York: John Wiley (1966).
3. Bergland, G. D. A guided tour of the fast fourier transform, I.E.E.E Spectrum 6:41-51 (1969).
4. Cooley, J. W., and Tukey, J. W. An algorithm for the machine calculation of complex Fourier series, Math. of Comput. 19:297-301 (1965).
5. D'Azzo, J. J. and Houpis, C. H. *Feedback Control System Analysis and Synthesis*, New York, McGraw-Hill (1966).
6. French, A. S. Comput. Prog. Biomed. 3, 45. (1973).
7. French, A. S., and Butz, E. G. Measuring the Wiener kernels of a non-linear system using the fast Fourier transform algorithm, Int. J. Control 17:529-539 (1973).
8. French, A. S., and Holden, A. V. Alias free sampling of neuronal spike trains, Kybernetic 8:165-171 (1971a).
9. French, A. S., and Holden, A. V. Comput. Prog. Biomed. 2:1-7 (1971b).
10. Koles, Z. J. *A Study of the Sensory Dynamics of a Muscle Spindle*, Thesis, University of Alberta (1970).
11. Lewis, P. A. W. Remarks on the theory, computation and application of the spectral analysis of series of events, J. Sound Vib. 12:353-375 (1970).
12. Shapley, R. Fluctuations in the response to light of visual neurons in limulus Nature 221:439 (1969).

13. Stein, R. B., French, A. S., and Holden, A. V. The frequency response, coherences and information capacity of two neuronal models, Biophysical J. 12:295 (1972).
14. Terzuolo, C. A., and Bayly, E. J., Data transmission between neurons, Kybernetic 5:83 (1968).
15. Watanabe, Y. Statistical measurement of signal transmission in the central nervous system of crayfish, Kybernetik 6:124-130 (1969).
16. Wiener, N. *Nonlinear Problems in Random Theory*, New York, John Wiley (1958).

chapter 29

SOFTWARE FOR SPECTRAL ANALYSIS OF NEUROPHYSIOLOGICAL DATA

A. S. FRENCH
Physiology Department
University of Alberta
Edmonton, Alberta, Canada

INTRODUCTION

Many neurophysiological systems may be modelled as single input-output systems with time varying input and output. In such situations linear systems theory provides a general method of obtaining a first order approximation to the system behaviour. Traditionally linear analysis has been carried out by determination of the autocorrelation and crosscorrelation functions of the input and output signals. These techniques are particularly easy to apply in neurophysiological experiments if the times of occurrence of action potentials are considered to be point events in time or Dirac delta functions. (Moore, Perkel and Segundo, 1966). More recently analysis in the frequency domain has become the established practice. The fast Fourier transform, rediscovered by Cooley and Tukey (1965), has made conversion between the time and frequency domains very efficient and this allows correlation and convolution operations to be performed via multiplications in the frequency domain. Another advantage is that the coherence function (Bendat and Piersol, 1966) is readily obtained from the frequency

---

Support for this work was provided by the Canadian Medical Research Council.

domain signals and gives a normalised measure of the linear correlation between input and output signals. It may also be used to estimate the information capacity of the system (Stein and French, 1970). A major problem in the application of frequency domain analysis to neurophysiological signals has been the high bandwidth of action potential signals. However, several techniques are now available which substantially overcome this problem (French and Holden, 1971a).

Although some neurophysiological processes may be completely modelled by linear systems, there are many which are clearly nonlinear. The response of many neurons to a sinusoidal current stimulus is not a sinusoid of identical frequency, as would result from a linear system, but a series of action potentials. Other well known nonlinearities are adaptation and accommodation to stimuli, and nonlinear summation of stimuli. One method which offers a general means of identifying such nonlinear behaviour is the Wiener functional analysis (Wiener, 1958). Although theoretically suitable for the analysis of a wide range of nonlinear systems, the method has not been widely applied because of its computational difficulties. However, the increasing power of digital computers has begun to make it a practical approach and it has recently been applied to neurological systems (Marmarelis and Naka, 1973; McCann, 1974). So far the kernels of the Wiener functionals have been measured almost entirely by the cross-correlation method of Lee and Schetzen (1965) but we have recently shown that the frequency domain again offers a more efficient route for system identification (French and Butz, 1973).

COMPUTATIONAL ROUTES

The system is assumed to have input and output signals which are functions of time, $x(t)$ and $y(t)$. For most purposes the input will be band limited white noise, since this provides the most suitable stimulus for the analysis of linear systems and is essential for computation of the Weiner kernels. The fast Fourier transform is then used to obtain the Fourier coefficients of the two signals:

$$x(t) \leftrightarrow X(f)$$
$$y(t) \leftrightarrow Y(f) \quad (1)$$

$X(f)$ and $Y(f)$ are both complex variables and so two values have to be stored for each frequency, f. However, the time domain signals are both real and so there is complex conjugate symmetry around zero frequency.

This means that as many useful frequency domain values are obtained as time domain values were supplied. Most fast Fourier transform algorithms can make use of this property to replace the original time series without using extra storage space. Using the Fourier coefficients the power spectral densities can be obtained:

$$S_{xx}(f) = <X(f) \cdot \overline{X}(f)>$$
$$S_{yy}(f) = <Y(f) \cdot \overline{Y}(f)> \quad (2)$$
$$S_{xy}(f) = <X(f) \cdot \overline{Y}(f)>$$

where $S_{xx}(f)$ is the input power spectrum, $S_{yy}(f)$ is the output spectrum and $S_{xy}(f)$ is the cross spectrum. The bar above the coefficients indicates the complex conjugate and the <> signifies an ensemble average. Figure 1 illustrates the above operations in the form of a flow-chart and also indicates some of the possible routes that can be followed once the spectra have been determined. The cross power spectrum can be used to determine the cross correlation function via the inverse fast Fourier transform. Similarly the input and output autocorrelations can be obtained by the inverse transform. The describing function $D(f)$, which may also be called the frequency response function, or if the system is linear, the transfer function, may be determined from:

$$D(f) = S_{xy}(f)/S_{xx}(f)$$
$$D(f) = S_{yy}(f)/S_{xy}(f) \quad (3)$$

The choice depends upon what is known about the system and on the properties of the input and output signals. Since the input signal is usually more reliable the former equation is usually used. The describing function

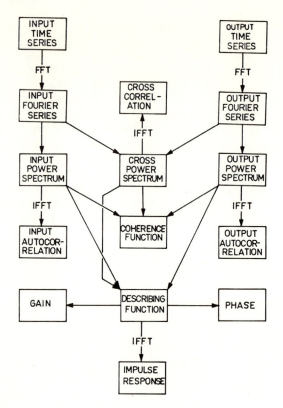

Figure 1. Processing paths which may be followed in the analysis of linear systems via the frequency domain. FFT indicates the fast Fourier transform and IFFT indicates the inverse fast Fourier transform.

is complex, since the cross spectrum is complex and the input spectrum is real, and so is normally dealt with in terms of two real functions, the gain and phase, $G(f)$ and $P(f)$. If the describing function is written explicitly in complex format,

$$D(f) = A(f) + iB(f) \quad (4)$$

where $i = \sqrt{-1}$. Then the gain and phase are obtained from:

$$G(f) = A(f)^2 + B(f)^2 \quad (5)$$

$$P(f) = \arctan(B(f)/A(f)) \quad (6)$$

Alternatively the describing function may be used to determine the impulse response of the system, $K(t)$, via the inverse transform. The duality of the time and frequency domain forms of the function reflect the fact that the system's behaviour may be modelled by multiplication of the input signal with the describing function in the frequency domain or by convolution of the input signal with the impulse response in the time domain:

$$Y(f) = X(f) \cdot D(f)$$

$$y(t) = \int_{-\infty}^{\infty} K(u)x(t-u)du \qquad (7)$$

$$K(u) \leftrightarrow D(f)$$

The coherence function is obtained from:

$$\gamma^2(f) = \frac{|S_{xy}(f)|^2}{S_{xx}(f) \cdot S_{yy}(f)} \qquad (8)$$

Since the coherence function is a normalised measure of correlation it has the value of one at all frequencies for a linear noise-free system. Values lower than 1 indicate the presence of inherent noise in the system or non-linear behaviour in the system. A value of zero indicates that there is no linear correlation between the input and output signals at that frequency. From the estimated coherence function and the number of degrees of freedom of the spectral estimates one can compute approximate confidence limits for the frequency response function estimate (Jenkins and Watts, 1968).

If the system being examined is non-linear, then the describing function is an approximation to the system behaviour, but it is the best linear approximation in the least mean square sense. The impulse response is the first time varying term in the Wiener functional expansion:

$$y(t) = \Sigma G_n[K_n, x(t)] \qquad (9)$$

where G_n are a complete set of functionals which are orthogonal to a white noise input signal. Identifying a non-linear system by this expansion involves measuring K_n, the kernels of the system. The functionals have the form:

$$y(t) = K_0 + \int K_1(u) x(t - u)\, du + \qquad (10)$$

$$\iint K_2(u, v) x(t - u) x(t - v)\, du\, dv - A \int K_2(u, u)\, du + \cdots$$

where A is the variance of the input noise. Since the functionals are orthogonal to white noise, measurement of the describing function by the method outlined above gives the first order kernel, $K_1(u)$, as the frequency response. The zero order kernel, K_0, is the average response to a white noise input, and may therefore be obtained from the DC component of the Fourier transform of the output signal. In the same manner that the most efficient estimation of the first order kernel is via the describing function in the frequency domain, it has been shown that the most efficient estimation of the higher order kernels is via multi-dimensional describing functions in the frequency domain (French and Butz, 1973). Using the Fourier transform pairs:

$$K_1(u) \leftrightarrow K_1^*(s)$$

$$K_2(u, v) \leftrightarrow K_2^*(r, s) \qquad (11)$$

$$K_3(u, v, w) \leftrightarrow K_3^*(q, r, s)$$

It may be shown that the Fourier transforms of the kernels can be estimated from:

$$K_1^*(s) = \frac{\langle Y(s) \cdot \overline{X}(s) \rangle}{A}$$

$$K_2^*(r, s) = \frac{\langle Y(r + s) \cdot \overline{X}(r) \cdot \overline{X}(s) \rangle}{2A^2} \qquad r \neq s \qquad (12)$$

$$K_3^*(q, r, s) = \frac{\langle Y(q + r + s) \overline{X}(q) \overline{X}(r) \overline{X}(s) \rangle}{6A^3} \qquad q \neq r \neq s$$

Since A may be estimated from:

$$A = <X(f) \cdot \overline{X}(f)> \qquad (13)$$

the first term in (12) is identical to our previous measurement of the describing function. The squared and cubed variance terms in the higher order descriptions may be obtained from:

$$A^2 = X(r) \cdot \overline{X}(r) \cdot X(s) \cdot \overline{X}(s)$$

$$A_3 = X(q) \cdot \overline{X}(q) \cdot X(r) \cdot \overline{X}(r) \cdot X(s) \cdot \overline{X}(s) \qquad (14)$$

PRACTICAL IMPLEMENTATION

We first began to use these techniques with relatively simple computing equipment, consisting of a PDP-8/I computer and a 32k word magnetic disc. We had insufficient storage space to digitalise all of our data from one experiment and the fast Fourier transform, although efficient, was not sufficiently fast to allow on-line processing. We were thus forced to take repetitive samples from data stored on analog magnetic tape and after obtaining estimates of the input, output and cross spectra, the spectra were added to an ensemble stored on the disc. Another section of data was then played back and the data digitalised, and so on until enough ensembles had been averaged. The process is shown diagramatically in figure 2. Once the spectral estimates are available the most suitable way to complete processing of the data is via some high level language which can easily handle arctangent and logarithm calculations. This process was quite satisfactory in terms of the accuracy of the results but it was rather time consuming and involved a great deal of operator attention (French and Holden, 1971b).

We subsequently obtained a magnetic tape storage unit, actually a DECtape, and this allowed us to store all of the available data on magnetic tape in digital format. The process was then similar to that outlined above, except that complete automation of the process was possible. Figure 3 illustrates the automated process and figure 4 illustrates the storage format which was used

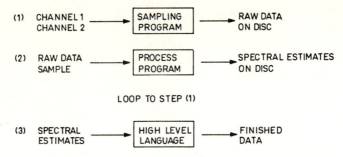

Figure 2. Implementation of linear spectral analysis using minimal computing hardware. Segments of experimental data are sampled, transformed into the frequency domain and then the spectral estimates are computed and added to a running average in memory or on a disc. The process is then repeated until the estimates are sufficiently smooth and then they are used to give the final data.

for the digitalized data on the magnetic tape. The format allowed four possible situations, with either input or output signals being either continuous analog signal or spike trains. Analog signals were stored as single word entries containing a 9-bit analog to digital conversion, while a 10 KHz crystal clock was used to store spike trains as single-word entries containing the number of crystal clock pulses since the preceding spike. One bit of each word was used to define it as belonging to the input or output signal and if spikes were so far

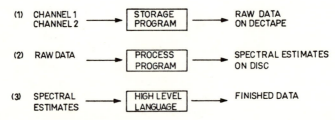

Figure 3. Implementation of linear spectral analysis using a large magnetic storage medium (in this case DECtape). The raw data is first stored entirely in digital format and then processed automatically to give the spectral estimates.

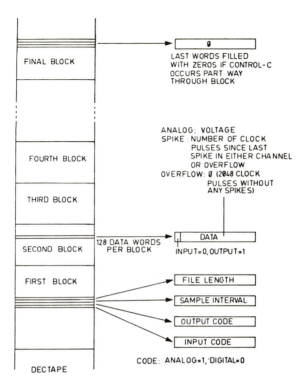

Figure 4. Storage format used for digitalization of data onto magnetic tape.

apart that the interval between them could not be contained in one word a zero word was used to indicate such a timing overflow. A four word header block defined the types of signal being stored, the sample interval used to store the analog signals and the total length of the file. Using this information succeeding programs could completely reproduce the original data to the resolution of the clock and analog to digital convertor. Processing programs computed the spectral estimates in floating point format and stored them on the disc. Spikes were sampled by the digital convolution technique described previously (French and Holden, 1971). A complete description of this package of programs, which includes visual and numerical inspection of the data at each stage of processing, has been published before (French, 1973).

The above system has now been duplicated with some modifications on a PDP-11/40 computer with a moving-head disc. The large storage space available on these discs makes it possible to store all of the experimental data which is normally required directly on the disc. Fortunately neurophysiological processes tend to be rather slow in engineering terms and very fast sampling rates are not normally required. Using the DECtape it was possible to store more than two thousand samples a second directly onto the tape, and with a moving head disc even faster rates are possible. Higher rates than this are rarely needed in physiological experiments. A number of changes have been made to the original process. First, a slightly different storage format for the raw data is used and this is illustrated in figure 5. Only three situations are catered for, both signals analog, both signals spikes and one analog-one spikes. Since the decision on which signal is input and which is output is not effectively made until the final processing stage, there is no point in defining them earlier. This allows a simpler header block of two words. Second, two bits are used for channel identification. This is possible because of the longer word of the PDP-11 and it makes decoding the data easier. Otherwise the storage format is similar. Figure 6 illustrates the complete set of operations and the principal difference is seen to be the storage of the individual Fourier coefficients, instead of immediately computing the spectral estimate ensembles.

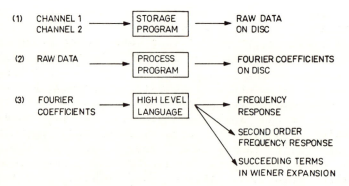

Figure 5. Implementation of general spectral analysis using a large storage medium. Instead of the spectral estimates the Fourier coeffecients are stored as a three dimensional array. They may then be used for estimation of linear and nonlinear terms in the Wiener functional expansion.

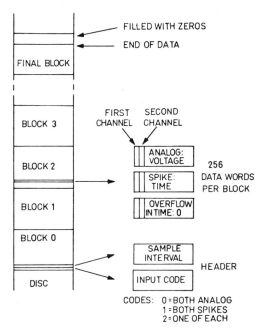

Figure 6. Simplified raw data storage format when a longer computer word is available.

The reason for this is made clear by inspection of equation (12) since although the Fourier coefficients may be used to compute all of the Wiener kernels, once the data has been converted into the spectra it can only be used to compute the first order kernel. The higher order kernels require cross frequency terms which can only be obtained from the coefficients. The Fourier coefficients are much less compact than the ensembled spectral estimates and occupy about the same amount of space as the raw data. The most convenient storage format for the coefficients is a three dimensional data file and this format is illustrated in figure 7. The three dimensions are the component elements (input real and imaginary, output real and imaginary), the frequency and the sample number. Figure 7 also shows how the three dimensional data file is stored as a linear array on the disc, in a conventional format varying in dimensions in order. All of the programs described for the PDP-11 are designed to run under our own operating system (French, 1975) and one of the features of this system is a version of the

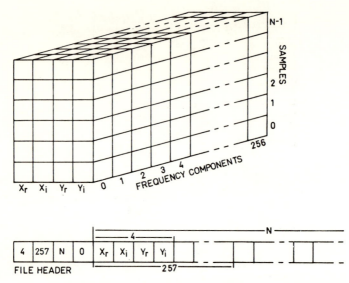

Figure 7. Format of the three dimensional file needed to store the array of Fourier coefficients. X_r, Y_r, X_i and Y_i are the real and imaginary components of the input and output signals.

Digital Equipment language FOCAL which is capable of accessing floating point data files in up to four dimensions. The limitation to four dimensions is not inherent in the system but is merely a feature of the versions implemented so far. The storage format illustrated in figure 7 with the four-word describing header is compatible with FOCAL and so the computation of the kernels from the Fourier coefficients may be carried out in this language if desired. In practice FOCAL is too slow if many samples are being ensembled and so machine language programs have been written to reduce the Fourier coeffients to one and two dimensional data files containing the first and second order kernels in the frequency domain. FOCAL is then used for final numerical output or plotting of the kernels. Figure 8 illustrates a second order Wiener kernel obtained by inverse Fourier transformation of a second order frequency domain kernel. This kernel was obtained by the analysis of less than ten seconds of data obtained by passing white noise through a first order linear filter and then a squaring element. The kernel has the expected form, although the small data sample means that it is not adequately smooth.

NEUROPHYSIOLOGICAL DATA   471

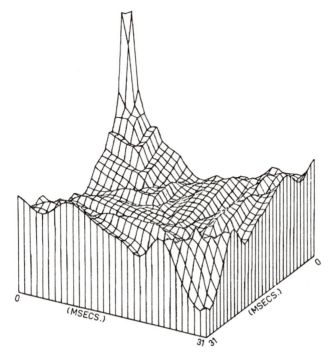

Figure 8. Second order Wiener Kernel obtained by the scheme described in figure 5 and the text.

Although most of the software described here has been implemented in machine language for efficiency, the relative cost of central processing units and fast memory is now sufficiently low that most of the operations could be carried out by use of compiled higher level languages such as FORTRAN. However, the basic steps in the analysis will probably remain the same as the techniques described here appear to present the most efficient ways of carrying out linear and nonlinear identification of single input-single output systems.

DISCUSSION

D.O. WALTER: I understand that as you go to higher-order kernels, the whiteness of the noise becomes more important. Have you encountered any difficulties with that?

FRENCH: Panos Marmarelis has done a lot of work with varying the input parameters to the Wiener kernel analysis, finding out a lot you can get away with. It turns out that the lower the bandwidth of your white noise, the faster the thing converges to a reasonable estimate, but of course the less information you get out of it. You don't have to worry about having very high bandwidth white noise to get a right answer.

D.O. WALTER: Is an estimated Wiener kernel very sensitive to non-stationarity of the data?

FRENCH: I suspect so.

RONKEN: What kinds of nonlinear systems are amenable to Wiener kernel analysis?

FRENCH: A very wide range. Only systems which go into spontaneous oscillations are almost impossible to deal with. But you can deal with "hard" nonlinearities as well as "soft" ones. It can handle things like rectification and hysteresis without any difficulty.

REFERENCES

1. Bendat, J. S. and Piersol, A. G. *Measurement and Analysis of Random Data*, New York, John Wiley (1966).
2. Cooley, J. W. and Tukey, J. W. Math. Comput. 19:297, (1965).
3. French, A. S. Comput. Prog. Biomed. 3:45, (1973).
4. French, A. S. Submitted for publication in Computer, (1975).
5. French, A. S., and Butz, E. G., Int. J. Control 17:529, (1973).
6. French, A. S. and Holden, A. V., Alias free sampling of neuronal spike trains, Kybernetik 8:165, (1971a).
7. French, A. S. and Holden, A. V., Comput. Prog. Biomed. 1:219, (1971b).
8. Jenkins, G. M., and Watts, D. G., *Spectral Analysis and its Applications*, San Francisco: Holden-Day, (1968).
9. Lee, Y. W., and Schetzen, M. Int. J. Control 2:237, (1965).
10. Marmarelis, P. Z., and Naka, K., J. Neurophysiol. 36:605, (1973).

11. McCann, G. D., J. Neurophysiol. 37:869, (1974).
12. Moore, G. P., Perkel, D. H., and Segundo, J. P., Ann. Rev. Physiol. 28:493, (1966).
13. Stein, R. B. and French, A. S., In: *Excitatory Synaptic Mechanisms*, eds. P. Anderson and J. K. S. Jansen. Oslo: Universitetsforlaget, (1970).
14. Wiener, N., *Nonlinear Problems in Random Theory*, New York: John Wiley, (1958).

chapter 30

REAL TIME PROGRAMMING CONTROL OF NEUROPHYSIOLOGICAL BEHAVIORAL EXPERIMENTS

T. MEDLIN

National Institute of Dental Research
Bethesda, Maryland

BACKGROUND

The National Institute of Dental Research established the Neurobiology and Anesthesiology Branch in 1974 in order to emphasize its research initiative in the area of pain and pain control. This branch is concerned with the elucidation of basic mechanisms of oral-facial sensation, with particular emphasis on pain sensation, and with the assessment and measurement of experimental and clinical pain in humans. The primary objective of this research group is to correlate behavioral responses to noxious and non-noxious stimuli in monkey with neuronal structure and function at various levels of the trigeminal system. The Scientific Applications Unit within the Data Processing Systems and Analysis Section is a technical development and support group within the Institute and this unit has provided the computer

---

I want to thank Ronald Dubner for whom this system was developed, Frederick Brown, who designed and built the necessary interface hardware, Merle Robinson who helped develop some of the processing software, Jon Rosenberg who implemented the histogram capability for the long range software, John Wilson, Chief of the Data Processing Systems and Analysis Section at NIDR, and Ralph Beitel who patiently reviewed this chapter.

software support for the projects to be described in this paper.

The first project is a behavioral study involving **pain and temperature discrimination in monkeys**. The second project, which uses the same hardware, but different software, is a study of first and second pain in humans. Both of these projects will be described from the vantage point of computer science, i.e., the systems analysis, software systems, and data base design will be emphasized.

COMPUTER SYSTEM

The current in-house installation has a Honeywell 316 and a Honeywell 516. These are configured into a dual processor arrangement with special multiplexed I/O channels and controllers in order to dynamically share a disk storage complex and to share peripheral equipment such as the card reader, line printer, and magnetic tape drive. This resource sharing concept has proven quite effective in terms of throughput, economics, and system availability.

The 516 is a 32k, 16-bit system with a 960 nanosecond cycle time. It has a 32 channel, 12-bit, multiplexer ADC, a DMC option, a mechanism for digital input and output (the Process Interface Controller, or PIC), high speed arithmetic option, and a 60 Hz real time clock (non-programmable). It does not support a floating point hardware option. Multiple teletype ports (10 cps) are supported as well as 1200 baud lines for interactive graphics terminals. The 516 is dedicated to data acquisition for real-time experiments and graphical analysis of data.

The 316 is a 32k, 16-bit machine with a 1.6 microsecond cycle time. It is not used for data acquisition except through a high speed paper tape reader (300 cps). It is used for program development and larger processing programs which are not suited for the 516 due to core limitations and the time constraints of real time programs.

The disk complex includes a moving head disk and a fixed head disk which combine to provide 4.5 million words of online storage capacity. Directories are kept on the fixed head disk since its average latency of 8.5

milliseconds (maximum of 17.5) is considerably less than for the moving head disk which has an average seek time of 87.5 milliseconds plus an average latency of about 12.5 milliseconds. At the user software level, the disk complex is simply one I/O unit since the operating system driver decides which disk(s) to use. The hardware switching mechanism and the software interleaving algorithms have been described in a separate paper (1).

Both systems run under a modified version of the OLERT (On-Line Executive for Real Time) operating system. This operating system requires roughly 16k in the 516 which leaves 16k for user programs. A variety of experiments can share this remainder since multi-programming is supported (2-5). Since the remaining core size is not large, all acquisition programs use a dynamic overlay technique (6-8) in order to reduce individual program size. This is even more critical since paging (9-10), swapping (11-12), reentrancy (13-14), position independent code (15), and other modern techniques are not supported by the hardware and software. The real-time requirements of the system obviously affect the ability to use certain of these concepts and the core residency requirements hamper efforts to provide adequate memory management.

In addition, CPU efficiency is enhanced by using multi-tasking concepts (16-18) within individual programs. Our real-time FORTRAN language supports enhancements for multi-tasking primitives (19), but these have not proven totally sufficient due to the inability to schedule various tasks within the same program at different priority levels. Priority levels for different programs can be different but there was no general mechanism to support varying priorities for different tasks within the same program. This problem was partially solved by developing a semaphore (20-22) mechanism for inter-process communication.

For future statements, it is important to realize that our system philosophy is to support multiple real-time experiments concurrently. Thus, there is the obvious requirement to develop software that is bug free and that will not degrade the ability of the operating system to handle other distinct software systems. Since the system runs 24 hours a day, 7 days a week with a crash on the average of once every 4 weeks, it is felt that this mechanism has worked well.

HARDWARE INTERFACE

The two software systems previously mentioned use the same hardware interface which was designed and fabricated internally.

To the computer software, this represents two priority interrupt lines (PIL's), two digital output lines, and one digital input line. The digital input and output goes through the PIC as mentioned earlier.

The signals are defined as:

PIL	RANK	FUNCTION
5	18	Trainer Buttons
7	20	Abnormal Stop Button

PIC ADDRESS	FUNCTION
4103	Digital input for PIL 5 to decode press or release of Trainer Buttons.
4105	Digital Output to turn training lights on or off and to initiate reward for correct responses
4167	Digital Output for controlling temperature of the contact thermode and for indicating stimulus beginning and termination
4105	Bit 16 - Light 1 15 - Light 2 14 - Light 3 13 - Reward
4167	Bit 7 - stimulus on/off (1=off) 8-16 - encoded temperature value
4103	Bit 16 = 1 when Button 1 is pressed 15 = 1 when Button 2 is pressed 14 = 1 when Button 3 is pressed value of 0 implies button release (only one button can be pressed at a time)

The digital output (4167) which controls the temperature of the thermode uses a 10-bit DAC; however, only 9 bits are used in actuality. The thermode was also designed and built internally. It is very linear as can be seen from the current calibration table:

CALIBRATION CHART

Value	Temperature	DELTA PIC
0	20.0 C.	--
65	25.0 C.	65
132	30.0 C.	67
197	35.0 C.	65
262	40.0 C.	65
328	45.0 C.	66
392	50.0 C.	64
454	55.0 C.	62
511	59.5 C.	57

When this data was fitted with a linear regression (23) on the function $f(x) = a*x + b$, the following results were obtained:

TERM	COEFFICIENT	STD. DEV.
b	-257.593	2.00647
a	12.9606	4.78269 E-2

Overall Standard Deviation = 1.84008
Correlation Coefficient = .999952

X	Y	Y-EST.	RESIDUAL
20	0	1.619	-1.619
25	65	66.422	-1.422
30	132	131.225	.774
35	197	196.028	.971
40	262	260.831	1.168
45	328	325.634	2.365
50	392	390.437	1.562
55	455	455.240	-1.240
59.5	511	513.563	-2.562

This linearity allows the software to use a linear interpolation scheme using this calibration chart to determine temperatures which are not exactly defined.

This scheme has worked very well. The inverse of the regression line could also have been used, but the linear interpolation scheme is clearly a better approach since it removes the effect of the higher degree terms which are more non-linear and which are rarely used in the current experiments.

The digital input from the training button is used to differentiate button presses or releases. At present, only one button is used but future paradigms will very possibly use up to three buttons. The abnormal stop button is used to handle unusual error conditions such as hardware malfunction or uncooperative monkeys who refuse to "work". Thus this stop button is basically a safety mechanism, but it can be used to alter parameters during an experiment if necessary. Presently, it is used to stop an experimental session in order to readjust the position of the contact thermode as needed.

EXPERIMENTAL SET-UP

A specially designed training chair was built to hold the monkey's head in a fixed position. This has been described elsewhere (24). The monkey, who has been previously trained according to specified paradigms, is placed in this chair. A contact thermode is attached to the chair and positioned to contact the monkey's upper hairy lip. The temperature of this contact thermode is controlled by the software but the contact thermode slope is determined by the experimenter. A tube located near the monkey's mouth allows a water or grape juice reward to be given to the monkey. The flow of liquid can be controlled by the software or manually. In front of the monkey is an illuminated button. The light associated with the button is controlled by the software. all button presses are monitored by the software.

Let us define a trial as the generation of one stimulus (change in thermode temperature) and the monkey's response to that stimulus. We shall define a set as a collection of trials. After the monkey is placed in the chair (and the software initialized), the first trial is initiated when the button is depressed. Depending on the paradigm (to be defined), the contact thermode temperature will eventually change. The monkey responds to this change by releasing the button. If he releases in a user-defined time window, he has responded correctly and is given a reward. The software can "run" a monkey through a series of sets automatically which can save many hours of researcher time per day.

After several sets are run on a given day, the data can be analyzed to see if parameters for "correctness" or "lock-out" need to be changed for the next day's work. Furthermore, data over several monkeys and many sets, in some cases representing a year or more of data, is added to existing data bases for additional analysis on a more long range basis. The software for this analysis will be described shortly.

Before the paradigms are described, we need to define some terminology:

MONKEY RESPONSE	MEANING
1. Reward	This is a correct response which means that the monkey responded to the stimulus within the user-defined time window.
2. Miss	This is a miss which means that the monkey responded to the stimulus after the correct response time window had terminated.
3. Early Release (TCA)	The monkey has released the (TCA) button before the stimulus was applied.
4. Catch	This was a catch trial. See discussion below.
5. Early Release (TCB)	The monkey has released (TCB) the button after the stimulus was applied but presumably before the change is suprathreshold.

TRIAL TYPES

1. Warming	The stimulus is one of increasing the thermode

	temperature to a non-noxious value.
2. Cooling	The stimulus is one of decreasing the thermode temperature to a non-noxious value.
3. Pain Critical	The stimulus is a large temperature shift into the noxious heat range. ($\geq$ 45 deg. C.)
4. Pain	The stimulus is a large, but non-noxious temperature shift (< 45 deg. C.).
5. Catch	This is a trial in which no stimulus is given and is used to detect possible monkey "timing" or anticipatory response strategies.

SET TYPES

1. Warming	Trial types of warming or catch are used.
2. Cooling	Trial types of cooling or catch are used.
3. Pain	Trial types of warming, pain, pain critical, and catch are used.

The mechanism for how the software knows the temperature sequence and how it determines monkey response is best explained after the paradigms are defined.

PARADIGM DEFINITIONS

The paradigm is the behavioral model that is used to analyze the monkey's responses. It is composed of timing cycles or phases, each of which has some time duration and functional meaning. There are currently two paradigms: onset and offset.

ONSET PARADIGM

The onset paradigm is used to study responses to the beginning of a stimulus. It has the following phases:

Timing Cycle	Meaning
FFA	Wait for monkey to initiate next trial
TCA	Trial has been initiated through the monkey's effort of depressing the trainer button.
TCB	The "lock-out" period defined as the period following the onset of stimulation in which the stimulus is presumably sub-threshold.
TCD	The reward period which is the phase during which the monkey will obtain a reward if he releases the button (i.e. denotes that he detects the stimulus).
TCE	This is the post stimulus response time (or post reward period).
TCF	This is the end of trial wait period. It can be short (TCFS) or long (TCFL).
TCG	This is a timing cycle used to allow all tasks within the program to finish prior to initiating the next trial.

When the monkey is ready to start the next trial, he depresses the button on the front panel of his chair. This initiates timing cycle TCA which is a random period of time between 2 and 8 seconds. The randomness of this phase prevents the monkey from timing. At the end of TCA, the stimulus is applied. This might mean increasing or decreasing the thermode base-line temperature by some amount (e.g., 0.4 to 3.0 degrees). The ending of TCA corresponds to the beginning of TCB or the "lock-out" phase. The duration of TCB is adjustable but is usually around 0.5 second. If the monkey releases during TCA or TCB, it is an "early release." The TCD or reward phase is entered at the completion of TCB (unless it is a catch trial). The duration of TCD is adjustable but is usually about 2.5 seconds. If the monkey releases during TCD it is a "correct" response and the monkey is given a reward (e.g., squirt of grape juice). If TCD terminates and the monkey has still not released, then the monkey has "missed" the opportunity for reward on that trial. Following the TCD phase, the TCE phase is used to eliminate stimulus changes in the immediate post-reward period. For exmple, the stimulus is not turned off until the end of TCE. This prevents the monkey from erroneously associating the button release with stimulus termination. If the monkey is still holding the button at the and of TCE, then the program will either terminate the stimulus or keep it on depending on a user-defined option. In the latter case, TCE is conceptually extended to handle this condition. This extended phase is called TCE-X. The TCF phase is, of course, last in line. This phase is always entered while some other phases may be bypassed (e.g., an early release in TCA by-passes TCB, TCD, and TCE). Early releases penalize the monkey by having a long TCF while a reward response will have a short TCF. A release in TCE also has a short TCF under the assumption that penalties should not be levied for honest misses. When TCF finishes, the software turns the button light back on to indicate to the monkey that he can start the next trial.

In the case of a catch trial, the TCB phase remains on until the monkey releases. Thus no reward can be given and none should be. This is another example of the fact that all trials do not use all timing cycles.

Typical values for the various phases are:

Cycle	Duration
FFA	undefined
TCA	2.0 - 8.0

TCB   0.5
TCD   2.5
TCE   2.0
TCFS   15.0
TCFL   35.0

STIMULUS CHART - ONSET

```
Trial      TCA   .  TCB   .  TCD   .  TCE   .  (TCE-X) .  TCF
                  .        .        .        .         .
                  _____
                 E.R.  |  E.R.     C.R.     MISS      MISS    |  MISS
Normal   _____|     .        .        .         .      |_____BASELINE
Warm               .        .        .        .         .
```

```
Trial      TCA   .  TCB   .  TCD   .  TCE   .  (TCE-X) .  TCF
                  .        .        .        .         .
Normal   _._____.        .        .        .         _____BASELINE
Cocl              |                                        |
                 E.R.  |  E.R.     C.R.     MISS      MISS    |  MISS
                  |_____|
                  .        .        .        .         .
```

```
Trial      TCA   .  TCB   .  TCF
                  .        .
Catch    _E.R._____CATCH_____
                  .        .
                  .        .
```

Note in the normal trials, the stimulus will be terminated in TCB if the monkey early releases. Furthermore the stimulus will be terminated at the end of TCE (and TCE-X not entered) if the monkey releases in TCD (correct) or TCE (miss). Also, if the monkey is still holding at the end of TCE and the user does not wish to use TCE-X, then the monkey will necessarily release in

TCF. This latter condition creates a penalty because TCF will be reinitiated when he finally does release.

OFFSET PARADIGM

The offset paradigm is considerably more complicated than the onset paradigm. This paradigm is used to study responses to the termination of a stimulus. It has the following timing cycles or phases.

Timing Cycle	Interpretation
FFA	Wait for monkey to start next trial by depressing the button
TCA	This is the stimulus time in this paradigm. The stimulus is turned on at the start of TCA and turned off at the end
TCB	This is the lock-out period for the offset case. The stimulus is terminated at the beginning of TCB, but the change in stimulus during this phase is presumably subthreshold.
TCD	This is the reward period. If the monkey releases during this phase, he gets a reward because of his correct response.
TCF	This is the inter-trial wait time which can again be either short (TCFS) or long (TCFL).
TCG	Same as onset

As before, a release in TCA or TCB is called an early release whereas a release in TCF is called a miss. Since the stimulus is off in TCD and therefore the monkey is responding to this off-condition, there is no need of a TCE phase as in the onset case. As in the onset case, if the monkey depresses or releases the button in TCF, then the entire TCF is restarted with the obvious result that the inter- trial period is lengthened. Thus the monkey is penalized for "playing around" or guessing.

As in onset, there is a catch trial sequence. In this case, the stimulus is not turned off as in the normal case. When the monkey releases, the software turns the stimulus off.

There is an additional complexity in the TCA phase when the monkey early releases. The software then checks to see if the stimulus is greater than a user-defined pain threshold. If it is greater than this threshold, the program turns off the stimulus immediately. If it is not greater than this threshold, the software checks another user-defined switch to see if it should turn off the stimulus or keep the stimulus on for the remainder of the TCA phase. This last "flag" allows for the software to help a monkey, accustomed to the onset paradigm, to adjust to the offset paradigm.

The normal offset trial corresponds pictorially to the ONSET normal cooling trial except that the base-line is the lower line rather than the upper. Obviously the catch trial is different and is depicted below:

```
     Trial       TCA   .   TCB   .   TCF
                        .       .
     Catch       _E.R._____CATCH___
                        .       |
                        .       |_____ Baseline
```

PARADIGM PARAMETERS

Timing/Temperature Tables

At this point, it is appropriate to discuss the parameters which the software uses to implement the complexities of these paradigms in the real-time situation. Some of the parameters are for actual control and some are of cosmetic value only. One multi-dimensional parameter is the calibration chart alluded to before. This table is stored on disk and is read into the program as required. This table can be edited very easily; thus, it can be changed without any software repercussions.

There are two basic features to the paradigms: times and temperatures. This extends to the development of "timing tables" and "temperature tables" which are established by individual programs prior to actually running a set. Many of these tables can be established and they can be edited in order to change items. Typically they are created in pairs, since the actual time durations are, of necessity, correlated with the temperature steps or deltas.

A timing table has the following parameters:

PARAMETER	MEANING
Base Temperature	Non-stimulus temperature level of the thermode.
Number of Trials	Number of trials in a set (maximum of 150)
Continue Stimulus Flag	In onset, if the monkey is still holding at the end of TCE, should the software terminate the stimulus or wait for the monkey to release before terminating.
TCA	Minimum time duration for TCA normally 2.0 seconds)
TCB	Lock-out time duration
TCD	Time window for correct response. (secs.)
TCE	Length of this phase (secs.)
TCFS	Short TCF length (secs.)
TCFL	Long TCF length (secs.)
Catch Temperature	For offset, the temperature to be used for catch trials in absolute form.
TCA fixed	For pain-critical, offset trials, this phase is a period of time fixed in length.

Pain-Critical Temperature	The temperature above which a trial is called pain critical rather than pain (in absolute form).
Number of Different	In offset, the TCB value depends on the temperature
Set Type	The type of set for which this timing table is designed
Abort or continue	The flag for the early release problem discussed in the offset TCA phase if an early release occurs on a non-pain critical trial.
Delta/TCB pairs	For offset, the TCB time durations for the various delta temperature steps
Threshold	Temperature above which we use TCFS rather than TCFL for early releases in TCA and TCB for offset, non- catch trials.

A temperature table is simply a list of temperatures for the sequence of trials in a set. This temperature list is another parameter to the program. Temperatures whose absolute value is between 0 and 20 are called deltas or delta temperatures while those between 20 and 60 are called absolute temperatures. Numbers in a temperature table have the following interpretation:

Paradigm	Value	Meaning
ONSET	$x = 0$	catch trial
ONSET	$0 < x < 20$	warming delta
ONSET	$-20 < x < 0$	cooling delta
ONSET	$20 <= x < 60$	absolute temperature
ONSET		warming or cooling
OFFSET	$x = 0$	catch trial
OFFSET	$0 < x < 20$	cooling trial
OFFSET	$20 < x <= P.C.T$	pain trial
OFFSET	$P.C.T. < X < 60$	pain critical trial

where P.C.T. = Pain Critical Temperature Parameter (in the timing table)

Thus the following diagram indicates the relationships discussed thus far:

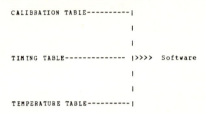

There are two other items to discuss: user input at the time the program is run and program output. The user defines the following items when he actually runs the program.

ID	Monkey identification
SET	Set number
TIMING TABLES	List of timing tables
TEMPERATURE TABLES	List of temperature tables
THERMODE SLOPE	User-adjusted thermode slope
COMMENTS	Textual information

Notice that a list of parameter tables is entered. Thus the program can run through several sets automatically which results in saving a large amount of researcher time. The cosmetic information is saved for the analysis software which will be described later.

As the program is running, it outputs the information it monitors on a trial by trial basis. The output mechanism is a thermal teletype in order to reduce noise which might distract the monkey. A sample of the programs output is shown is figure 1. The following columns appear:

Column	Meaning
TRIAL	Sequential trial number
TYPE	Type of trial WA = warming CO = cooling PA = pain PC = pain critical CA = catch

MR	Monkey Response
	    EA = TCA early release
	    EB = TCB early release
	    RE = Reward
	    MI = Miss
	    CA = Catch

BASE	Base temperature
DELTA	Change in temperature
LATENCY	Response time in seconds
TC	Timing Cycle of release A = TCA , etc.
HOLD	Hold time in seconds
PLAY	Number of times monkey releases after his first release in a trial but before the start of the next trial.
RAN/A	Random part of TCA in seconds.
TCB	Length of TCB in clock tics, e.g., 30 = 0.5 seconds.

TCF	Which type of TCF
	    FS = TCFS
	    FL = TCFL

For each trial, the program saves the following information in an output disk file:

    Trial Number
    Monkey Response
    Trial Type
    Stimulus Temperature
    Latency
    Release Timing Cycle
    TCA Length
    TCB Length

This information is augmented by a table header in which cosmetic information such as time, data, and comments are

TRIAL	TYPE	MR	BASE	DELTA	LATENCY	TC	HOLD	PLAY	RAM/A	TCB	TCF
1	CO	EA	35.0	2.0	0.00	A	11.13	0	2.18	30	FL
2	CO	RE	35.0	6.0	1.13	D	6.85	1	3.72	42	FS
3	CO	RE	35.0	2.0	0.68	D	8.58	0	5.90	30	FS
4	CO	RE	35.0	4.0	0.65	D	5.75	0	3.10	36	FS
5	CO	MI	35.0	2.0	6.98	F	9.20	0	0.22	30	FS
6	CO	RE	35.0	6.0	2.78	D	5.40	0	0.62	42	FS
7	CO	RE	35.0	4.0	0.72	D	2.88	0	0.17	36	FS
8	CO	RE	35.0	4.0	0.63	D	7.90	0	5.27	36	FS
9	CO	EA	35.0	2.0	0.00	A	1.88	0	2.75	30	FL
10	CO	RE	35.0	6.0	0.77	D	4.78	0	2.02	42	FS
11	CO	RE	35.0	2.0	0.57	D	7.52	0	4.95	30	FS
12	CO	RE	35.0	6.0	0.72	D	8.63	0	5.92	42	FS
13	CO	RE	35.0	4.0	0.62	D	6.40	0	3.78	36	FS
14	CO	RE	35.0	6.0	1.75	D	5.68	0	1.93	42	FS
15	CO	EA	35.0	2.0	0.00	A	3.47	0	2.23	30	FL
16	CO	RE	35.0	4.0	0.67	D	7.73	0	5.07	36	FS
17	CO	RE	35.0	2.0	0.55	D	5.87	0	3.32	30	FS
18	CO	RE	35.0	4.0	0.75	D	3.03	0	0.33	36	FS
19	CO	RE	35.0	4.0	0.68	D	6.70	0	4.02	36	FS
20	CO	RE	35.0	6.0	0.33	D	5.57	0	2.73	42	FS
21	CA	CA	35.0	4.0	8.10	B	11.45	0	1.35	1800	FS

Figure 1. Sample logging output.

maintained. This output table is then used for further analysis.

The updated diagramatic representation is thus:

The latency time is defined as the amount of time from the start of TCD until the monkey releases the button. Response time is latency time plus the lock-out time. Hold time is response time plus the length of TCA. These parameters are calculated by using the real-time clock of the computer and by monitoring every aspect of the monkey's action. Since the clock has a 60 Hz. frequency, this limits the resolution of such calculation to 16.67 milliseconds. This was deemed adequate for the work. Furthermore, it was felt that a 10 millisecond response was needed to turn off the stimulus when the monkey releases and that a 100 millisecond response was needed to generate a reward pulse for correct responses. These constraints represented no problem. The unknown was how fast the monkey could press and release the button. It was felt that he could not perform the two events in less than 5 milliseconds. Since we operate a multiprogramming system with many interrupts, this time is not large in comparison to an interrupt service time of 750 microseconds. Therefore the software monitors events and detects when it misses them. This has aided debugging and represents a significant point of the philosophy which the software uses, namely, to try to detect as many error conditions as possible. Currently, seventeen such error conditions are monitored. Errors rarely appear, but when they do, mechanisms exist to handle them. This allows the software perforace to exhibit the characteristic of "graceful degradation".

DAILY PROCESSING

The output table from any given set can be used as input to other programs to draw histograms and to analyze the data statistically. The histogram programs allow the analysis of hold times on early releases, latencies of correct responses and early releases (in TCB), and latencies of correct and early release response in non-critical trials in the offset paradigm.

The statistical analysis is used to obtain counts, means, medians, ranges, grand means, and percentages on a variety of different data stratifications based on type of response, delta temperatures, response times, etc. Examples appear in figures 2 and 3.

On a weekly or as needed basis, these output tables are dumped to magnetic tape and then added to data bases on a DEC-SYSTEM 10 computer facility which is available through the Division of Computer Research and Technology (DCRT) at the NIH. These data bases will continue to grow and the design has allowed for this expandability. The DEC SYSTEM-10 was decided upon due to its large online storage capacity and the existence of a sophisticated graphics package which allows clipping, windowing, and blanking (25). The internal system simply did not have sufficient storage capacity to handle these large data bases. Thus the chart on the top of page 496 applies.

LATENCY OF CORRECT TRIALS

DELT=	0.8	MEDIAN=	1.39	MIN=	1.22	MAX=	2.24	N= 7
DELT=	1.6	MEDIAN=	1.04	MIN=	0.94	MAX=	1.10	N= 10
DELT=	3.0	MEDIAN=	1.03	MIN=	0.98	MAX=	1.11	N= 10
DELT=	5.0	MEDIAN=	1.03	MIN=	0.96	MAX=	1.16	N= 10

LATENCY FOR DELT MORE OR EQUAL TO 3

DELT=	0.8	MEDIAN=	0.00	MIN=	0.00	MAX=	0.00	N= 0
DELT=	1.6	MEDIAN=	1.19	MIN=	0.98	MAX=	1.43	N= 20
DELT=	3.0	MEDIAN=	0.00	MIN=	0.00	MAX=	0.00	N= 0
DELT=	5.0	MEDIAN=	0.00	MIN=	0.00	MAX=	0.00	N= 0

LATENCY FOR DELT LESS THAN 3

DELT=	0.8	MEDIAN=	1.39	MIN=	1.22	MAX=	2.24	N= 7
DELT=	1.6	MEDIAN=	1.04	MIN=	0.94	MAX=	1.10	N= 10
DELT=	3.0	MEDIAN=	1.03	MIN=	0.98	MAX=	1.11	N= 10
DELT=	5.0	MEDIAN=	1.03	MIN=	0.96	MAX=	1.16	N= 10

MED AND RANGE LAT FOR CORRECT & MISSED TRIALS

DELT=	0.8	MED=	2.140	(1.91,	2.37)	MIN=	1.22	MAX= 3.01	N= 10
DELT=	1.6	MED=	1.045	(1.04,	1.05)	MIN=	0.94	MAX= 1.10	N= 10
DELT=	3.0	MED=	1.035	(1.03,	1.04)	MIN=	0.98	MAX= 1.11	N= 10
DELT=	5.0	MED=	1.035	(1.02,	1.05)	MIN=	0.96	MAX= 1.16	N= 10

NO. TRLS	NORMAL	CATCHES	MISSES	REWARDS	EARLY R.
63	60	3	3	57	0

Figure 2. Daily processing for onset.

TRIALS	WARM	COLD	PAIN	PAIN-CRT
78	0	37	13	25

	NONCRITICAL	CRITICAL
CORRCET	46	20
MISS	1	0
EARLY R.	3	5
CATCH	3	0

PERCENT OF EARLY RELEASES FOR NON-CRITICAL TRIALS AT TEMP 2.0 = 0.0
PERCENT OF EARLY RELEASES FOR NON-CRITICAL TRIALS AT TEMP 4.0 = 11.1
PERCENT OF EARLY RELEASES FOR NON-CRITICAL TRIALS AT TEMP4 1.0 = 0.0
PERCENT OF EARLY RELEASES FOR NON-CRITICAL TRIALS AT TEMP4 3.0 = 16.7

PERCENT OF EARLY RELEASES FOR CRITICAL TRIALS AT TEMP 45.0 = 0.0
PERCENT OF EARLY RELEASES FOR CRITICAL TRIALS AT TEMP 47.0 = 0.0
PERCENT OF EARLY RELEASES FOR CRITICAL TRIALS AT TEMP 49.0 = 33.3
PERCENT OF EARLY RELEASES FOR CRITICAL TRIALS AT TEMP 51.0 = 37.5

Figure 3. Daily processing for offset.

EARLY RELEASES FOR PAIN CRITICAL TRIALS

TEMP	TCA	TCB
45.0	0	0
47.0	0	0
49.0	1	1
51.0	1	2

MEDIAN AND RANGE OF LATENCIES FOR ALL CORRECT TRIALS

TEMP	MEDIAN	INTERP	MIN	MAX	N
2.0	1.020	0.00-0.00	0.77	2.72	19
4.0	1.150	0.00-0.00	0.91	1.62	15
41.0	1.490	0.00-0.00	1.03	1.78	7
43.0	1.730	0.00-0.00	1.47	1.89	5
45.0	1.945	1.94-1.95	1.80	2.07	8
47.0	2.2120	0.00-0.00	2.02	2.21	3
49.0	2.690	2.55-2.83	1.95	2.89	4
51.0	2.660	0.00-0.00	2.34	3.01	5

MEDIAN AND RANGE OF LATENCIES FOR ALL CORRECT AND MISS TRIALS

TEMP	MEDIAN	INTERP	MIN	MAX	N
2.0	1.020	0.00-0.00	0.77	2.72	19
4.0	1.160	1.15-1.17	0.91	3.80	16
41.0	1.490	0.00-0.00	1.03	1.78	7
43.0	1.730	0.00-0.00	1.47	1.89	5
45.0	1.945	1.94-1.95	1.80	2.07	8
47.0	2.2120	0.00-0.00	2.02	2.21	3
49.0	2.690	2.55-2.83	1.95	2.89	4
51.0	2.660	0.00-0.00	2.34	3.01	5

MEDIAN AND RANGE OF PAIN CRITICAL EARLY R.

TEMP	MEDIAN	MIN	MAX	N
45.0	0.000	0.00	0.00	0
47.0	0.000	0.00	0.00	0
49.0	4.200	3.20	3.20	2
51.0	6.500	4.50	6.70	3

MEDIAN AND RANGE OF CRITICAL HOLD TIMES FOR COR. AND E.R.

TEMP	MEDIAN	INTERP	MIN	MAX	N
45.0	6.900	6.90-6.90	6.80	7.00	8
47.0	7.100	0.00-0.00	7.00	7.20	3
49.0	7.200	6.90-7.50	3.20	7.90	6
51.0	7.400	7.30-7.50	4.50	8.00	8

Figure 3. (*continued*) Daily processing for offset.

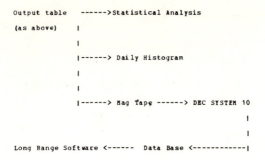

These data bases can then be further analyzed in order to compare sets and to detect trends. The program that performs this phase of the analysis is rather complicated since it allows for a variety of analyses with exclusions and arbitrary data combinations. A histogram facility is included. This aspect will be covered after the data base design is described.

The magnetic tapes which are created as described above, are used to create or extend data bases. The software which reads the tape can be used to pick out certain records or selected files from his tape. It can also be used associatively in order to pick out data based on the contents of the data. The following parameters can be used to select out portions of a magnetic tape:

Name	Abbreviation	Meaning
First File	FF	first file on tape to process
Last File	LF	last file on tape to process
First Record	FR	first record in the first file to use
Last Record	LR	last record in the last file to use
Search ID	ID	the desired monkey
Search Set	SET	the desired set for the given ID

Base Temp.	BASE	desired baseline temperature.
Thermode Slope	SLOPE	desired thermode slope
Type	TYPE	desired set type

The program uses the following defaults:

FF	2
LF	E.O.T.
FR	1
LR	E.O.F.
ID	all
SET	all
BASE	all
SLOPE	all
TYPE	all

The default for the first file is 2 rather than 1 since the first file on the tape is a header and does not contain actual data. These data bases are placed on the DEC SYSTEM 10 disk. This mechanism has proved exceedingly convenient and easy to use.

The data bases are created on a monkey by monkey basis, i.e. each monkey has its own data base. Furthermore, each monkey has a separate data base for each type of paradigm (currently ONSET and OFFSET).

Each data base is composed of logical (and physical) records which are 128 words in length. The file (i.e. data base) has a directory for all sets in the file which allows sets to be accessed directly. Further, there are four threads or links which are used to aid the software in searching and pattern recognition activities. One thread links all "set header" records in a one way linked list. The other thread links all sets in the same "class" where a class is a collection of sets with the same set type, base temperature, and thermode slope. The third thread is a two way chain for all "table of contents" records which contain cosmetic information as well as the "class blocks". The last thread is a pointer from a set header record to the first trial data record of that set.

The first logical record of the file is the first "table of contents" record. Records 2 - 9 are initially reserved for the set directory. Records 10 through the current size contain set header records, data records, or additional table of contents records.

The format of record one or the first table of contents (T.O.C.) record is as follows:

WORD	CONTENTS
1	Number of records in file
2	Backward Link (= 0)
3	Forward Link to next T.O.C.
4	Set Link Word
5	Number of sets in file
6	Number of Classes in file
7 - 28	Unused (= 0)
29 - 33	Class block 1
34 - 38	Class block 2
39 - 43	Class block 3
44 - 48	Class block 4
49 - 53	Class block 5
54 - 58	Class block 6
59 - 63	Class block 7
64 - 68	Class block 8
69 - 73	Class block 9
74 - 78	Class block 10
79 - 83	Class block 11
84 - 88	Class block 12
89 - 93	Class block 13
94 - 98	Class block 14
99 - 103	Class block 15
104 - 108	Class block 16
109 - 113	Class block 17
114 - 118	Class block 18
119 - 123	Class block 19
124 - 128	Class block 20

The format of a class block is:

Relative word		
	1	Set type (integer)
	2	Base temp. (real)
	3	Slope (real)
	4	Class link (integer)
	5	# sets in this class (integer)

All links are record numbers. Thus word 4 contains the record number of a set header record. Word 3 contains the record number of the next table of contents record and relative word 4 of a class block points at the first set header record in that particular class. A set is in one, and only one, class.

The directory records are reserved initially to allow for 1000 sets. This can be expanded if needed. Each word in the record contains the record number of a set header record. Only the first 125 words (of 128) are actually used. The remaining 3 words are reserved for possible expansion.

The "additional" T.O.C. records, i.e. other than the first, have the following format.

WORD	MEANING
1	Number of classes in this record
2	Backward T.O.C. link
3	Forward T.O.C. link
24 - 128	Class block information in 5 word groups

A set header record has the following set-up.

WORD	CONTENTS
1	set number
2	month
3	day
4	year
5	number of trials
6	set type
7	base temperature (real)
8	slope (real)
9	class link field
10	TCA length (real)
11	TCB length (real)
12	TCD length (real)
13 - 23	Comments (onset)
13 - 16	Comments (offset)
17 - 23	Zero - unused (offset)
24 - 33	TCB array (offset)
24 - 33	Zero - unused (onset)
34	set link field
35	trial data link field
36 - 128	Zero - unused

The trial data, at which the set header record points, is the compressed data from the tape. It is stored in consecutive physical blocks with monkey responses first, then delta values, then latency values, and finally hold times. The number of values in each area is equal to the number of trials in the set.

This design has proven to be very useful. The inverted orientation allowed by the class block scheme is very useful for statistically comparing sets with the same characteristics. Furthermore, it is clear that the design can be extended to handle any number of sets and any number of classes.

The long range program is run in a time-sharing mode, as opposed to batch, which allows the researcher to set the results immediately. This is very inconvenient because it reduces the delay to obtain answers to questions and it allows previous calculations to help the researcher determine what to do next. The existing statistical packages, such as BMD, SPSS, SAS, and PSTAT (26), are batch oriented and were thus not appealing to our interactive design. Furthermore the necessity for histograms on a CRT also disallowed selection of these packages.

The software is command oriented, i.e. one tells the program what to do on a step-by-step basis. The program commands are:

PARA	<word>	
	ON	for ONSET
	OFF	for OFFSET
ID	<2-chars>	
		for monkey ID
SETS	Forms 2 and 3 can be combined by separating them with semicolons, e.g., 100-103; 107/20 33;203-205	
	CLEAR	to scratch previous requests
	#1 - #2	process sets 1 through 2 e.g., 100-199
	#1 / #	list process set 1 and exclude the trial(s) in # list e.g., 100/1 9 27

CROSS   <word>

This parameter pertains to the question, "Should the calculations for a set be done if the block size is smaller than normal?" By using YES and an appropriate BLKSIZE, sets can be combined for calculation purposes.

YES   implies calculations will only be done when blocks are filled which allows for sets to be combined.

NO   implies trials from one set will not be calculated with trials from another set.

TRANS   <word>

NONE   implies no transformations
LOG   implies log transforZation
REC   implies reciprocal transformation
ALL   implies both LOG and REC

CALCULATE   <what-list>

A list of words are typed to determine which calculations are performed. This list can be one or more of the following:

MED   median and mean medians
MEAN   means and grand means
PER   counts and percents
RANDOM   foreperiod/stimulus duration
HOLD   means and medians of hold time of early releases and catch trials.
ALL   all of the above

RESPONSE   <what-list>

A list of words are typed to indicate which responses are to be included in the calculations. This list can be one or more of the following:

CR	Use correct responses only for calculating means and medians.
CM	Use correct responses and misses for calcuating means and medians.
ER	Use only early releases for hold time calculations
CT	Use only catches for hold calculations.
ALL	Same as CM ER CT

BLKSIZE   <number> or <word>

This command is used to set the block size length for calculations

SET   implies grouping by set
ALL   implies one group
(CROSS should be YES)

When this number of trials has occurred for a particular delta bin, then the requested calculations are performed.

At the end of all sets, the means for the blocks are performed. When the mean of medians is calculated, only those of the blocksize will be included in the calculation.

HOLDBLKSIZE   <number> or <word>

Same as for BLKSIZE except this grouping is for the hold calculation

EXDELT   <relation> <number>

This command is used to divide trials into two groups depending on whether the previous delta falls above or below the specified number. It is used to separate filler and non-filler trials. The relations are:

GT	greater than
GE	greater than or equal
EQ	equal
LT	less than
LE	less than or equal

EXHTIME   <relation> <number>

This command is used for dividing trials into two groups according to hold time of the trial. See discussion for EXDELT above.

USEDELTAS   <word> or <number-list>

This command is used to indicate which delta values (in the specified sets) are to be used. Furthermore, deltas can be collapsed into larger groups for calculation purposes.

     ALL   implies that each delta will be in a separate bin

     #   implies that this one delta will be in a separate bin by itself

     (#1 #2...)   all delta listed will be collapsed into one bin

     #1/#2   trials with delta in this range will be analyzed in one bin

     #1-#2   delta in this range will be analyzed in separate bins.

If we type the following: USEDELTAS .2 (.2 .6 1.6) 1.6/3.0 1.6-3.0 then the following bins will be analyzed:

Bin

1	all trials with delta = .2
2	all trials with deltas = .2, .6, or 1.6
3	all trials with deltas between 1.6 and 3.0
4	bin for delta 1.6
5	bins for deltas between 1.6 and 3.0
.	
.	
.	bin for delta = 3.0

DEL   <number>

This command is used to delete the indicated set from subsequent calculations.

COMBINE   <number1> <number2>

This command is used to specify which 1 second foreperiods should be combined and analyzed together

DUM   <number1>-<number2>

This command dumps the class criteria for the indicated range of sets.

GO   Indicates that all parameters are set and the program can begin the calculations

KEEP   This command creates a disk file of the raw trial data. This file can then be looked at with the text editor.

STOP   This terminates the program.

The program remembers all parameters and reuses them to alleviate the problem of the researcher having to reenter them prior to every calculation.

The usage of SAIL, Stanford Artificial Intelligence Language, greatly reduced the development time of the long range software due to the presence of syntactic constructs which allowed structured programming techniques to be used (27). SAIL is an enhanced version of ALGOL-60 which allows IF-THEN-ELSE, CASE, DO-WHILE, and DO-UNTIL statements, among others, to be utilized. Furthermore, the existence of more core eliminated the need to develop program overlays. This contribution was also a significant aid in program development.

The algorithms used by the software are fairly straightforward. Formulae for means, standard deviations, and standard error are the unbiased versions. The program performs quite a lot of searching activity. This is aided by the data base design. In addition, the binary tree search technique(28) is also used to reduce core search time from an average of N/2 accesses to log N. This gain is not insignificant due to the size of some of the data arrays which are used for complex calculations over many sets.

Perhaps the biggest time saving comes in calculating medians. The typical approach is to sort the entire array and take the middle element (or the average of the middle two if the size is even). A vastly improved technique is to use the quick-sort algorithm and utilize the enhancement suggested by Knuth (29) to allow finding the element in a specified cell WITHOUT sorting the entire array. This procedure runs about thirty to fifty times as fast as the "old" bubble sort and is especially effective for large arrays in which the elements are in random order. It is less effective if the list is partially sorted but is still enormously better than bubble sort or shell sort. This latter problem can be alleviated by using a median estimate other than the first component.

The histogram facility provided by this software is fairly typical in that it allows user or automatic scaling, user-defined or automatic bin widths, and user-defined or automatic graph type (i.e. bars or line graph). In addition, a mechanism has been established to generate copies of the histogram on a CALCOMP plotter for higher quality needs or to generate plots that are to be included in a paper. This addition is expected to be heavily utilized.

PAIN SOFTWARE

In an effort to facilitate the efforts of researchers who were interested in studying the effects of noxious heat pulses on first and second pain, a program was developed to generate a pulse train with the following characteristics:

The various phases are described as:

PHASE	MEANING
1	Base-line temperature established
2	Optional conditioning stimulus can be applied
3	Pre-stimulus time
4	Warming part of stimulus
5	Cooling part of stimulus
6	Post-stimulus time

The time duration of each phase is user definable and the stimulus temperature is selectable in the range of 20 - 60 degrees C. The diagram above represents a typical usage although inverted combinations can also be used by appropriate selection of baseline and other temperatures. Researchers are using this software to see if noxious heat pulses can evoke first and second pain (30).

The slope of the temperature change is selected by the researcher at the hardware interface and is thus not under program control.

As in the other software system, the computer's real time clock is used as the mechanism for beginning and terminating the various phases. The 16.67 millisecond resolution of this clock was deemed adequate.

In addition to controlling the pulse parameters, the software also monitors the reaction time of the human subjects who respond to the stimulus. The output force of a dynamometer was used to determine the degree of pain. This aspect was not monitored by the computer. A total of four subjects have been run. Using the ANOVA procedure in SPSS, a 3 way analysis of variance based on subject, inter-pulse interval, and sequential pulse number has indicated that significant reliability in evoking first and second pain can be obtained.

REFERENCES

1. Wood, Wilson, and Brown, *ACM/NBS Fourteenth Annual Technical Symposium* (1975).
2. Wirth, "On Multiprogramming, Machine Cooling, and Computer Organization", CACM 12, 9, (1969).
3. Kleinrock, "A Continuum of Scheduling Algorithms", Proc. AFIPS 36, SJCC, (1970).
4. Lampson, "A Scheduling Philosophy for Multiprocessor Systems", CACM 11, 5, (1968).
5. Coffman and Kleinrock, Computer scheduling methods and their countermeasures", Proc. AFIPS, SJCC, (1968).   6. Pankhurst, "Program Overlay Techniques", CACM 11, 2, (1968).
7. Spacek, "A Survey of Overlay Techniques", University of Maryland Tutorial, CSC 007, (1970).
8. Presser and White, "Linkers and Loaders", Computing Surveys, September (1970).
9. Denning, "Virtual Memory", Computing Surveys 2, 3, (1970).
10. Denning, "The Working Set Model of Program Behavior", CACM 11, 5, (1968).
11. Flores, "Swapping vs. Paging", Modern Data, April, 1970.
12. Mcgee, "On Dynamic Program Relocation", IBM Systems Inc., Vol 4, No. 3, (1965).
13. Watson, *Timesharing System Design Concepts* McGraw-Hill:New York (1970).
14. Wegner, *Programming Languages, Information Structures*

and Machine Organization, McGraw-Hill:New York (1968).
15. Bell and Hewell, *Computer Structures : Readings and Examples*, McGraw-Hill:New York (1971).
16. Dijkstra, "Cooperating Sequential Processes in Programming Languages", ed. Gennys, Nats Advanced Study Institute.
17. Saltzer, "Traffic Control in a Multiplexed Computer System", MIT Thesis, MAC-TR-30.
18. Horning and Randell, "Process Structuring", Computing Surveys 5:1, (1973).
19. Dennis and Van Horn, "Programming Semantics for Multiprogrammed Computation", CACM 9:3, (1966).
20. Dijkstra, "The Structure of the Multiprogramming System", CACM 11:5, (1968).
21. Dijkstra, "Solution of a Problem in Concurrent Programming Control", CACM 8, (1965).
22. Hansen, "Concurrent Programming Concepts", Computing Surveys 5:4, (1973).
23. Draper and Smith, *Applied Regression Analysis*, Wiley:New York, (1966).
24. Dubner, Beitel, and Brown, "A behavioral Animal Model for the Study of Pain Mechanisms in Primates", In *Pain: New Therapeutic Approaches and Frontiers for Immediate Research*, (In Press) ed. M. Weisenberg and B. Tursky Plenum Press, (1976).
25. Newman and Sproull, *¼Principles of Interactive Computer Graphics*, McGraw-Hill:New York (1973).
26. Schucany, Shannon, and Minton, "A Survey of Statistical Packages", 4:2, (1972)
27. Dahl, Dijkstra, and Hoare, *Structured Programming* , Academic Press, (1972).
28. Price, "The Lookup Techniques", Computing Surveys 3:2, (1971).
29. Knuth, *Sorting and Searching*, Addison-Wesley,(1973).
30. Price, Hu, Dubner, and Gracely, "Peripheral and Central Neural Mechanisms That Modify First and Second Pain Evoked By Noxious Heat Pulses. (To be published).

chapter 31

DADTA IV: A COMPUTER BASED VIDEO DISPLAY CONTROL AND DATA COLLECTION SYSTEM FOR BEHAVIORAL TESTING

K. J. DRAKE
Poughkeepsie, New York

K. H. PRIBRAM
Psychology Department
Stanford University
Stanford, California

In the past five years the small, general purpose digital computer has become a common feature in many neurobiology laboratories, but to a considerable extent it has remained a tool of the electrophysiologist. Application of the minicomputer to the problems of the experimental analysis of behavior has been relatively limited and real-time computer systems for stimulus control and data collection are still somewhat uncommon.

The authors are deeply indebted to Jon Glick, who designed and constructed much of the hardware, and to R. Bruce Rule and James Bright, who are reponsible for the extensive library of useful software available. We also wish to thank Dr. Robert Phelps for frequent and timely advice, Reese Cutler for aid in maintaining the hardware, and Dr. Carol Christensen for assisting with the preparation of the manuscript and figures. This work was supported in part by NIMH Grant MH 12970.

Precise, automated control of the contingencies which guide or produce behavior is almost universal in behavioral studies, as is some method for the unequivocal recording of behavioral effects of interest. The control systems employed are generally composed of hard-wired logic modules in combination with cumulative event counters or graphical output devices for the collection and recording of data. This approach is eminently satisfactory for the great variety of operant techniques characterized by a simple logical structure, and where the behaviors of interest are accurately reflected. However, in many discrimination paradigms the fine grain of behavior, such as reaction time, is of interest and the relevant data can only be obtained by detailed analysis on a trial by trial basis. The implementation of the necessary computational capabilities and output facilities with logic modules can become quite complex and it is often difficult to retain the desired degree of flexibility. These limitations and the use of increasingly complex discrimination paradigms provided the original impetus to seek more versatile alternatives to hard-wired logic systems and the rationale remains much the same today.

The computer based video display and control system described in this chapter is a descendent of a special purpose digital processor christened DADTA, for Discrimination Apparatus for Discrete Trial Analysis, designed and constructed some fifteen years ago in this laboratory (Pribram, et al., 1962) and resembles in many respects its immediate predecessor, DADTA III (see Pribram, 1969). The development of DADTA IV epresents an attempt to take advantage of the then current (1970) display and minicomputer technologies to produce a general purpose system for the analysis of behavior. Although our primary applications relate to visual discrimination performance in non-human primates, the intent was to achieve sufficient generality and simplicity that DADTA IV would be useful in other laboratories and easily duplicated.

Moderate cost (about $10,000 for a minimal configuration) was a prime design objective, but wherever possible and cost effective, commercially available system elements were employed. Particular emphasis was placed on developing a flexible method for stimulus presentation and minimizing the complexity of user constructed devices.

DESIGN CONSIDERATIONS

A variety of processors meeting the requirements of the DADTA IV were evaluated and the decision to select the PDP-8/e was based partially on considerations of our own experience with the PDP-8 family, compatibility with other computers in the laboratory, and our library of applications software. However, availability of peripherals, system software, and service facilities also figured heavily in the decision process. For similar reasons DECtape magnetic tape was chosen for mass storage of programs and data.

The selection of a general method for the presentation of visual stimuli presents many problems that are beyond the scope of this discussion, but some of the factors will be briefly outlined.

Probably the most common method of generating patterns for visual discrimination experiments involves the use of one or more individual display readouts. Typically these devices employ rear projection of a film mask onto a single-plane viewing surface and incorporate multiple bulbs permitting programmed selection of a small stimulus set, typically ten to twelve different patterns. If it is necessary to change the stimulus set the entire readout unit or the film mask must be changed -- at best a time consuming process that cannot, of course, be done under program control. Another serious drawback is that the readout is located at a fixed position in space. If a paradigm calls for the presentation of stimuli in a variety of spatial positions, it is necessary to employ multiple readouts. This was the approach taken with DADTA III which employed a 4x4 square array of 12 symbol projection units for display purposes (Pribram, 1969). The multiple readout method requires a fair amount of decoding and power driving hardware, cabling and maintenance.

Sequential or random access 35 mm slide projectors represent another alternative method for programmed display of complex spatial patterns of arbitrary nature, but the unreliability and slowness of the electromechanical components, the need to photographically prepare all stimulus configurations, and the inability to easily create dynamic patterns limit the usefulness of the technique.

It is clear that the ideal device for this application would offer a large surface area on which complex spatial patterns could be generated with good, programmed control over visually important parameters such as luminance, contrast, rise time, and chromaticity. This ideal can be only distantly approached at the present, but a number of CRT and plasma graphics devices exhibit some of these desirable characteristics.

Large screen electrostatically deflected CRTs directly refreshed by the CPU offer good spatial resolution, but are expensive and the generation of even moderately complex patterns places excessive demands on the processor to the point where display flicker becomes a limiting factor. Most self refreshed CRT displays incorporating internal memories are character and text oriented and do not offer the desired degree of flexibility.*

Large screen storage tube displays (e.g., the Tektronix 611) offer economy, freedom from refresh requirements and good graphics capabilities but the stored image has poor luminance and contrast qualities and control over these important parameters is limited.

The use of a scan conversion technique to generate a television video signal from the information written on a special storage CRT was selected for DADTA IV, thus retaining several of the advantages offered by the storage CRT while gaining the desirable contrast and luminance characteristics of a television monitor.

The low cost of the actual display device ($100 - $500 for a typical video monitor) is a distinct advantage because it is economically feasible to locate parallel display monitors in a number of locations for monitoring of the stimulus display during experiments or program development. Additionally the critical and expensive display hardware can be kept out of the harsh environment of the animal testing chamber. Finally the ability to modify the computer originated display by mixing the scan converter output with external video sources (eg. a

---

* Extremely fast, internally refreshed high resolution displays have been built, but the expense is prohibitive for most applications. Continued advances in memory technology will hopefully bring the cost to a reasonable level in the not too distant future.

spatial white noise signal or a masking pattern) proves useful in some applications.

It is recognized that the performance of television raster scan displays is only marginally suitable for many visual experiments because of the limited spatial and temporal resolution. The scan converter employed in DADTA IV can resolve approximately 100 lines vertical and 125 lines horizontal and the "frame" rate is 30 per second. For virtually all behavioral studies and for some types of electrophysiological stimulation these characteristics are acceptable.

SYSTEM ORGANIZATION

Physically the DADTA IV system is constructed in two enclosures.

1) The computer and peripheral rack contains the PDP-8/e processor and internal options, the Tektronix Type 4501 Scan Converter unit, DECtape transports, power supplies and A/D converter input panel.

2) The testing chamber consists of a sound attenuated enclosure for the subject, subject response panels, the video display monitor and reinforcement dispensers. A small interface box on the rear of the chamber contains the connectors that receive cables from the computer rack, the drivers and power supply to control and provide the current for operating testing chamber devices (e.g., chamber lighting and reinforcement dispenser) and the signal conditioning buffers for the subject response switches.

In the usual configuration for discrimination testing, the face of the CRT is overlayed with a 3x3 matrix of transparent plastic disks, each mechanically coupled to a microswitch. Discriminanda are displayed in positions behind the disks and depression of a panel constitutes the subject's response. In effect the video display is being used to simulate a set of nine individual display devices. The overlay is easily removed for paradigms not requiring a response spatially associated with the stimulus or the number and geometry of response regions can be quickly altered by substituting a different switch panel (see figure 1).

The processor enclosure and testing chamber are linked by two flat ribbon cables and a single coaxial

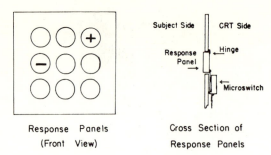

Figure 1. Configuration for discrimination testing.

cable carrying the video signal. The cables are limited to a maximum of ten meters in length.

Figure 2 indicates the overall organization of the DADTA IV system presently in use. Devices not essential to the operation of a minimal configuration are indicated with a star. Processor options, I/O channels, the real time clock and controllers for all the peripherals except the scan converter are mounted on the PDP-8/e processor which is equipped with a total of 12k words of memory. Although most applications can be written to run in 4k, the very useful DEC operating system, OS/8, requires a minimum 12k configuration (or 8k and a read-only-memory) when operated with the TD8-E DECtape system.

Operator interaction with the DADTA IV is via a teletype which also serves as a hard copy output device. In the minimal system, not including magnetic tape storage, the paper tape reader is used for program loading. The TD8-E DECtape control and TU56 dual tape transports provide the medium for program and mass data storage. Generally DADTA IV is operated under OS/8 or X/SYS software operating systems which provide convenient tape library facilities. The TD8-E is a rather primitive non-interrupt, non-data break controller with the saving grace that it is an order of magnitude less expensive than the more elegant TCO-8 controller. Since the primary role of the magnetic tape in the DADTA IV is program storage, the slow, CPU intensive data transfers are not a serious problem.

The time base for applications programs is provided by the DK8-E real time clock which is programmed to interrupt the processor at 1 msec. intervals, at which time a software clock is updated.

The major communication channel between the processor and the external world is the DR8-EA static I/O

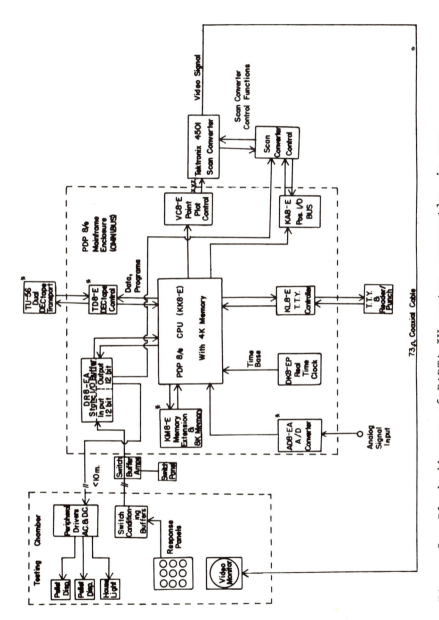

Figure 2. Block diagram of DADTA IV system presently in use.

buffer. The output portion of the buffer is a 12-bit register that can be loaded, cleared or read by the processor providing simple switching control of low speed peripherals. Eight bits are allocated for control of testing chamber functions and four are reserved for selection of display options. The 12-bit input buffer is identical to the output buffer except that the register bits are normally set by external events and only the input buffer is connected to the interrupt line. All subject responses are input via the DR8-EA. The positive I/O bus interface is included on the Omnibus to permit the implementation of input-output transfer (IOT) instructions for peripheral controllers constructed with Digital Equipment Corporation (DEC) logic modules. It was felt that the use of the external positive I/O bus rather than direct interfacing with the Omnibus would reduce the complexity of the scan converter control and facilitate system expansion by other users.

TESTING CHAMBER INTERFACE

The testing chamber interface is constructed from DEC K-series modules mounted in a K724 interface shell and communicates with the static I/O buffer via a flat ribbon cable (maximum length of 10 meters). The K-series logic exhibits excellent noise immunity.

1) Output Section (shown in figure 3): Bits 04 to 09 of the DR8-EA, after inversion, drive the gated inputs of the peripheral device current drivers. Bits 04 to 07 control isolated A.C. switches (rated at 500 V.A.) whereas bits 08 and 09 control D.C. switches (1A. at 55V.). Bits 10 and 11 are one-shot coupled to the D.C. drivers for triggering pulse operated devices. In the usual discrimination configuratin of DADTA IV the pulsed D.C. outputs operate reinforcement dispensers, one A.C. driver contols testing chamber illumination and the remaining switches are uncommitted and available for special requirements.

2) Input Section (shown in figure 4): Twelve slowed K123 buffer gates provide switch conditioning and line drive to transmit switch closure information to the static input buffer. The lines are also buffered at the receiving end where a parallel set of switch inputs is available. As shown, nine of the testing chamber inputs are committed to the subject response panels. The

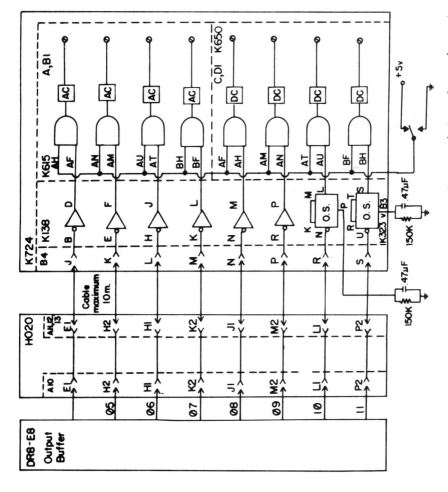

Figure 3. Output buffer and drivers for peripheral equipment.

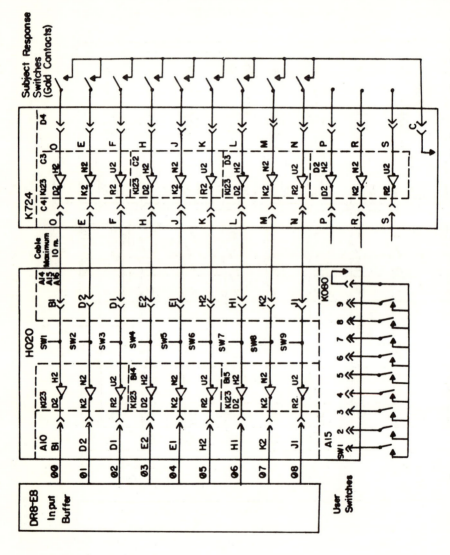

Figure 4. Output drivers and buffers for user switches and subject response switches.

parallel switch set at the computer end is particularly useful during program checkout if the testing chamber is located remotely.

DISPLAY HARDWARE

The Tektronix Type 4501 scan converter serves both as an image storage medium for non-dynamic displays and as an interface element between the processor and the video display monitors. Write-in is accomplished by supplying x and y deflection information from the VR8-E point plot control (high speed dual channel D/A converter) along with the appropriate intensity modulation (z axis). Although the resolution of the VR8-E is 10 bits on both the x and y axes the scan converter coordinate system has a 4x3 aspect ratio corresponding to the dimensional characteristics of a conventional television frame. The storage principles of the scan converter are similar to those of the conventional storage oscilloscope but the CRT storage target is constructed in a manner that allows read-out of the stored information by scanning the target with a "read beam" in a systematic manner. The Type 4501 scans in a pattern corresponding to a television raster and develops the necessary synchronization and blanking pulses to produce an E.I.A. standard 525 line composite video signal. A signal beam is used for both reading and writing and it must be time shared in situations where it is necessary to update the display without interrupting the video output or if displays are being generated with the scan converter in the "non-store" mode. The beam time sharing dictates certain timing requirements for data transfer from the CPU and these will be discussed when the scan converter control hardware is described.

There are several operational modes for the 4501 which in DADTA IV are selectable either manually or by program instructions. The options are:

> 1) Dark/light background: The written area of the CRT may be selected to appear as white on a black level background or black on a white level.
> 2) Read only: In "read only" mode the storage target is scanned and a video signal is produced from the stored information, but no new information can be written.

3) Write only: In "write only" mode, read raster scanning is disabled and the video output is held at background level. New information may be written on the target and stored for subsequent readout.

4) Store/non-store: In store mode information written on the tube face is retained for periods of up to 15 minutes. In non-store mode information is not retained beyond the normal target persistance time.

SCAN CONVERTER CONTROL

Option mode selection is controlled by bits 00 to 03 of the DR8-EA static output buffer (figure 5). Setting a bit in the buffer enables the corresponding buffer gate which activates the 4501 control input line. The static buffer bit allocations are summarized in Table 1. If neither "read only" nor "write only" is selected, the mode of operation will be "read and write".

Stored information is removed from the display by selecting the scan converter erase function. Selective data erasure is not possible. The entire screen must be erased, modified and rewritten to delete a portion of the stored image. The erase function is implemented as an I.O.T. instruction via the KA8-E positive I/O bus and appropriate device selector. A complete 4501 erase cycle requires 174 msec., during which time the video output is gated off. However, if an erase instruction is issued while read scanning is in progress, a transient flash of light and loss of monitor synchronization may occur. This problem can be avoided by only initiating erase cycles during the vertical blanking interval. A negative going vertical drive level with the leading edge coincident with the start of vertical blanking is available and is used to provide programmed detection of the start of the vertical interval. The "erase cycle in progress" level from the scan converter is OR'd with the level shifted and inverted vertical drive to furnish the I.O.P. gating signal. The resulting I.O.T. instruction, "skip on vertical drive on or erase cycle in progress" minimizes hardware requirements while providing all information necessary for proper erase timing. It would, of course, be straightforward to handle all erase cycle timing directly with hardware.

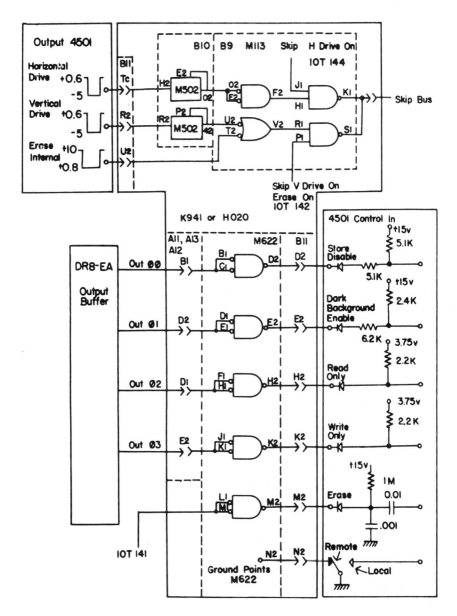

Figure 5. Control for Tektronix 4501 Scan Converter.

Table 1. Scan Converter Control Options.

Option	Output Buffer Bit
store/non-store	00
dark/light background	01
read only*	02
write only*	03

Scan Converter Control I.O.T. Instructions

I.O.T.	Instruction
6141	Erase display
6142	Skip if vertical drive on <u>or</u> erase cycle in progress
6144	Skip if horizontal drive on

* If neither "read only" nor "write only" is asserted, the default condition is "read and write".

The scan converter control includes a flag indicating the start of the interval between successive horizontal scan lines. The horizontal drive level from the 4501, after level shifting and buffering, gates I.O.P. 4 to give a "skip on horizontal drive on" I.O.T. instruction. This instruction is used for proper synchronization of data transfer from the VC8-E point plot control when the scan converter is operated in the "read and write" mode. In this mode, read scanning and video output are maintained continuously, but the beam is available for writing during the 9.2 μsec between the end of each horizontal line and the beginning of the next (see figures 6a and b). Generally only one point can be written in each line interval since the scan converter requires 8 μsec (maximum) to write a point in "store" mode. This limits the maximum writing rate to 15,725 points per second. The programmer must be somewhat careful to pre-load the D/A converter registers to ensure that sufficient time will be available to deflect the write beam and intensify the point within the write time slice. The "read and write" mode can be used in non-store mode for generation of dynamic displays as the target persistence is sufficient to maintain a point until the next line scan. In the current controller, the scan converter flags are not connected to the interrupt line. However, in some applications where the demands on the CPU are heavy it might be desirable to do so.

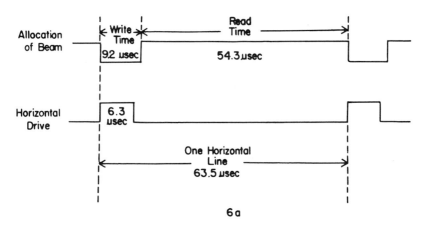

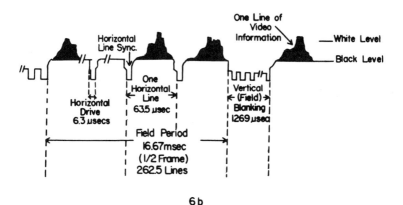

Figure 6. Timing diagrams for scan converter. (a) Scan converter beam share timing in "Read and Write" mode. (b) Video signal (simplified).

SYSTEM PERFORMANCE

DADTA IV has been operational for approximately three years (more than a dozen neurobehavioral experiments have been run and completed) and has proven flexible and reliable in an environment where a variety of experiments with unique stimulus requirements are in progress simultaneously. The scan conversion technique produces moderately complex displays of good quality. Reliability of the entire system has been excellent and there have been no significant hardware difficulties.

A library of assembly-level subroutines including programs for certain display functions, time-keeping and response processing has been developed and these are integrated with control programs written in a real-time modified version of FORTRAN II. The use of compiler level coding greatly simplifies the task of handling computation, report generation, tape I/O, and operator interactions. The display software allows specification of stimuli in a variety of formats depending on the degree of flexibility required for a particular application. In a simple case, for example, only a panel position, a size parameter and a character from the standard set might be necessary to specify a stimulus, whereas a more complex display might require furnishing a set of stimulus descriptors for use by the more general graphics routines.

SYSTEM EXPANSION

A shortcoming of DADTA IV in its present form is that the processor is dedicated to the operation of a single testing station at any one time. Expansion of the system to handle multiple stations simultaneously is not difficult from either a hardware or software point of view, but requires the addition of a scan converter, scan converter control, and static I/O buffer for each additional independent display. This represents a hardware cost in excess of $3000. If more than a few independent stations are contemplated, it becomes economical to use a single time shared scan converter for production of video signals and a multiple channel video disk recorder for refreshing the individual monitors. In multiple channel systems of either type the generation of multicolor displays becomes feasible by allocating two or more channels to a single color monitor.*

FINAL COMMENTS

DADTA IV is primarily intended for behavioral testing applications, but the display system has been used for stimulus presentation in several electrophysiological investigations. It should be mentioned that some preliminary data from this laboratory suggest that there may be some problems associated with the use of television displays in single unit studies. Although the displayed images are well defined and flicker free to the human observer at usual intensity levels, the emitted light contains a large modulation component at the 60 Hz. scanning rate (i. e., the television field rate). Over small vertical screen distances the modulation components are nominally in phase and it apears that many visual system neurons are driven at the scanning rate when the animal is exposed to a bright spot stimulus within the unit's receptive field ** (Phelps, personal communication). Even with a modified scan converter and monitor operating at a 120 Hz scan rate, some visual units continue to respond to the scanning modulation although direct following does not occur. Regardless of whether this phenomenon produces alteration of normal visual information processing, the introduction of unintended noise within the bandpass of the system under study is definitely undesirable. Additionally there is some tentative evidence that under certain conditions intensity modulation of a visual stimulus at frequencies above the flicker fusion point may affect visual discrimination performance in the monkey (Christensen, 1973).

This type of problem is not unique to DADTA IV; all systems for stimulus presentation are less than ideal in

---

*The use of a color monitor in a single channel system will provide the capability to vary the chromaticity of the entire display. A very limited form of multicolor display could be obtained with some additional hardware.

**Lateral Geniculate and Area 17 of the unanesthetized cat.

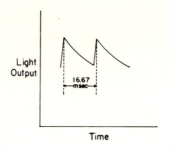

Figure 7. Modulation of light output from television screen.

some respects. Display and memory technologies are developing at a rapid rate and it is likely that the near future will bring a variety of low cost, self refreshed graphics displays potentially suitable for stimulus presentation in vision research. However, it should always be kept in mind that subtle differences between displays may exist that can influence experimental results in unpredictable ways. Stimuli generated by different methods may be perceptually equivalent or similar to the adult human observer, but this does not imply that they are equivalent to an animal or immature subject, particularly when we are dealing with a damaged or modified central nervous system. Furthermore, stimuli that are indistinguishable to the subject may not produce identical electrophysiological responses at all points in the sensory system. These obvious principles are often given too little consideration in the design of stimulus generation equipment for behavioral and even electrophysiological studies.

DISCUSSION

D.O. WALTER: Has this been exported?
DRAKE: In a few cases.
MCINTYRE: The decision about the I/O bus structure is a little odd. The cable costs and module costs could make your individual interfaces awfully expensive in comparison to the quad board or the alternative of using

TTL logic directly on the quad board, by a factor of 5 or 10.

DRAKE: That's certainly true, although it was less true at the time. We hoped this would appeal to psychologists who were used to working with logic modules but had no experience with computers.

MCINTYRE: The trick is to tell them the little ICs are miniature logic modules.

CLARKE: If you were putting this system together today, would you have used a 4501?

DRAKE: Tektronix doesn't make them anymore.

REFERENCES

1. Christensen, C. A. The Effects of Inferotemporal and Fovial Prestriate Lesions on Discrimination Performance in the Monkey. Unpublished doctoral dissertation. Stanford University, (1973).
2. Pribram, K. H., Gardner, K. W., Pressman, G. L., and Bagshaw, M. H. An automated discrimination apparatus for discrete trial analysis (DADTA). Psychol. Rev. 11:247-250 (1962).
3. Pribram K. H. DADTA III: An on-line computerized system for the experimental analysis of behavior. Percep. and Motor Skills 29:599-608 (1969).

For detailed information on Digital Equipment Corp. devices and systems the reader is referred to the following publications available from Digital Equipment Corp., Maynard Mass:

Small Computer Handbook. 1972
Control Handbook. 1971
Logic Handbook. 1972
Logic System Design Handbook. 1972

For detailed information on the Tektronix Type 4501 Scan Converter refer to:

Tektronix Type 4501 Scan Converter Unit Instruction Manual 1969. Available from Tektronix, Inc., Beaverton, Oregon.

APPENDIX I

MINIMAL DADTA IV SYSTEM

I. Computer Enclosure

    D.E.C. PDP-8/e Processor with 4k words of memory
    D.E.C. ASR 33 Teletype with paper tape punch/reader
    D.E.C. KL8-E T.T.Y. Controller
    D.E.C. DR8-EA 12-bit buffered I/O
    D.E.C. VC8-E Display (point plot) Control
    D.E.C. KA8-E Positive I/O Bus
    Scan Converter Control (modules available from D.E.C.)*

II. Testing Chamber

    Television Monitor
    Input Response Interface and Peripheral Output Drivers (modules available from D.E.C.)*
    5v. Power Supply for above
    Davis PD 109A pellet dispensers

EXTENDED DADTA IV SYSTEM

In addition to the minimal system, the following are suggested:

    8k memory and extended memory control
    D.E.C. DE8-E Extended Arithmetic Option
    D.E.C. TD8/e DECtape Controller
    D.E.C. TU-56 Dual DECtape Transports
    D.E.C. AD8-EA A/D Converter
    D.E.C. DK8-EP Real Time Clock

---

* Not commercially available

chapter 32

AUTOMATED SYSTEMS FOR BEHAVIORAL NEUROPHYSIOLOGY–THREE STRATEGIES FOR INTERACTION WITH NEUROPHYSIOLOGICAL EXPERIMENTS

S. L. MOISE, JR.
Center for Health Sciences
University of California
Los Angeles, California

My use of computers in neurobiology has been in an area I call behavioral neurophysiology. That is, the collection, evaluation, and manipulation of neurological parameters in the intact, behaving organism where the behavior of the organism is often as important as its neurophysiological responses.

In view of the goal of this project to develop a computer language for neurobiologists, I felt it appropriate to briefly desribe three systems with which I am familiar. These represent three strategies for biological data collection specifically designed to satisfy somewhat different experimental requirements. All three are typical of many laboratory situations that could benefit from the use of computers and as such deserve consideration in the planning of a language which may someday be used in similar circumstances.

This research was supported in part by the Public Health Service Research Grant GM-16058-08 and in part by the AF Office of Scientific Research of the Office of Aerospace Research under Contract F44620-70-C-0017.

The three systems are: 1) EVE - A monolithic assembly language data collection only scheme for a fixed behavior paradigm. No control over environmental conditions is required, but the experimental variables to be collected may vary considerably from experiment to experiment. 2) FOCAL SB - A FOCAL (Dec. 1972) based data collection, computation, and experimental control scheme. A fixed behavioral paradigm is used in which several environmental parameters must be controlled on the basis of results from real-time calculations made on the collected data. 3) ASPRIN - A modular assembly level monitor with user-callable functions for data collection and experimental control. This system is capable of controlling many environmental conditions in a wide variety of experimental paradigms while collecting behavioral and neurobiological data.

EVE

This data collection application is in the Women's Hospital at the University of Southern California. It is part of a research unit headed by Dr. Hahn with Michael Murphy as the principal software architect. This system is being used to monitor several physiological indices from mother and fetus during labor. The data is then being analyzed for the eventual computation of a risk index (prediction of birth outcome) for the fetus. This index is ultimately envisioned as a function of many variables including the parents' medical history. At this early stage of the research, the variables of greatest interest are those being collected during labor. These presently include: 1) uterine pressure; 2) maternal temperature; 3) maternal blood pressure; 4) maternal ECG and 5) fetal ECG (from which fetal heart rate is calculated). The system requires capability for high data input rates for examination of the fine structure of the ECG and the eventual inclusion of EEG data. Current emphasis in this program is centered on maternal and fetal heart rates.

The hardware for this system is shown in figure 1. The PDP - 8/I has 12k of core and the laboratory peripherals include A/D converters, a programmable clock, display scope and controller and Schmitt triggers. The special hardware devices were built by Mr. John Kracik and include: 1) A 36-bit, microsecond resolution,

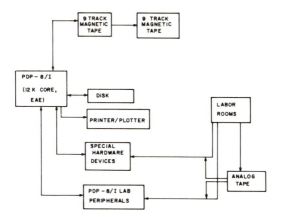

Figure 1. EVE Hardware System

free-running clock. This clock does not generate interrupts but may be read "on the fly" by the program. 2) Multiplexed Schmitt triggers: Up to 256 triggers can be multiplexed to provide only one interrupt instead of requiring separate interrupts (and IOTs) for each trigger. An 8-bit code informs the program which trigger generated the interrupt.

The software is organized in three parts: acquisition, preliminary data reduction, and conversion to EBCDIC format on magnetic tape for analyses more easily done on a large computer.

The heart of the system is the acquisition program. It is a very general, event-driven analog and digital data acquisition program written in assembly language. Since data is often collected by nurses or paramedics with little computer experience, this program was designed for simple interaction with the experimenter so that these people can correctly set up experiments with a minimum of training.

When an event occurs, the program records the 36-bit time of the event, what the event was, and associated analog variables. These variables are selected by the user before the start of each experimental run. Examples of events which might trigger data collection are: 1) The firing of a peak detector (Schmitt trigger) on the ECG to give the time of each fetal heart beat. 2) A key on the system console device: The experimenter may strike a key to indicate special events or artifacts which may be particularly difficult to recognize in the data. 3) The overflow of a programmable clock for doing clocked A/D conversions. 4) Digital stop/start signals from a time code generator.

Considerable effort was invested in writing this program to make it general and efficient. The program currently handles up to 16 analog inputs and 8 Schmitt triggers. An indefinite number of analog channels and up to 256 Schmitt triggers may theoretically be monitored with a maximum throughput of approximately 10k events/sec.

Data is stored in binary using the device independent facilites of OS/8 (2). Any standard OS/8 device driver (such as disk or DECtape) that will run with interrupts on can be used. Since standard OS/8 handlers sit and wait on a transfer done flag, no furthur output may be started until previous putput is complete. Interrupts are left on at all times, however, so data collection continues under all conditions. This I/O scheme is effectively a background, single-buffered, data output task (DMA transfer with little CPU intervention required) with a foreground, cyclic-buffered data input task.

After the experiment has terminated, a special program reads the data back and converts it to OS/8 ASCII. At this point, preliminary data reduction is carried out using OS/8 FORTRAN IV and OS/8 BASIC. A higher level language is necessary at this stage as the investigator often wishes to manipulate (transform, plot, reorganize) his data extensively before proceeding to further analysis on a large computer.

To date, most of the acquisition has been done off-line from analog tapes. All data collection will be done on-line in the future when the interfacing and software for a newly acquired PDP-11/40 is complete. The PDP-11 will also perform all analyses now being done on larger machines.

FOCAL SB

This system is located in the laboratory of Dr. J. Weldon Belleville and Dr. George Swanson at UCLA. This group is doing studies in human respiratory physiology that require precise regulation of inspired CO_2 concentration with measurement of respiratory response data on a breath-to-breath basis. The mechanical system developed in this laboratory is reportedly unique in that it is flexible and fast enough, for example, to force a subject's end-tidal CO_2 to follow a specific wave-form independent of his ventilation or mixed venous return. This ensures that the respiratory controller stimulus is independent of the associated ventilation response and allows respiratory controller behavior to be studied independently of gas exchange and body-tissue effects. Further details of the physiology and mechanics of this system may be found in the papers by Swanson, Carpenter, Snider and Belleville (1971) and Ward, Michels, Swanson and Belleville (1974).

The computer system used in this research has to act as a controller for the breathing apparatus, a data collector for various respiratory variables, and a real-time computation device to determine parameters for modifying the breathing apparatus. The hardware assembled for this purpose is shown in figure 2. The computer is a PDP-8e with 12k of core, disk, laboratory peripherals (programmable clock, scope controller, A/D converter and D/A converters), high speed paper tape, and a Tektronix 4010 graphics terminal as the system device.

In a typical experiment, the subject breathes from the chamber through a two-way valve. His exhaled gas flow is monitored by a pneumotachograph and CO_2 level is monitored by an infrared CO_2 monitor. These signals go to an analog computer which integrates the flow,

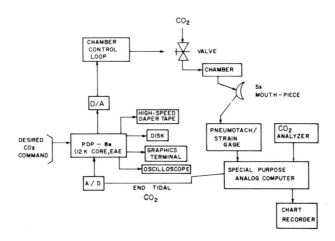

Figure 2. FOCAL SB Hardware System.

determines breath cycle timing, and sends the end-tidal CO_2 level to the digital computer. The algorithm in the digital computer calculates the chamber command signal based on the desired end-tidal CO_2 and the S's present end-tidal CO_2 concentration. The command is delivered to the chamber loop at the start of S's exhalation. The chamber loop maintains this concentration during the S's subsequent inhalation.

A typical experimental paradigm is shown in figure 3. Each breath may be considered to be an inhalation followed by exhalation. The inhaled CO_2 concentration for the Nth breath must be set up in the chamber during the exhalation of breath $N-1$. The control algorithm currently used calculates this command during the N-1st inhalation based on the error between the desired and actual end-tidal CO_2 concentration on the $(N-2)$nd breath.

The software to run these experiments consists of a highly modified version of DEC's FOCAL (modified by Jim Crapuchettes of Freeland Associates) for data collection, and programs in OS/8 FORTRAN IV and OS/8 BASIC for data analysis. Modifications to FOCAL permit it to use the extended arithmetic element for calculation, access disk blocks, use additional memory, and control the breathing chamber apparatus.

FOCAL was chosen to run these experiments since it allowed relatively unsophisticated computer users to generate experiments quickly. Perhaps even more importantly, FOCAL allows considerable ease and facility for generating and changing computational algorithms.

Several equations must be solved during time $N-1$ (figure 3) to generate command signals for the next

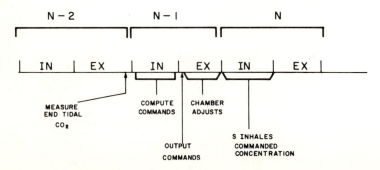

Figure 3. Typical experimental paradigm for repiratory response experiment. IN = INHALE, EX = EXHALE.

breath at time N. These equations must sometimes be quickly modified to eliminate problems of patient well-being, compensate for changes in the subject's breathing behavior over time, and the like. Since data acquisition requirements are modest (6 channels of data at less than 100 Hz), FOCAL, which is relatively slow for real-time operations, has been ideally suited to this application. However, on-line computational demands in these experiments have become sufficiently complex to push FOCAL to its speed limits. The next logical sequence of experiments will require conversion to a faster computational procedure on the acquisition of additional hardware (such as a floating point processor).

ASPRIN

This is the Automated System for Primate Instruction which I have developed in the Environmental Neurobiology Laboratory at UCLA (Dr. W. Ross Adey, Director). Our requirements were for a system that would: 1) Act as a task controller for a very large number of experimental paradigms in which any organism capable of pushing buttons could be tested. It was also desirable to permit the use of other response manipulanda as well. 2) Collect detailed behavioral data in a user specified format for real-time and subsequent analysis. 3) Monitor and record biological data (EEG, ECG, EMG, etc.) for real-time and subsequent analysis. 4) Permit control of a potentially large number of experimental parameters.

Real-time versions of higher level languages did not exist or were not suitable for our purposes when development began. The real-time version of DEC's FORTRAN IV is very difficult to adapt to non-standard peripherals as the run-time system is integrated around characteristics of these peripherals. DEC's real-time BASIC and the many versions of FOCAL are too restrictive in terms of the rates at which data may be collected and processed.

Our needs were too variable in terms of experimental design to be able to create a monolithic assembly-level program full of options such as EVE and it seemed that assembly language programming was still a necessary evil for the users in our laboratory environment. Therefore,

I developed a general purpose monitor designed to facilitate writing new experiments by relieving the need for the the programmer to deal with hardware dependent functions such as interrupt handling and input/output. Equally important, it provides a wide array of user callable functions commonly needed for controlling behavioral tasks and collecting biological data. The user program typically consists of a series of monitor calls with appropriate logic for presenting stimuli, determining appropriateness of response, recording elapsed time, processing biological data, etc. Details of hardware and software components of an earlier version of this system have been described by Moise, Olsen and Huston (1974).

The hardware for this system is shown in figure 4. The system was designed around an 8k PDP-8/I with peripherals as shown. The PDP-8 has since acquired 24k of core. The test apparatus is a primate test chamber with a behavioral test panel. This panel has a 3x3 array of stimulus response windows which are transparent Plexiglass response switches. Behind each switch is a 1 plane digital readout unit that can project any combination of 7 symbols and 5 colors through the switches. This panel may be removed from the chamber and placed on a table for use with man (adults or children). A pellet feeder which can be outfitted for monkey pellets, M&M's or other rewards is also attached to the panel.

In order to be able to select any combination of the 12 stimulus lights on any of the 9 display windows at any given time, the user must be able to select any of 108 unique output lines. In addition we wanted to be able to control numerous other devices such as house lights, brain stimulation, etc. Since minicomputers generally only provide 12 to 24 bits for output (depending on word size) we built a simple, relatively inexpensive decoder with memory (in the system console of figure 4). Eight of the PDP-8's 12 output bits are decoded in 2 groups of 4 to provide 256 output possibilities. The remaining 4 bits are not presently decoded, but could be if necessary.

The monitor provides the user with debugging and program modification facilities similar to those in DEC's Octal Debugging Technique. All internal calculations are done in floating point using a modified 27-bit floating point package. This package, or any other arithmetic

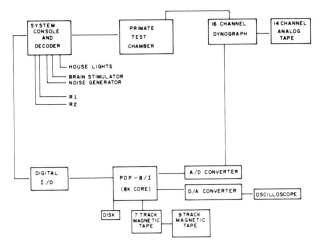

Figure 4. ASPRIN Hardware System. R1 = relay 1, R2 = relay 2

package, may be appended to the user's program for user computations. The monitor does not presently provide computational algorithms for the user.

User callable functions available in the monitor include the following: 1) CLOCK: Initialize the clock counter 2) CLOCK1: Read the clock counter. Provides elapsed time in msec. since the last call to CLOCK. 3) WAIT: Wait a specified interval (with or without servicing external inputs such as button presses). 4) MONITOR: Transfer control to the monitor to wait for an external input or end of WAIT. 5) ENTER: Provide a location where the monitor is to transfer when an external input is received. 6) ROOM: Output control signals to the decoder to drive external devices. 7) FPOUT Output a floating point number to mass storage. 8) INTEGER: Output an integer to mass storage. 9) OUT: Output a character to mass storage. 10) STRING: Output a string of characters to mass storage. 11) READ: Input from the console device. 12) CONSOLE: Output to the console device. 13) OPEN: Open an OS/8 output file. 14) CLOSE: Close an OS/8 output file. 15) SAMPLE: Begin analog data collection. 16) SAMP1: Stop analog data collection. 17) RAND: Generate a random number on a specified range. 18) PERM: Permute a specified array of items. 19) SCOPE: Output via the D/A converter.

The heart of the monitor is the clock interrupt

service routine. Interrupts from other devices lead to acquisition of data points or identification of external events. Initiation of actions to service these events is managed by the clock handler which is activated every millisecond. All timing and response servicing is accurate to 1 msec.

Analog data collection is initiated by a call to the monitor routine SAMPLE. Arguments for this call include the address of a ring buffer where samples are to be placed, the size of this buffer, the sampling rate, and the number of channels to sample. A pointer is available to the user so he knows where the latest sample has been put in the buffer.

When the user passes control to the monitor to wait for an external response, or end of a timed interval, the clock interrupt routine determines if sampling is in progress and exits to a user specified routine to process these samples if there are no external responses or end of interval to service. The user specified routine must remove samples from the ring buffer, process them, and return to the monitor. During involved processing, the user may still be computing when a response occurs or a timed interval ends. Therefore two flags are provided which indicate the occurrence of these events. These flags may be interrogated periodically during computation, and when the user is ready to service these events, he can transfer control back to the monitor. Collection of analog data is continuous until the user calls routine SAMPL1. One must be careful that processing of the analog data is on the average complete before the next sample is taken or data will be lost.

Data is output via standard OS/8 device handlers. As in the EVE system, any handlers that will run with interrupts on may be used. For experiments in which data collection and computations require as much CPU time as possible, data is output in binary. After the experiment a separate program converts the binary to OS/8 ASCII. For experiments with less demanding processing needs, the data may be converted to ASCII before output. Once in OS/8 ASCII format, analyses are performed on the data via OS/8 FORTRAN IV programs. Conversion from ASCII to EBCDIC on magnetic tape permits the use of larger computers for more involved computations on large quantities of data.

In our laboratory, as in many, the demand for

computer time now exceeds the available hours in a day. We run experiments 6-10 hours each day and need to process the large volume of data thus generated as soon as possible. It would be extremely useful to be able to do data analysis as a background task while experiments run in the foreground. In some experiments there are times during which the CPU is utilized fully for real-time computation, thereby eliminating the possiblility of a background task. Not all experiments are of this nature and those that are generally are so only in bursts.

At present, the monitor runs a null background task when waiting for an event. Given at least 12k of core, this could be replaced by some more complex task such as FOCAL. A substantial effort would have to be expended to modify FOCAL to run in this environment and to be able to assess data via OS/8 files.

A far more powerful solution would seem to be to implement DEC's RTS-8 (real-time system for the PDP-8 and PDP-12 family of computers) (3). We have recently acquired timesharing hardware. With this inexpensive modification (four modules and a few wire deletions) and our present 24k of core, we should be able to run foreground real time tasks and OS/8 in the background. While a second teletype is recommended for OS/8 in the background it is possible to run the system without this feature.

Much of the ASPRIN monitor can be directly converted to run as tasks under RTS-8. User programs would have to modify their calling sequences and means of passing parameters to these tasks. Then, while the experiment executes in the foreground, a single user could run OS/8 (including FORTRAN IV) in the background.

As mentioned, several experiments which run under the ASPRIN monitor require real-time calculations on collected data. At present, computation algorithms must be coded as calls to a floating point package. Unlike higher-level language systems such as FOCAL SB it is difficult to change algorithms and parameters. It would be useful to have an incremental compiler built into the monitor that would interpret a simple syntax language into floating point processor calls. The user could then type in the algorithms he wished to try, and change it as often as necessary by halting the experiment, retyping the algorithm and continuing the experiment.

CONCLUSION

In considering the design of a language for neurobiology, characteristics of the systems described here must be kept in mind if this language is to be used by the behavioral neurophysiologist. 1) It must permit complete freedom in the use of existing hardware for whatever experimental designs that hardware is capable of. 2) It must be efficient for optimal throughput. Data collection rates vary considerably over the systems described here, but these are modest compared to the requirements of some spike train data collection experiments (e.g., see chapters by D. K. Hartline and J. J. Capowski in this volume). 3) The user should have some means of quickly and easily generating computational algorithms for handling data collected in real time. 4) Data collected should be stored in a format compatible with the system software of the computer being used. In particular the data should be easily available for analysis by a higher level computational language such as FORTRAN or BASIC.

DISCUSSION

KEHL: How much core does the monitor require?

MOISE: The monitor takes 4k, so user programs reside in Field 1 and higher.

BROWN: Is this exportable?

MOISE: No, although I suppose much of the software might be.

MCCORMICK: How's this distinguishable from any other real-time operating system for process control?

MOISE: There isn't much difference.

KEHL: Aside from the fact that it works.

MOISE: I haven't really found any other system that would do what I want. I would have been delighted to use some other system instead of wasting so much of my time making my own. One of the impressive feelings emerging from this symposium is one of despair that we each have to do these things over again for our own labs.

REFERENCES

1. *Programming Languages, Volume II.* Digital Equipment Corp., Maynard, Mass. (1972).
2. *OS/8 User's Manual*. Manual DEC-S8-OSHBA-A-D, Digital Equipment Corp., Maynard Mass. (1974).
3. *RTS/8 User's Manual*. Manual DEC-08-ORTMA-A-D, (1974).
4. Moise, S. L., Jr., Olsen, D. E. and Huston, S. W. An automated system for primate instruction for behavioral neurophysiology. Behav. Res. Math. and Instr. 6:557-564 (1974).
5. Swanson, G. D., Carpenter, Theodore M., Jr., Snider, D. E. and Belleville, J. Weldon. An on-line hybrid computing system for dynamic respiratory response studies. Computers and Biomed. Res., 4:205-215 (1971).
6. Ward, Denham S., Michels, David B., Swanson, George D., and Belleville, J. Weldon. Automatic control of end-expiratory carbon dioxide. *Proc. of Eleventh Ann. Rocky Mountain Bioeng. Symposium and Eleventh Int. ISA Biomed. Science Instrumentation Symposium* (1974).

chapter 33

A DIGITAL SYSTEM FOR AUDITORY NEUROPHYSIOLOGICAL RESEARCH

W. S. RHODE
Laboratory Computer Facility
University of Wisconsin
Madison, Wisconsin

INTRODUCTION

Neurophysiology studies of the auditory nervous system have been performed in our laboratory for over 20 years. Averaged evoked response recording and the determination of the vibratory characteristics of cochlear structures, along with microelectrode studies of all-or-none neural spike activity have benefited from the

---

This research is supported in part by grant NS 06225 from the National Institute of Neurological Diseases and Stroke and in part by grant RR0249 from the Biotechnology Resources Branch of the Division of Research Resources.
The realization of the equipment designed for this project is due to collaborative efforts with several individuals. Mr. R. Olson, of the U.W. Medical Electronics Laboratory, is principally responsible for the implementation and much of the design of the Digital Stimulus System, the event timer and the histogram binners. Mr. P. Wilson, also of the Medical Electronics Laboratory, is responsible for the programmable A/D. Dr. R. Borchers of the U.W. Physical Sciences Laboratory is responsible for the graphic processor and the formatter/controller for the Datacord tape cartridge drive.

availability of a small laboratory computer (1). The computer system, which is based on a Classic LINC computer, has been used since 1963 for control and monitoring of stimulus parameters and data collection. While the LINC continues to be a useful tool, it nevertheless has some limiting features such as small memory (2k), 12-bit word length, lack of direct memory access (DMA) capability and the lack of standard peripherals. In order to overcome these limitations, a new computer system (Harris 6024/5) has been purchased. A major effort is under way to enhance the experimental facilities with new and possibly unique stimulus generation and data collection equipment.

The stimulus waveforms used to conduct research on the auditory system include sinusoids, trapezoids, pseudo-random noise, amplitude- and frequency-modulated sinusoids (AM and FM), swept frequencies, tone complexes, and biologically meaningful sounds such as speech. Digital stimulus generating equipment has been designed to produce these classes of stimuli and to be completely computer controlled.

For data collection, an event timer, a programmable analog-to-digital converter, and a digital spectrum analyzer have been designed to meet the demanding needs of the research. Each device is designed for ease of use and includes data buffers which relax the requirements for rapid response by the computer system.

Besides stimulus generation and data collection, the data storage and analysis aspects of experimentations are equally important. A graphic display processor has been designed which includes complete buffering of the display yet provides dynamic capability to handle rapidly changing displays. An electrostatic printer/plotter provides hard copy graphic output which is essential in data analysis. The primary data storage device is a tape cartridge unit with a 100 Mbit capacity, which is sufficient to handle most storage problems. The computer system has a Disc Monitor System which supports multiprogramming. Nearly all programming is done in FORTRAN, eliminating the tedium of machine language coding and speeding program development.

STIMULUS GENERATION AND CONTROL

The primary goal of the stimulus generation and control apparatus is to produce a wide class of auditory stimuli with proper fidelity and with complete computer control of the sequence of presentations. The apparatus includes a digital stimulus system, a pseudo-random noise generator and a variable length circulating buffer for the presentation of arbitrary waveforms. Each of these devices is discussed below.

Digital Stimulus System

This system grew out of the need for a programmable sine-wave oscillator with digital control of stimulus presentation. While a number of programmable frequency synthesizers appeared on the market in the early 1970's, their principal limitation for auditory research is the harmonic content of their output (greater than 0.1%). In 1972 we began the design of an all-digital technique for waveform synthesis combined with electronic gating of the signal, digital timers, and a stimulus counter for stimulus sequence control. The device is called the Digital Stimulus System (2) or DSS.

The front panel of the DSS is illustrated in figure 1. LED displays are used to indicate either the current or the preset value for each of the three timers and the stimulus counter. The frequency is displayed as a seven-digit number, two of which are for the fractional part of the frequency. The phase is displayed as a fraction of the period in 3 digits. The display of the fractions is generated by using a PROM or programmable-ROM which decodes the binary numbers into two or three digits. LEDs are also used to indicate operating modes and the rise/fall time. Most functions are computer or manually settable. The integer part of the frequency can be swept using a manual lever, spring-loaded to return to the zero-position. The rate at which the frequency changes using the manual control is dependent both on the displacement of the lever from its resting position and the current frequency.

A simplified hypothetical example of the technique for synthesizing a sine wave (3,4) using table look-up is shown in figure 2A. The contents, F, of a 4-bit frequency register are added to θ, the contents $C(SAR)$ of the sine address register, which at time T equals $F \cdot T$ modulo $(2)^6$ where $T = 0,1...,\infty$, and $2^6 =$ the clock frequency. A table of 64 values of the sine function is stored for each of the 64 possible values of θ. The

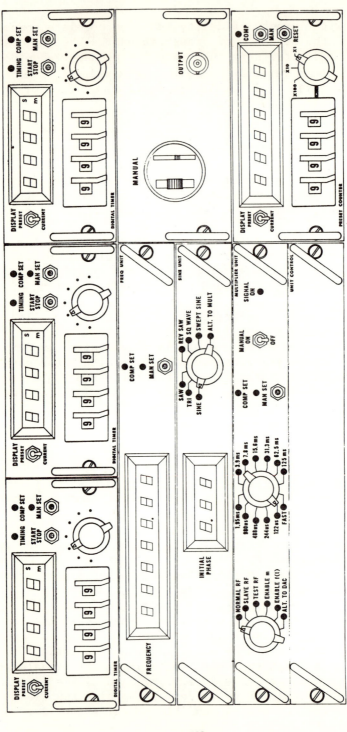

Figure 1. The front panel of the Digital Stimulus System. See text for detailed description.

AUDITORY RESEARCH 547

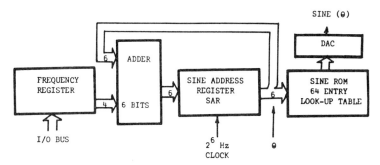

Figure 2A. Simplified diagram of a hypothetical sine-wave generator.

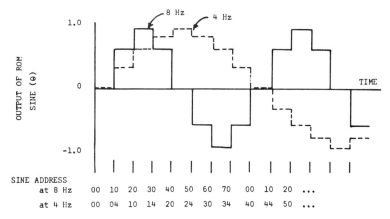

Figure 2B. An illustration of the output of the sine-wave generator for a frequency of 8 Hz (solid line) and for a 4 Hz (dashed line) frequency. Larger steps in output occur for higher frequencies.

value of F (1 to 15) determines the size of the step through the sine table. If $F = 1$, then each of the 64 values of the sine (θ) are read out each second, whereas if $F = 2$ then every second value of the sine function is read out twice a second. In general it can be seen that the value of F is the frequency of the synthesized sinewave. The result of synthesizing 4 to 8 Hz sinewaves is shown in figure 2B. The effect of quantizing a signal in time and amplitude is obvious and can be reduced by shortening the sample time (higher clock rate) and using more bits to represent the signal.

The actual DSS has a 16-bit frequency register and a 19-bit sine address register; it therefore has a frequency range of 64,000 Hz and a clock rate of 2^{19} or 524,000 Hz. If the sine (θ) was stored for each of the 2^{19} values of the sine address register at an accuracy of 16 bits, an 8,000,000 bit memory would be necessary. Only one quarter of the sine function need be stored and trigonometric identities can be used to further reduce the size of the ROM to 16,000 bits. A small sacrifice in accuracy is made to accomplish this saving in memory size; the table is accurate to 1 part in 2^{15} for 2^{17} arguments of θ. The distortion of the DSS has been determined to be less than 0.01% for frequencies below 10 KHz; this distortion is due to the digital-to-analog converter. The DSS can also generate triangular waves, squarewaves, sawtooths and reverse sawtooths. The frequency can vary from 2^{-16} to 2^{+16}, that is, any frequency up to 65,000 Hz can be generated. The frequency is specified in two parts, a 16-bit integer and a 16-bit fraction, and the initial phase angle of the signal can be specified with 16-bit accuracy.

In order to reduce the amount of energy at frequencies other than the one being generated, the signal is turned on gradually. That is, it is multiplied by a trapezoidal waveform which has a programmable rise/fall time. A 16 x 16 bit digital multiplier is used to perform the multiplication. The advantage of this approach over the use of an analog multiplier is that it doesn't introduce any distortion when the signal is fully on since the signal waveform is merely multiplied by a constant value. Our previous electronic switch used a Hall effect multiplier and had a distortion level of about 0.2%. The rise/fall time of the DSS can be varied from 0 to 125,000 microseconds; there are 14 settings which change by a factor of 2 except for the 0 μsec rise/fall time. The trapezoidal gate signal is generated by a pair of 16-bit counters as shown in figure 3. (The first counter generates the rise-time delay and the second counter generates the gate signal, the trapezoidal waveform, after the expiration of the risetime delay. When the enabling logic level goes to zero, counter 2 begins counting down to 0). This ensures that the duration of the gate signal is identical to the stimulus duration. The waveforms are shown with dotted lines to emphasize that they are digital until they pass through the digital-to-analog converter. The stimulus duration logic level is generated by the stimulus sequencer.

The third major subsystem is the stimulus sequencer consisting of three digital timers, each with 4 presettable digits and a timing range of 1 μsec. to 9999 sec., plus a 6-digit presettable counter which permits 1 to 9999 repetitions of a timing sequence. The timers and counters are interconnected so that sequences of stimuli can be generated as shown in figure 4. A command from the computer starts the delay timer. When the delay time has expired the repetition interval timer and stimulus duration timers are retriggered if the stimulus counter is not at its final value when the repetition interval timer and the risetime delay counter reach their preset counts.

The final step in the synthesis process is the digital-to-analog conversion (figure 3). This is the step which limits the accuracy of waveform production. Theoretically, a 16-bit system should result in a system which has harmonic distortion in the neighborhood of 2^{-16} or -96 dB. The present system achieves about -80 dB distortion, which is quite acceptable for sound production. The limiting system component is the digital-to-analog converter or DAC. Any DAC can produce 'glitches' or large undesirable steps in output when undergoing transitions in input addressing which involve a change in the state of a large number of bits, e.g., 00111111 to 01000000. A deglitcher amplifier is used to suppress these transients; this is a fast sample-and-hold

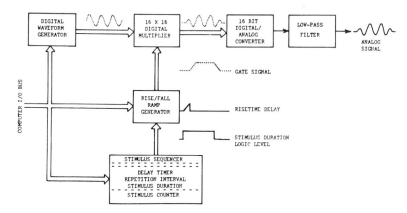

Figure 3. The relationship between the enabling logic level and the trapezoidal gate signal when the rise time is specified to be t_1 seconds.

circuit which maintains the output of the DAC at its previous value until the input address has had time to change and the output of the DAC has stabilized.

The philosophy incorporated into the design is to make most of the functions of the system capable of both manual and programmable control. This allows initial exploration of neural unit responsivity without the need for computer interaction. It is also useful for maintenance of the system.

Pseudorandom Noise Generator

Pseudorandom numbers are generated using a maximal length sequence generator (5). This device is useful because repeatable sequences are generated and the length of the sequence can be varied. This permits a reduction in the amount of computations necessary to recover the transfer function of the system being analyzed.

The basic idea is that a system's transfer function, the relation of its output to input, is recoverable by merely measuring its output, $Y(jW)$, when the input, $X(jW)$, to the system is white noise. When the input is white noise, $X(jW)$ is equal to constant, K, and since $Y(jW) = H(jW) \cdot X(jW)$ it is obvious that the transfer function is $H(jW) = Y(jW)/K$.

The design of the noise generator for our system has not yet been completely specified. A method of implementing a programmable bandwidth with arbitrarily sharp slopes in a bandpass filter characteristic is being explored.

Waveform Generator

In certain auditory experiments it is desirable to combine 3, 4, or more harmonics of a given frequency with a specifiable amplitude and phase. The cost of building a separate sinewave synthesizer for each harmonic becomes prohibitive and has resulted in alternate methods of synthesizing harmonic complexes (6). One method of accomplishing this is to have a variable length circulating buffer which stores one cycle of the desired waveform. For example, if the 9th, 10th, 11th, and 13th harmonics of 100 Hz were to be combined, the basic or fundamental period of the complex wave would be

1/100 sec. or 10 msec. Therefore, a 10 msec. sample of this waveform must be stored and the desired stimulus duration achieved by repeating the 10 msec waveform. The maximum size of the buffer is to be 16,000 words. Therefore, at a sample rate of 64k, the maximum period can be 250 milliseconds. This period can be increased by increasing the length of the buffer or by decreasing the sample rate. This device can also be used as a simple buffer for delivering stimuli of arbitrary duration by merely transferring data from a disk file to the buffer such as speech, phonemes, or animal sounds. The output of the buffer replaces the output of the waveform generator as the input to the digital multiplier in the DSS.

Multiple System Interaction

A common experiment involves the presentation of two tones which are harmonically related (7). If both the frequencies f_1 and f_2 are integers there is no problem since the two synthesizers have a common clock. If, however, one wishes a 3:4 ratio with f_1 = 1000Hz, then f_2

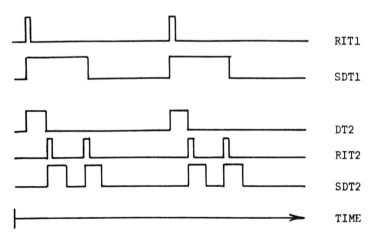

Figure 4. An illustration of one of the timing sequences possible with two Digital Stimulus Systems. RIT: repetition interval timer; SDT: stimulus duration timer; DT: delay timer. RIT1 triggers DT2 which initiates a sub-sequence fo DSS2 (the second digital stimulus system) consisting of two repetitions of stimulus 2.

Table I. The different synthesized signals, $g(t)$, when the second input to the frequency register is $MX(t)$, where $MX(t)$ is the output of a second DSS, f_0 is the carrier frequency and M is a fraction settable in the digital multiplier.

With three Digital Stimulus Systems it will be possible to generate an AM signal, that is $g(t) = 1/2(1+M \sin W_m t) \sin (W_c t)$, W_m is the modulating frequency and W_c is the carrier frequency, and M is the modulation coefficient, $0<M<1$.

Other types of interaction can occur when pseudo-random noise or a complex wave form is combined with a sinusoid as a probe tone. These combinations of stimuli are frequently used in psychoacoustic masking experiments.

$x(t)$ Modulating Waveform	$g(t) = \sin 2\pi (f_0 + MX(t))$ Synthesized Waveform
sawtooth	sweep frequencies from f_0 to f_0+M (or f_0+M to f_0)
sinusoid	frequency modulation (FM)
square wave	alternately present f_0+M and f_0-M
triangular wave	linearly sweep from f_0-M to f_0+M

= 1333.33 ... Hz. This is the reason a 16-bit fraction is included for specifying frequency, permitting the presentation of such a ratio with less than 1% phase drift for a 100 second stimulus duration by virtue of a very accurate representation of the second frequency.

It will be possible to synthesize several other complex signals. In order to accomplish this, the frequency register of one of the DSSs is replaced by a frequency register and second input which are added which allows the second input is varied in an appropriate manner, AM, FM, swept frequencies or an alternating frequency can be produced. These possibilities are listed in Table I.

DATA COLLECTION FACILITIES

The principal data collected in our laboratory consist of trains of all-or-none unit spikes. Other frequently encountered data types include averaged evoked responses and continuous physiological signals such as EEG recordings, respiratory variables, etc. Several devices have been designed which are used to collect these separate data types including a multiple event timer, histogram binners, a programmable A/D system and a digital spectrum analyzer.

Event Timer

The conversion of the all-or-none unit activity to a discrete sequence of times of occurrence is by far the most important activity in the lab. Each time the voltage recorded from a microelectrode exceeds a specified threshold value the time is recorded as indicated in figure 5. The event timer designed to perform this task resolves events by means of a time base which can be varied from 1 μsec to 999 μsec under program control. An 8.3 sec period can be timed with 1 μsec resolution without overflow in the 24-bit counter. The event timer can time up to 16 events and includes a 32-word buffer (a FIFO or First-In-First-Out buffer) to

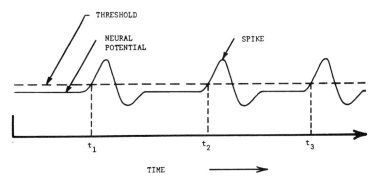

Figure 5. Whenever the electrical potential recorded with a microelectrode passes a predetermined threshold value, the time, t_1 is recorded. The sweep trigger circuit of an oscilloscope is used to discriminate the all-or-none spikes from the background noise.

store event times. This relaxes the need for rapid response to timer events by the computer system. Whenever the FIFO contains any information, an interrupt is generated for the computer.

The diagram in figure 6 illustrates the use of a 16-event scanner. Only events which are selected will be timed. Each time an event occurs a 'scan' is made of the 'event occurred register' for every possible event where the 'event occurred register' is a set of 16 flip-flops which are set by an event occurrence. If an event did occur, the count (time) is recorded and saved in the FIFO buffer. Simultaneously, the identity of the event is saved in the status register FIFO. The event timer can be turned on and off either by computer command or by a selectable sync (synchronization) pulse. The sync can arise in any of the three DSSs or an external source; one of the sources must be selected. The sync occurs whenever the signal is first turned on. The event timer is stopped by a terminate pulse which occurs whenever all the stimuli have been presented and the proper time has

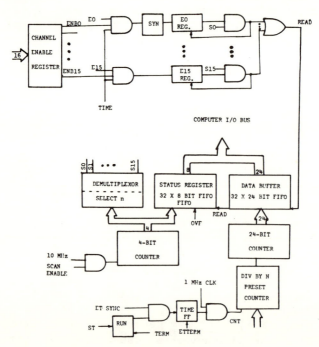

Figure 6. The event timer.

elapsed. Because of the use of the scanner the events are ordered. That is, the occurrence of simultaneous events is arbitrated by the timer. No decision about event order need be made. With a time base of 10 µsec, up to 9 events could occur at the same time and be recorded as such.

The event timer can be used to determine the frequency of an oscillator in two different ways. the first is by using the positive zero-crossings as an input to the timer. The difference in two successive event times is then an approximation to the period of the wave form. At higher frequencies, e.g., 10,000 interrupts/sec would be generated and 100 cycles would have to be timed if the accuracy is to be 0.01% at 1 µsec resolution. This interrupt rate is far too high for any computer's programmed I/O; therefore a preprocessor is included in the event timer to count cycles of an oscillator before the event timer is read. This timer sync is taken from this on the N+1st zero crossing. When the event timer sync is taken from this device the time corresponding to N cycles is read upon the occurrence of the terminate pulse and therefore frequency is simply determined by the relation $F = N/\text{Time}$.

The timer is reset to zero upon occurrence of the sync and read whenever one of the enabled events occurs. The status register is then read to determine whether an error occurred or what event occurred or what event caused the counter to be stored. Then the timer FIFO is read and the time stored in the proper memory location.

Analog-to-Digital Conversion System

The design objectives of the analog/digital converter system are simplicity of use and flexibility. It is a 16-channel system having sample rates from 0.001 Hz to 160 KHz with 12-bit accuracy. The A/D inputs can be used single-ended, differential or pseudodifferential. The channels to be sampled are program selectable, as is the sample rate and number of samples to be taken (figure 7). The device is connected to te computer on a DMA channel so that the high sample rates can be accommodated by the system. No handling of the data is necessary except to move it out of its buffer in the computer's memory. Programmed I/O use of the A/D is also possible if it is necessary to make a decision about the sampled

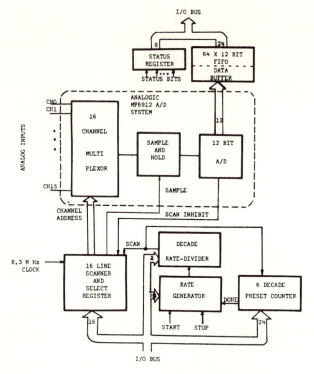

Figure 7. The A/D subsystem.

process in 'real time'. A 64-word buffer (FIFO) is included in the output of the A/D which relaxes the time constraints of the computer in responding to I/O requests by the A/D system when using programmed I/O.

LED displays indicate the sampling rate, the number of samples to be taken (or the current number of samples left to taken), the mode the system is in and the status of the A/D. The technique of generating the sample rate is identical to that used in the Digital Stimulus System. The A/D system is an Analogic MP6912 which includes a 16-channel multiplexer, a sample and hold amplifier and a 12-bit A/D converter.

Three sampling techniques are available in the A/D system: 1) Quick-Scan, in which conversion of each channel occurs as soon as the previous channel is converted until all selected channels have been sampled; the A/D then waits for the next clock pulse, 2) Uniform Interval, in which the next channel is converted after a clock pulse occurs; and 3) Rapid Sample, in which only channel is sampled, permitting samples to be taken at a 160 KHz rate.

The Quick-Scan technique allows one to sample all channels at the same instant in time. This is especially advantageous for low sample rates although it is desirable in general. In the Uniform Interval mode, the rate is the channel sample rate multiplied by the number of channels to be sampled. Sampled data is often analyzed using Fourier techniques, especially the Fast Fourier Transform (FFT). It is desirable to be able to collect 2^m samples in a given epoch so that the frequency resolution is a 'nice' number. For example, if the frequency resolution is to be 0.2Hz, the epoch is 1/0.2 or 5 seconds and, if the bandwidth is 30 Hz, the sample rate would have to be 102.4 Hz. This can be determined since the minimum sampling rate must be at least 2 x 30 Hz or 60 Hz and the minumum number of samples is 5 x 60 or 300. The actual number of samples, N, must be equal to $2^m \geq 300$, therefore $m = 9$ and $N = 512$. To obtain a non-integer frequency, the A/D system permits a 10, 100 or 1000 divisor to be specified or the sample rate. In this example the sample rate is set to 1024 and the divisor to 10.

Histogram Binners

One of the fundamental operations in auditory neurophysiology is the creation of phase or period histograms. This is a cross-correlation between the unit discharges and the positive (or negative) zero crossing of the stimulus waveform and has been called a cycle histogram. The abscissa varies from 0 to 2π radians or 0 to T sec, corresponding to one period of the applied waveform.

The auditory system generally behaves in a nonlinear fashion. The generation of harmonics and intermodulation frequencies has been studied by using two tones and examining intermodulation products. The IM product studied most extensively (7) is $2f_1-f_2$ where f_1 and f_2 are the frequencies of the primary tones. Other 'distortion' products can occur at the frequencies f_2-f_1, $2f_1+f_2$ and, in general at $(n+1)f_1 \pm nf_2$. It is desirable to determine the energy at each of these frequencies in the unit's response. In the past the tone-pair had to be presented each time a component was to be analyzed using a period histogram. With the Digital Stimulus System a very simple method is used to determine the bin in which a spike occurred since a register in the DSS contains the instantaneous phase of the wave form. This register can be read when a spike occurs. This register is updated every 0.12 sec so that the error (time jitter due to quantization) contributed by the measurement system is negligible. Each DSS can be used to compute a separate cycle histogram. In the case where the tones f_1 and f_2 are harmonically related, all the intermodulation products, primaries and their harmonics are multiples of a fundamental frequency. Therefore, if the third DSS is set to the fundamental frequency, all the information regarding the frequency content of the response can be recovered from the fundamental cycle histogram either by spectral analysis or curve fitting.

The intermediate phase register provides a 12-bit number which corresponds to a fraction of the cycle. Therefore, a 4096 bin histogram can be created. This fraction can also be multiplied by the desired total number of bins to obtain the number of the bin whose contents are to be incremented by 1. Expansion capability has been built into the system. If a second event (such as a second spike in a multiunit experiment) is to be binned a second buffer register can easily be added to each DSS. The identity of each histogram binner is encoded in the status word and is the first item decoded in a data transfer.

Digital Spectrum Analyzer

There are a number of signal analysis problems which require the determination of spectral energy at the harmonics of the stimulus frequency. While a device such as a wave analyzer can perform this task it is slow and cumbersome. A wave analyzer is a very narrow-band filter with a center frequency which can be varied, often under the control of a motor. Phase-locked amplifiers are also feasible for this task but they too are somewhat cumbersome for determining spectra. These devices are not particularly inexpensive either.

A device has been designed which will not only permit spectra to be determined (both amplitude and phase) but will also act as an average response computer (ARC) and a post-stimulus-time (PST) histogram computer.

It is relatively inexpensive since a programmable A/D system exists. Therefore, the only additional equipment needed is a 24-bit adder and a 4k, 24-bit memory system. A diagram is shown in figure 8. When the DSA is in the ARC (averaged response computer) mode, each sample that is taken is added to the contents of the current location and the address register is incremented. Up to 4096 samples can be averaged without danger of overflow. The DSA can be single- or multi-channel. The exact configuration of channels to be sampled is determined by the A/D system. In the histogram mode the input address is derived from the intermediate phase register of the DSS; the contents of the device are modified only if an event occurred. A sync pulse must be derived from the DSS. A count of the number of cycles is kept using the four decade preset counter.

In general one would like to use sample frequencies which result in samples per epoch so that FFT analysis can be used. As the stimulus frequency increases to greater than 20 KHz one can take only a few samples/cycle. It will then be advantageous to switch to a straight Fourier analysis so that $3f$, $5f$, $6f$, and $7f$ sampling frequencies can be used when appropriate. The highest sample rate obtainable from the A/D system is 160

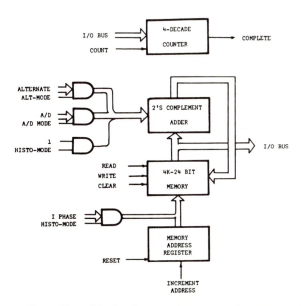

Figure 8. The digital spectrum analyzer or DSA.

KHz in the single-channel mode with 12-bit accuracy. Therefore information can be recovered for frequencies up to (160 KHz)/3 or 53 KHz, which is adequate for nearly all of the types of experiments we perform.

In using the system as a transient averager, the sampling process is initiated upon the occurrence of some synchronizing event. As many channels can be sampled as desired as long as the total number of samples is less than the available memory. The sampling process is stopped during a sweep as soon as the end of memory is encountered or another sync occurs. This may result in more data than is desired but one needs to read in only that data which is of interest. A variation of the ARC mode is to use the device to create PST histograms. The sync is the same but now a 1 is added to the proper location whenever a 'spike' event occurs. A maximum resolution of 4 μsec/bin can be obtained in ARC mode while 0.12 μsec/bin can be obtained in PST mode.

Graphic Display Subsystem

Much of the analysis of neurophysiological unit data is graphical and requires a reasonably fast display with which the user can interact during an experiment. While many graphic systems were available in 1972 when the design of this system was begun, all seemed to be too expensive or else they utilized storage CRTs and hence were not useful for dynamic displays.

The system that has been designed uses a 512 x 512 x 1 bit memory constructed out of 4k RAMs; it has a hardware vector generator with multiple line types, and a cursor which is both manual and programmable. It is an add-on to any alphanumeric terminal which uses standard TV format video. The vector generator is a hardware implementation of an interpolation algorithm described by Bresenham (8). This allow a dot to be written every memory cycle or every 2 μsec. This speed is fast enough to allow the display of rapidly changing data while continually refreshing the display. About 100 inches of vectors can be displayed every 10 milliseconds assuming a 50 point/inch resolution. The block diagram is shown in figure 9. Note that the alphanumeric terminal communicates with the computer via its own bit-serial channel.

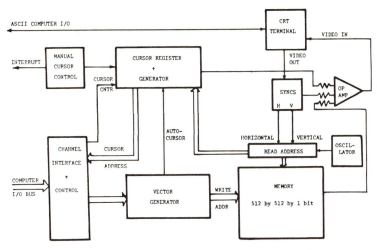

Figure 9. The graphic display is intended to provide dynamic display capability.

The unit provides multiple line types including solid, dotted and dashed with the length of the dashes being programmable. A non-storing cursor is included which can be positioned with a 4 key pad. The graphic unit is connected to the computer via a DMA channel. This allows the entire string of commands necessary to generate a complete display to be sent to the graphic processor without CPU intervention. The graphic unit interprets each word to determine whether it is a command or data.

The graphics processor could easily be expanded for greater resolution and intensity or color coding.

Digital Attenuators

The amplitude of the output of the Digital Spectrum analyzer is controlled with the use of a relay operated attenuator pad. The device has a range of 1 to 110 dB in steps of 1 dB. One attenuator is dedicated to each DSS. The device is programmable. A simple modification of the DSS will permit automatic control of the attenuator when the frequency is changed so that a constant SPL (Sound Pressure Level) can be maintained.

Digital-to-Analog Converters

These DACs provide voltages for the control of an external device for use in feedback circuits.

Computer System

The computer, a Harris 6024/5, has a 24-bit word length which permits the frequently occurring data types in our lab to be represented in a single word rather than using double word operands. This influenced our selection of the computer. Up to 14 I/O chanels can be connected; each I/O channel can have up to 16 units attached. And, for economy's sake, one can design a unit to contain an unlimited number of sub-units. This has cost-saving features due to the high standards that must be maintained when interfacing units to a Harris I/O channel. This approach was used for the channel with the Digital Stimulus System (figure 10). The limit of sub-units/unit was set at 16 in this system. The overall system organization is shown in figure 11.

Like most computer systems, two kinds of I/O are available: 1) programmed I/O and 2) DMA (which Harris calls ABC for Automatic Block Control). This latter mode allows the specification of memory address and the word count; the ABC channel will then transfer the block of data and generate an interrupt when the transfer is complete.

The Harris computer has a Disc Monitor System which supports multi-programming and foreground/background operations. It has several programming languages including FORTRAN, BASIC, RPG II, SNOBAL, Macro Assembler and FORGO, a diagnostic FORTRAN. It also has a reentrant editor called ACRONIM which has extensive file maintenance features.

The supporting peripherals are extremely important to the total system. The graphic display unit has been described. It is hoped that several of these units will be connected to the system. the Versatec 1100A electrostatic printer/plotter provides hard copy graphic output, a very important capability. An elaborate modification of the Versatec plot package was undertaken which resulted in an improvement in its performance by a factor of more than 100 in speed with many extensions being made to the package (9). A future expectation is to use the same graphic processor unit developed for the

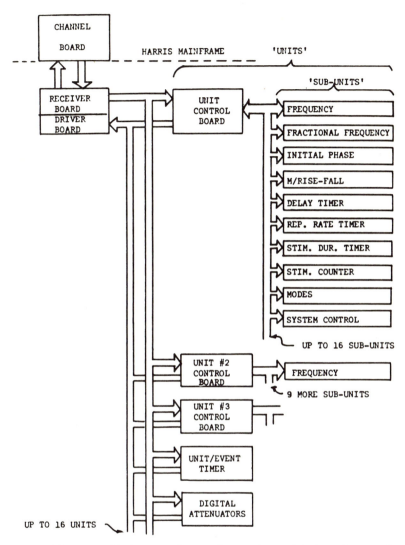

Figure 10. A 24-bit I/O channel which has been implemented using TTL logic. All the units reside in a single chassis so that the bus can be easily extended without incurring cabling costs. Each Digital Stimulus System consists of 10 subunits each of which can be addressed.

display as an interface to the printer/plotter. The buffer memory in this case would have to be either 1 Mbit or possibly 2 Mbit. The latter would allow a 10" x 20" plot to be handled. Given the plot speed of the graphic

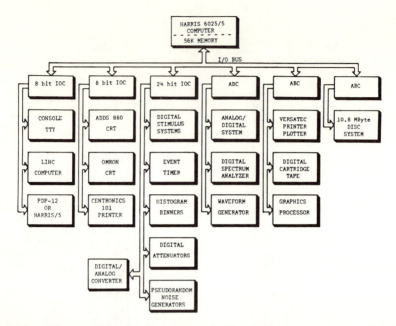

Figure 11. A block diagram of the Harris computer system with the 6 I/O channels. Three 24-bit ABC (Automatic Block Controller) channels and three programmed I/O channels are used. Each of a potential 14 channels can handle 16 units. Most of the devices in the system are self evident.

processor, the limiting feature would be the Versatec plotter. This embellishment would save time and free up core space and disk file space. It awaits lower memory prices.

Another important part of the computer system is the mass storage. While the disk cartridge drive (10.8 Mbytes) is a key feature of the system it is of limited utility for data storage due to the large amounts of data collected. Each user of the system is likely to require a method of storing and retrieving his data. LINC tapes (1.5 Mbit capacity) were the mainstay previously with 2500 tapes already in use in our lab. Recent advances in tape cartridge drives have considerably improved their performance. The Datacord drives (10) purchased for our system use a 3200 bpi - 8 track serial format. Each cartridge can store up to 150 Mbit (unformatted). Using a reasonable format, 60-70 LINC tapes can be stored on one of these cartridges. A Read/Write speed of 30 ips is standard with a Read/Search speed of 240 ips, also standard. The formatter controller we have specified will treat the device as a random access device. Each tape block will have an address written in its preamble which the controller uses to locate blocks of data.

Serial interfaces have been constructed between the Harris computer and a second Harris computer, between our LINC computer and the Harris, and between a PDP-12 and the Harris. These permit data transfers to take place when necessary.

SUMMARY

The digital equipment necessary for the performance of a wide variety of auditory neurophysiological experiments has been described. Many of the devices have a utility beyond our own lab. Much of this equipment has been designed and built within our own facilities to meet the specifications we developed.

It is expected that this system will continue to evolve indefinitely as unanticipated needs arise. Facilities for the design and construction of sophisticated equipment have proven to be indispensible to our research operations.

DISCUSSION

BROWN: Does your interval timer have a DMA mode?

RHODE: No, just accumulator transfer.

MCINTYRE: Is your display really cheaper than the GT-40?

RHODE: The parts cost would run around $3,000.

MORAFF: All of the plans I've heard over the past few years for hardware development have justified the development in terms of "when we planned to do this, it was the only way we could do it", either because of lack of capital equipment money, or because the manufacturers didn't provide quite the right thing, or even because someone wanted to learn how to do it by building one. I wonder how much longer this can go on.

RHODE: It depends entirely on what you need. The manufacturers don't make what I need. A commercial company could make something like this for $1000, and market it for $3000. But they just haven't done it.

When the chips get down to one dollar per 1k, you'll only have to spend maybe $250 for the memory. The vector generator could go on one chip, and there's not much left to the system after that.

MORAFF: How does your average neurophysiologist implement all this? We're a bunch of hotshots who know how to do all this -- we're not loners sitting out there in our labs with no electronics or computer training -- what about the unwashed masses? What do they do?

RHODE: If they're smart, they stay away from this stuff.

KEHL: You know, the answer is partly what George Baruch said in Aspen: he said to the biomedical community, "You ain't a market". When a manufacturer builds something, he has to sell at least 10,000 units to make a profit. You're just a market of a few hundred.

MORAFF: But wait until next year.

KEHL: Next year they'll need a market of 50-100,000. We don't even keep up, and our funding capabilities are going down instead of up.

FRENCH: I think waveform synthesis could be done much more easily with analog techniques like phase-locked loops.

RHODE: We looked at that. We wanted complete programmability of frequency and phase.

FRENCH: You can do that with phase-locked loops.

BROWN: How good is spectral purity of phase-locked loop generated sine waves?

FRENCH: As good as the number of diodes and resistors you want to put in your shaping circuit.

BROWN: Eighty decibel?

RHODE: Ours is 1:10,000, which is one of the determining factors. Triangle/sine-wave conversion gives about 1% distortion, so ours is about 100 times better. Anyway, a table look-up is about as simple as you can get. You can buy a memory for $40 if you don't want 16-bit output.

D. O. WALTER: Why do you need such high quality of sine wave?

RHODE: The best transducers available will give about -80 dB distortion, so we match the transducer. The ear is very sensitive, and we want to look at distortion products. So we don't want to put in the distortion products with the signal.

BROWN: To give non-auditory people some perspective, let's say we were stimulating an auditory neuron one octave below best frequency. What is the attenuating factor per octave of the tuning curve?

RHODE: Some of the slopes of our iso-rate intensity curves are as much as 60dB/octave. So a stimulus one octave down could present a suprathreshold input via the second harmonic and still be subthreshold for the fundamental. We aren't compulsive without good reason.

REFERENCES

1. Hind, J. E. and Rhode, W. S. Computer-based auditory 33:2351-2355 (1974).
2. Rhode, W. S. and Olson, R. E. A digital stimulus system. Monograph No. 2, Laboratory Computer Facility, Madison, Wisconsin (1975).
3. Tierney, J., Rader, C. M. and Gold, B. A digital frequency synthesizer. IEEE. Trans. Audio Electroacoust. AU-19:48-56 (1971).
4. Cooper, H. W. Why complicate frequency synthesis, Electronic Design 15:80-84 (1974).
5. Anderson, G. C., Finnie, B. W. and Roberts, G. T. Pseudo-random and random test signals. Hewlett-Packard Journal, (Sept. 1967).
6. Wolf, V. R. and Bilger, R. C. Generating complex waveforms. Behav. Res. Meth. and Instr. 4(5):250-256 (1972).
7. Goldstein, J. L. and Kiang, N. Y. S. Neural correlates of the aural combination tone $2f_1-f_2$. Proc. IEEE 56:981-992 (1968).
8. Bresenham, J. E. Algorithm for computer control of a digital plotter. IBM Systems Journal, 4(1):25-30 (1965).
9. Mansfield, M. GRAPH-PAC, Monograph No. 3, Laboratory Computer Facility (1975).
10. Blakeslee, T. Powering up a cartridge tape drive. Digital Design, February pp. 28-38 (1975).

chapter 34

THREE COMPUTER INTERFACES FOR NEUROPHYSIOLOGISTS

P. B. BROWN, D. DUFFY, AND T. W. McINTYRE

Department of Physiology and Biophysics
West Virginia University
Morgantown, West Virginia

In order to run multiple tasks simultaneously using a minicomputer like the PDP-8, it eventually becomes necessary to devise moderately intelligent peripherals which can run by themselves with a minimum of programming. In our own research, it is often necessary to simultaneously present complex patterns of multiple stimuli via multiple channels, sample interval data via several channels, and average voltage data via several channels. In addition, the CPU must often be free to perform online analyses on the incoming data. This requires three peripheral devices which can be preprogrammmed by the computer and which will run essentially autonomously once they have been programmed.

These three devices are the Pulse Interval Timer (PIT), Analog Digitizer/Averager (ADA), and Digital Stimulator (DS).

The PIT is a modified version of similar devices built for the PDP-8 by Dr. Jay Goldberg, and for the IBM 1800 and PDP-15 by Dr. Paul Brown. The work reported here was supported by USPHS grant NS09970 and NSF grant DCR74-24765.

THE PIT

The PIT consists of an interval timer, capable of digitizing intervals between input events (TTL pulses on any of six channels) with a resolution of 10 μsec.

The device operates in accumulator transfer or direct memory access (DMA, data break) mode, and has an internal First-In-First-Out (FIFO) buffer with maximum capacity of 32 words. Each output word consists of two twelve-bit PDP-12 words, with the format illustrated in figure 1. Each data word consists of six bits of input channel code (any combination of input channels may thus have simultaneous inputs) and 18 bits for interval code. Timer overflow is indicated by a zero interval code equivalent to 1,000,000(8) timer intervals; channel code for an overflow word may be zero or not, depending on which , if any, input channels were active during the clock cycle in which the timer overflowed.

The PIT is controlled using eight IOT instructions, listed in Table 1.

The IOT instructions are used to read or load registers in the PIT, the Preset Word Count (PWC), Word Count (WC), Timer (T), Preset Memory Address (PMA), memory address (MA) and Control (X) registers, and to skip on the DONE flag, which is bit 0 of the control register.

The T register is 24 bits long, and actually consists of an 18 bit counter, a six-bit channel input register and synchronization register, and a 32-word FIFO.

The 24 bits of the FIFO output word are read by two consecutive PTRA commands in accumulator transfer mode. In DMA mode, two 12-bit words are transmitted to memory on each PIT output.

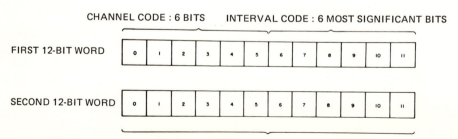

Figure 1. Format of 24-bit PIT output word.

Table 1. PIT IOT Instructions.

Instruction	Function
PTLA	Load PMA from accumulator by jam transfer. If PIT not running (RUN flag of control word = 0), loads PMA and MA. If PIT is running (RUN = 1), loads PMA only.
PTLX	Load X. Loads control word from accumulator by jam transfer.
PTLC	Load PWCC from accumulator by jam transfer. If RUN = 0, loads PWC and WC. If RUN = 1, loads PWC only.
PTRA	Read MA into accumulator in Data Break mode, read T in accumulator transfer mode (2 PTRAs needed for 24 bits of T register, 1 PTRA for 12 bits of MA).
PTRX	Read X (control) register into accumulator.
PTRC	Read WC into accumulator.

The remaining registers are all 12 bits long. The PMA contains the initial memory address for the next memory buffer to be filled after the currently active buffer is filled. The MA is the next address to be filled in the currently active bufffer. The PWC is the twos complement of the number of words to be filled in the next buffer, and the WC is the twos complement of the number of words remaining unfilled in the currently active buffer. This permits double-buffering with a maximum of time for the program to refresh address and word count values.

The PMA and PWC are jammed into the WC and MA each time the current buffer is filled, at which time the DONE flag of the X register is set. The PWC and PMA are loaded with the PTLC and PTLA commands. If the RUN bit of the X buffer is zero (PIT not running) the PWC and PMA drop through to WC and MA. If RUN=1, WC and MA are not loaded from the PWC and PMA until the next buffer overflow. Thus, PWC and PMA may be set up any time during the filling of the currently active buffer. The functions of the different registers are listed in Table 2.

Table 2. PIT Registers.

Register	Function
Preset Word Count (PWC)	In Data Break mode, used to (a) load WC register when PIT is not running (b) hold next value to be loaded into WC automatically on occurrence of DONE status. Loaded with PTLC instruction.
Word Count (WC)	In Data Break mode, contains the twos complement of the number of 12-bit PDP-12 words (e.g., twice the number of 24-bit T output words) lift to be filled in the current PIT input buffer in core. Loaded by PTLC via the PWC when PIT is not running, or by automatic transfer from PWC when DONE status occurs. Read with PTRC instruction. The WC is incremented after each 12-bit data transfer via Data Break (i.e., twice per 24-bit T output word). When the count goes to zero, the DONE flag is set.
Preset Memory Address (PMA)	In Data Break mode, used to (a) load the MA when PIT is not running; (b) hold the next value to be loaded into MA automatically on occurrence of DONE status. Loaded with PTLA instruction.
Memory Address (MA)	In Data Break mode, contains the next address to be filled in the current PIT input buffer. Loaded by

Table 2 (*continued*). PIT Registers.

Register	Function
Memory Address (MA), *continued*	PTLA via the PMA when PIT is not running, or by automatic transfer from PWC when DONE occurs. Read with PTRA (in Data Break mode only). Incremented after each 12-bit transfer via Data Break.
Control (X)	Used as a combined status and control register, and to transmit the Memory Field to the Data Break control. Some bits can be modified by the PIT. All bits can be modified under program control, jam transfered by PTLX. They are read by PTRX. To modify individual bits, PTRX is followed by appropriate bit modification, followed by PTLX.
Timer (T)	A 24-bit, 32-word FIFO and input registers, used to buffer up to 32 PIT output words. As each word enters the FIFO, it ripples through to the last unoccupied level. When the output level is occupied, DONE flag is set in accumulator transfer mode, or Data Break request is initiated in Data break mode. See figure 1 for bit configuration. Can be read by two consecutive PTRA instructions in accumulator transfer mode.

The functions of the 12 X register bits are listed in Table 3. Generally the PIT is initialized before sampling, and the DONE bit is cleared by the program after servicing each DONE event (output word in accumulator transfer mode, buffer overflow in DMA mode). The DONE bit is used to generate interrupts in the Interrupt Enabled mode. Note that the PIT can be paused, the timer can be cleared, and all internal flags can be cleared, by using appropriate X register bits. The ARM bit permits arming the TRIGGER input channel, a seventh channel which sets the RUN bit and starts the PIT when asserted. An external clock can be selected with the EXT CLK bit, permitting (a) a resolution of 10 μsec or longer, equal to the EXT CLK square wave period X 2; and (b) synchronization with external devices. The PIT has an INT CLK oscillator of 200 kHz, with an output which may be used to synchronize external devices.

Table 3. Control (X) Register Bits.

Bit	Function
0 = DONE	Set by PIT on WC ← 0 in Data Break mode (AC MODE = 0) or appearance of an output word in T register in accumulator transfer mode (AC Mode = 1), or by PTLX instruction with AC bit 0 = 1. Forces jam transfer of ᴾWC into WC, and PMA into MA, ᴏn PTLC or PTLA, respectively. Initiates interrupt if INT EN = 1. Cleared only via PTLA instruction, or BA INIT L.
1 = CLEAR PIT	When asserted via PTLX, disables PIT inputs and clears T register. Does not stop the timer or clear the trigger flipflop, although timer overflow will not be detected by the T register. The output word control WH II which controls which half of the T register output is read, is

Table 3 (*continued*). Control (X) Register Bits.

Bit	Function
1 = CLEAR PIT, *continued*	cleared so next PTRA in accumulator transfer mode or Data Break transfer will read the first 12 bits (channel code and upper 6 bits of interval code). None of the other control register bits are affected. Cleared only via PTLX, with AC bit 1 = 0 or BA INIT L.
2 = AC MODE	When asserted, via PTLX, accumulator transfer mode is selected. When cleared, via PTLX or BA INIT L, Data Break mode is asserted.
3 = EXT CLK	When asserted, via PTLX, selects external clock input. When cleared, via PTLX or BA INIT L, selects internal 200 KHz crystal oscillator.
4 = not used.	May be used by program as a program flag.
5 = INT EN	When asserted, via PTLX, causes interrupt request on DONE = 1. DONE must be cleared by interrupt service routine before interrupts are turned back on. When cleared, via PTLX or BA INIT L, interrupts cannot be initiated by DONE = 1.
6 = RUN	When asserted, by PTLX or by TRIGGER input when ARM = 1, PIT timer will increment on clock pulses. MA and WC are refreshed from PMA and PWC on (DONE = 1) and (AC MODE = 0), PWC and PMA

Table 3 (*continued*). Control (X) Register Bits.

Bit	Function
6 = RUN, *continued*	may be loaded without affecting WC and MA when RUN = 1. Contents of T register are unaffected by RUN bit. Cleared only via PTLX, or BA INIT L.
7 = ARM	Set only via PTLX. If RUN = 0 and ARM = 1, TRIGGER input will set RUN flag and start timer. Cleared by PTLX or BA INIT L.
8 = CLEAR	Set only via PTLX. When asserted, 18-bit timer (counter) register is held cleared, although PIT may still be running and generating output words. Cleared by PTLX or BA INIT.
9 - 11 MEM FLD	Set only via PTLX. Specifies memory field in core to which PIT output data is transfered via Data Break. Cleared by PTLX or BA INIT H.

The "works" of the PIT are diagrammed in figures 2-9. In figure 2, generation of timing pulse and operation of the internal timer is illustrated. Either the internal or external clock is used to generate four square waves, of half the clock frequency, at one quarter-cycle delays (C, C1/4, C1/2, C3/4). These occur only if RUN=1. These trigger 1 μsec. pulses from their positive transitions (Q1, Q2, Q3, Q4), at 2.5 μsec. intervals if the internal clock is used, or at other intervals if the external clock is used. These pulses control the sequence of events within the PIT. The timer consists of five 4-bit binary counters used to produce the 18-bit interval code. On cycling to zero, the TMR OVF H is set, signalling timer overflow. The overflow flipflop can be cleared by Q2 H or by TMR RESET H. Timer increments occur on trailing edges of Q1 H pulses. The timer is cleared on TMR RESET H, which can occur under

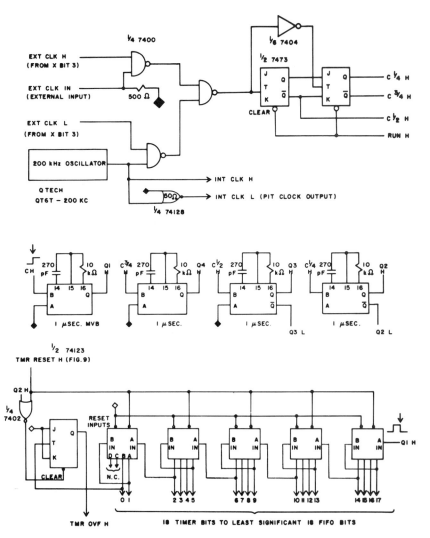

Figure 2. Generation of timing pulses and operation of interval timer.

PIT control whenever Q4 H is asserted and there is a set flipflop in the synchronization level. It can also occur on CLEAR COUNTER=1 (bit 8 of the X register), CLEAR PIT=1 (bit 1 of the X register), or the I/O clear pulse generated the the CPU at power on or issuance of appropriate program control instructions. These latter conditions are synchronized with CPU operation, not with the PIT timing pulses. See figure 9 for generation of TMR RESET H.

In figure 3 a timing diagram illustrates the temporal relations between events in the PIT. There are four "event times" synchronized with the PIT timing pulses: (a) the synchronization level flipflops, where input channel bits enter the synchronized portion of the T register, are loaded from the input level flipflops, which are in turn cleared by feedback from the synch level flipflops (figure 4), on the trailing edge of Q2 H. (b) On the trailing edge of Q3 H the FIFO is loaded with the current timer contents and the synch level flipflop bits, if any of the synch level flipflops are set. (c) Then the timer is cleared, on the leading edge of Q4, if any of the synch level flipflops are on, via TMR RESET H. (d) Finally, the synch level flipflops are cleared on the leading edge of Q1 H and the timer is incremented on the trailing edge of Q1H. Note that the timer goes to

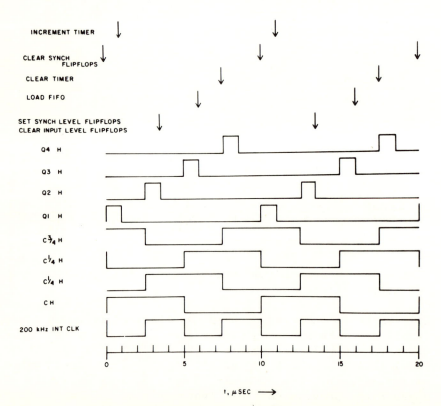

Figure 3. PIT timing diagram. Timing signals and the timing of internal events are shown.

000001(8) on the first Q1 H̄ after the synch level is entered on any channel, whether one of the 6 external event channels or the timer overflow.

In figure 4, the input synchronization is diagrammed. An input level flipflop can be set via the Schmitt-trigger NAND gate any time CLR PIT (X bit 1 = 0), except during the 6.5 μsec. the corresponding synch level flipflop is on. This only occurs during the 10 μsec. cycle following the previous arrival of a channel input, during which time a second arrival at the same channel would be considered simultaneous anyway. The input level

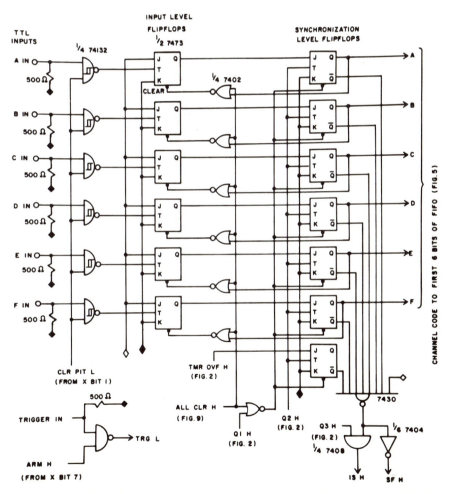

Figure 4. PIT input synchronization.

flipflop contents are loaded into the synch level flipflops on the next Q2 H trailing edge, and are cleared by feedback from the synch level flipflops which are turned on (other input level flipflops are not cleared, permitting input to them during the next clock cycle), or by ALL CLR H (figure 9). If any synch level flipflop is on, SF H goes high, and IS H goes on for the duration of Q3 H. Q1 H or ALL CLR will then clear the synch level flipflops.

In figure 5, the FIFO and output multiplexer are diagrammed. Contents of the timer and synch level

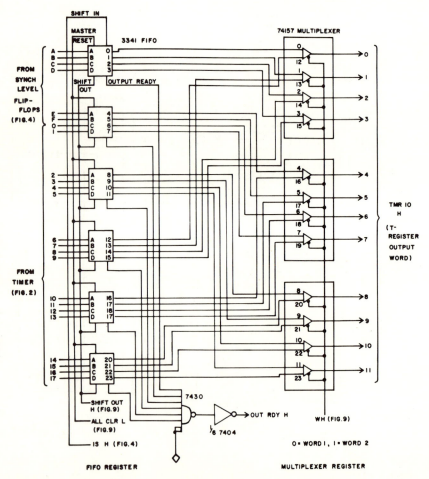

Figure 5. FIFO and output multiplexer.

flipflops are shifted into the FIFO on IS H (figure 4) and ripple down to the last unoccupied FIFO level. Presence of an output word in the FIFO causes OUT RDY H (X bit 0) to be asserted, signalling presence of an output word. WH = 0 selects word 1, WH = 1 selects word 2 for TMR IO H, the 12-bit word which will go to the PDP-12. Upon transfer of the 24-bit FIFO output words (two Data Break transfers to memory, or two PTRA instructions in accumulator transfer mode), the PIT logic (figure 9) generates SHIFT OUT H, which shifts the next FIFO word into the output. If there are no FIFO words besides the output word, OUT RDY H goes low and stays low until the next FIFO word is input. If there are more FIFO words, the next one is shifted into the output. In this case OUT RDY H goes low during SHIFT OUT H, and goes high again once all the 4-bit FIFO registers have settled (just after SHIFT OUT H goes low). The FIFOs, OUT RDY H, and WH are all cleared by ALL CLR L (figure 9).

In figure 6, the PWC, WC, PMA, and MA registers are illustrated. They are loaded via BAC H signals (accumulator output) into D registers (PMA and PWC) on the PTLA and PTLC instructions, respectively. The PMA and PWC are loaded into the MA and WC on word count overflow, or whenever PTLA or PRLC occur with RUN H off. The MA and WC are incremented at the end of each Data Break cycle initiated by the PIT, by ADDR ACC H (See figure 9). Word count overflow is signaled by WCO L and WCO H, a 1 μsec pulse triggered whenever PWC bit 0 goes to 0.

In figure 7, the X register is illustrated. The register is jam-loaded by the PTLX instruction, and cleared by BA INIT L, the CPU IO clear signal. In addition, DONE is asserted by a) word count overflow, or b) output ready status in the T register, in accumulator transfer mode. The RUN bit can be set by TRG L, generated by the TRIGGER input when ARM = 1. This permits the user to start the PIT with a TTL pulse into the TRIGGER input.

In figure 8, the generation of output words to the computer is illustrated. Open-collector NAND gates are used to produce low levels on the IO lines. The line to the accumulator is multiplexed, to provide the control word (PRTX), the word count (PRTC), the T register output words (PTRA in accumulator transfer mode) and the memory address (PTRA in direct memory access mode).

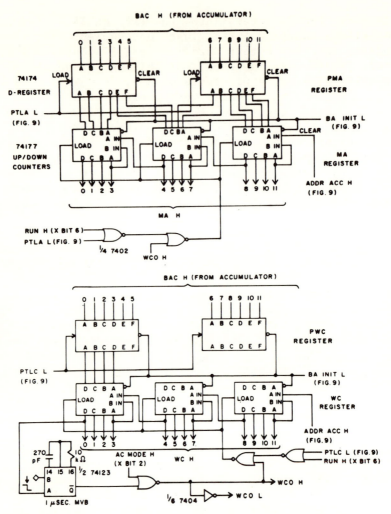

Figure 6. PWC, WC, PMA, and MA registers.

Figure 9 illustrates the generation of signals used to communicate between the CPU and the PIT. Interrupt requests result from DONE H (1) AND INT EN H (1); pulses occur on PTSK H (1) AND DONE H (1). Data Break requests occur on AC TR L(0) AND DONE L (1) AND OUT RDY H (i.e., data break mode, no word count overflow, and output data in FIFO register). W H, which controls which half of the 24-bit output is transmitted, is initially 0, and goes to 1 after transfer of the first 12 bits, controlling the multiplexer of figure 5. Initiation of DMA transfers is signalled by the PDP-12 with ADDR ACCEPTED L, and completion of transfer by BTS 5 H. On BTS 5 H, W H toggles, and when W H toggles a second time (back to zero), SHIFT OUT H is generated, shifting the FIFO contents forward. In accumulator transfer mode, AC MODE

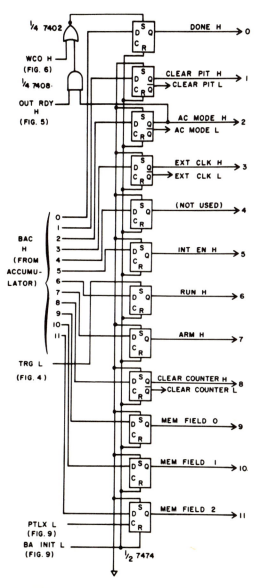

Figure 7. X register.

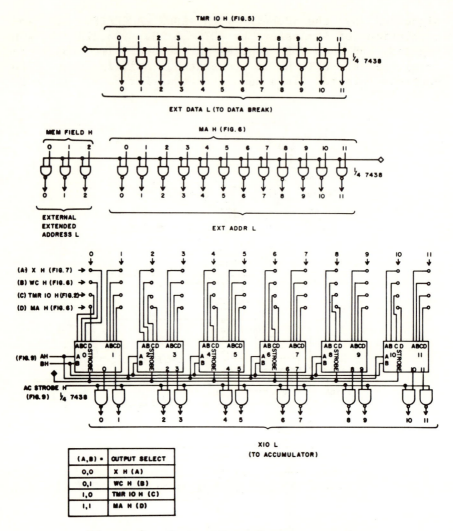

Figure 8. Generation of output words.

H (1) AND PTRA H (1) are used to toggle W H and produce SHIFT OUT H. The device-to-accumulator commands, PTRC L, PTRA L, and PTRX L are all used to generate AC STROBE H, which enables the output gates of figure 8.

TMR RESET is generated by the CLEAR COUNTER bit of the X register, Q4 H AND SF H (data enters synch level figure 4), or the PDP-12 hardware initialize signal, BA INIT H. This signal clears the timer and TMR OVF flag, without modifying anything else in the PIT.

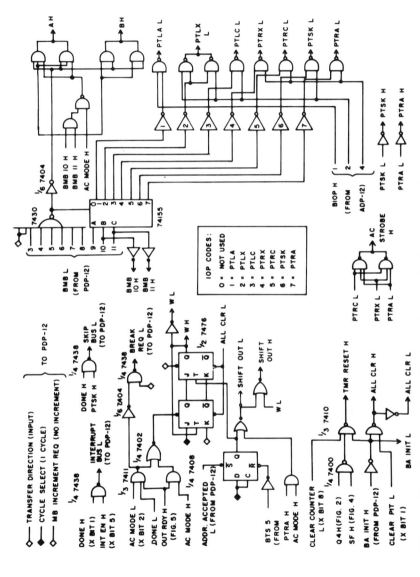

Figure 9. Generation of signals used for communication between PIT and PDP-12.

ALL CLR L and H are generated when BA INIT H or the CLEAR PIT bit of the X register is asserted. This signal has no effect on the PIT clock or the timer, but it clears the input and synch levels, blocks the input gates and clears the FIFO. The X-register is not affected, nor are the PMA, MA, PWC, or WC, by the CLEAR PIT bit of the X-register. However, all these registers are cleared by the PDP-12 hardware initialization pulse BA INIT.

The PIT was constructed on four double-length, double-width ("quad") DEC prototype boards. The entire device, not including labor or connector cables to the laboratory or the PDP-12, cost about $500, $300 of which went for the very costly DEC boards. We believe the design to be a very flexible one, permitting a wide variety of modes of operation, and easily modified.

THE ADA

The ADA is still in the design stage. As illustrated in figure 10, it will consist of an A/D input section, controlled by control registers similar in many ways to the PIT X register. The device will free-run, initiating a sampling "sweep" on each trigger input, sampling the specified channels at the specified sampling rate. An internal 8,000 sample memory will buffer output. Accumulator transfer and Data Break modes will be available, with and without interrupts. Data logging and averaging modes will be available, using an internal arithmetic unit for averaging. Three DAC outputs will provide a dynamic display of the contents of the current

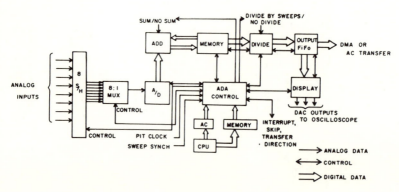

Figure 10. Block diagram of ADA.

sweep or the average, with channels vertically displaced from each other as separate "beams". Internal control registers will include PWC, PMA, WC, and MA, similar to those in the PIT. A clock derived from the PIT will be used, with an internal preset count register to control sampling rate during a sweep. The trigger input will be from any of the 7 PIT inputs, under program control, from an ADA trigger line, or from the DS lines, depending on programmed selection.

Due to the large number of internal control registers, the device will be programmable either by loading individual registers from the accumulator or by filling a program buffer in core and transferring it via the DMA facility.

The cost of the device will depend primarily on the cost of memory. The device can be built for $2000 by early 1976.

THE DS

We decided, after considering the use of D/A converters for analog stimulus generation, to generate pulsatile (TTL) outputs on 12 digital channels, and to use these outputs to control laboratory - resident devices for generation of analog waveforms. Thus, with appropriate control signals from these 12 lines, such analog signals as variable amplitude, gain, and offset analog pulses, trapezoids, pseudo-random noise, and even sinusoids, can be generated moderately easily under computer control.

The basic devices, of which there are 12, are the type illustrated in figure 11. We presently use them in the laboratory for stimulus control. To place them under computer control, they will be modified as diagrammed in figure 12. With the use of control registers, the 12 devices can generate a 12-bit output word. The 12 devices can interact with each other, with a trigger dispatch control and an external clock dispatch control (each one 12 bits X 12 bits), routing outputs to conditioning and count inputs. Each counter is programmed by loading its count, multiplier, and mode registers. The entire device, like the ADA, may be programmed via the DMA facility, to speed setup.

We estimate the device will cost less than $1200 to build.

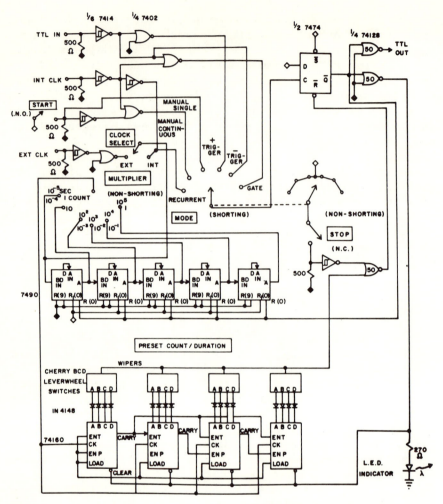

Figure 11. Preset counters presently used for stimulus programming.

These three devices, simple in concept but extremely versatile in their operation, will be fully documented, eventually with printed circuit masks made available from us. Construction costs would be minimal. We hope that by adoption of devices similar to these as standard laboratory hardware, neurobiologists will be able to quickly implement computer-aided research in their laboratories, with a minimum of computer science expertise. The adoption of such devices, and associated

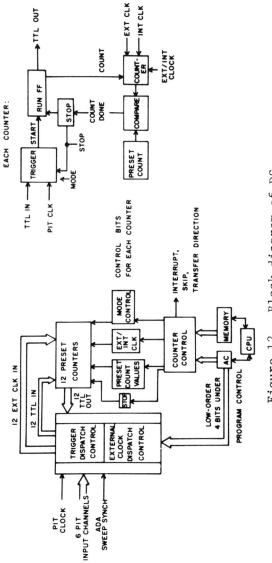

Figure 12. Block diagram of DS.

input data formats, is in our opinion an important step in the specification of a general-purpose computer operating system for neuobiologists.

DISCUSSION

L. PUBOLS: Will one of the analog inputs be the stimulus monitor?
BROWN: There's no reason why it couldn't be.

REFERENCE

1. Brown, Paul B., Maxfield, B. W., and Moraff, H. *Electronics for Neurobiologists*. MIT Press: Cambridge, Massachussetts.

chapter 35

USES OF THE LM² IN NEUROBIOLOGY

T. H. KEHL AND L. DUNKEL

Department of Physiology and Biophysics
University of Washington
Seattle, Washington

LOGIC MACHINES

A Logic Machine consists of a microprogrammable control processor, one or more bi-directional data buses, several functional units, and a microprogram, all of which, taken together, perform a digital algorithm. Logic machines have been built for a variety of digital algorithms: graphics display terminals, floating point processors, auxiliary processors and, in this case, a 16-bit minicomputer (Logic Machine Mini-computer: LM^2).

A Logic Machine is designed and constructed with the aid of the Logic Machine Design System. In this system, functional units are stored in a software library. When needed in a special application, these virtual functional units are collected together, and a paper tape used to control a semi-automatic wire-wrap machine is punched. Hence, we have no inventory of hardware modules. With the MSI-LSI (Medium and Large Scale Integration) which is now commonly available, an inventory works against flexibility. Little, if any, time is necessary to produce a Logic Machine as compared to other modular design systems, notably macromodules (1) or RTM's (2). In any event, we have found the Logic Machine Design System to be a rapid response design/construction system which accumulates designs as functional units and allows experimentation in the interaction between microprogram and hardware; that is, by our definition, firmware.

FIRMWARE APPROACHES TO NEUROBIOLOGY

A system allowing the rapid, easy design and construction of hardware optimized to a specific task has long been a goal of hardware designers. Clark and Molnar (1) introduced macromodules in 1965. Since then they have been able to build a variety of special purpose, high-performance devices. A macromodule system consists of a group of modules arranged so that a data path is formed and each module in sequence performs some operation on this data (figure 1a). Module A, the first module in the data path, would perform its operation, A_0. At completion Module A signals the next module, B, in sequence to take the data and perform operation B_0, and so C_0, D_0, etc. Signals performing the sequencing of data flow are also sent in parallel with the data. Alteration of an algorithm consists of changing the cabling for both data and control paths. Note that two or more parallel sequences of data can be obtained because control flows with data.

Bell, et al (2) produced the Register Transfer

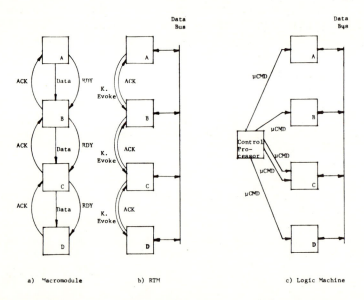

Figure 1. Different approaches to modularization: (a) Macromodules; (b) Register Transfer Modules; (c) Logic Machine.

Module (RTM) system, consisting of a bi-directional data bus forming a single data path. Signals to sequence data are passed from module to module by K (evoke) commands, just as with macromodules, but the single data bus prevents any parallelism. (figure 1b).

Programming either of these systems consists of resequencing the data flow; that is, rewiring the system. The Logic Machine substitutes a microprogrammable control processor to control the sequencing (figure 1c). Consequently, we are able to make major alterations to the data flow by remicroprogramming. Our control processor executes microinstructions in 100 ns (8-bit words) and 200 ns (double 8-bit words); these speeds are competitive with hardwired devices, and, for the most part, faster than macromodules and RTM. (Our higher speed really results from the fact that we are able to employ newer, faster circuits. Architecturally, one would expect a microprogrammable machine to be somewhat slower, given a constant set of modules.)

Microprogramming has another "quiet beauty" coming from the fact that, as computor hardware has become very inexpensive, large numbers of scientists and technicians have learned to program. A microprogrammable system is at home, i.e., familiar, in today's scientific world, whereas rewiring patch panels seems to most of us a return to older, more cumbersome techniques.

Why, then, do we not simply restrict ourselves to programming? Why incorporate a functional unit construction capability? Two major reasons justify the inclusion of special purpose functional units.

First, optimization of hardware gains maximum efficiency at low cost. For example, take the problem of acquiring multi-electrode data at reasonably high rates (10 KHz sample rate per electrode, for example). Few mini-computers are able to sustain sample rates much over 30 KHz. Of course no analysis can be done on incoming data; the computer can barely acquire the data in the first place.

Secondly, functional units can be specialized for character/string manipulation in such a way as to enhance and standardize compilation techniques without "forcing" language conformity. Software is easily the most expensive part of a system; consequently simplifying and speeding software development and execution has major advantages.

A/D CONVERSION WITH NON-TERMINATING BUFFER

LM^2 uses two 16-word scratch pads and two counters which act as DMA channels. When a memory request is made by a peripheral, the scratch pads at that level are simultaneously addressed. One counter is the Memory Address register and the other is the Word Count register. Thus, as the A/D is connected to level 6, word 6 of Memory Address Scratch Pad and of Word Count Scratch Pad are loaded into their respective registers. At "decode" time (3) the control processor cooperatively permits, under priority discipline, a branch to a special A/D microroutine.

A typical microroutine might be:

		Nsec.
ADMDR	A/D to Memory Data Register	100
WCSPMAR	Word Count to Memory Address Register	
CNTDN	Count Word Count Down (by one)	100
LDWCSP	Back to C Scratch Pad	100
MASPMAR	Memory Address to MAR	100
WRITE	Start Write Cycle	100
CNTUP	Count Up Memory Address	100
LDMASP	Back to MA Scratch Pad	100
DECODE	Return to Macroprocessing	200

Additional macroinstructions allow a scanning of a preset number of analog multiplexer channels at the maximum scan rate. At completion a clock is enabled and the next scan is started when the clock is done.

Non-terminating buffer operation is useful when one wishes to have a pre-stimulus period. The following commands are given to the A/D channel:

First Address of Memory Buffer
Length of Buffer
Number of Channels to scan
Location to store Trigger Memory Address
Start

At Start the channel begins filling the buffer. After the last location of the buffer is filled the next A/D conversion is automatically sent into the first location of the buffer. As long as no interrupts are sent into the system, non-terminating action continues.

An Interrupt (trigger) will cause a channel trap to pre-programmed location, starting an interrupt service routine. In this service routine an instruction will read the contents of the A/D's Memory Address register (actually in the MAR Scratch Pad) into memory. A simple computation determines how many post-stimulus samples are still needed and the A/D's MAR is jam-set to that value.

Both the MAR's value at trigger time and the run-out factor are stored with the data when it is sent to disk or tape. It is then easy to present pre- and post-stimulus recording in the proper order.

Note that the additional logic necessary to make a non-terminating buffer operation consists of adding microroutines. The advantage of this approach is quite obvious; given other requirements from another experiment, we could again change the microprogram to accomodate. For example, it could quite easily sample at a high rate of speed until some number (N) of samples were taken and then shift to averaging over groups of K samples, storing the averages. Alternatively, we could compute signal averages, make interspike interval histograms, or nearly any function used in neurobiology -- just by re-microprogramming. Firmware, in our opinion, provides the speed, flexibility and low cost to greatly simplify these problems.

CHARACTER/STREAM FUNCTIONAL UNITS FOR COMPILATION LEXICAL ANALYSIS

Recently a graduate student (Kenneth Burkhardt) completed a dissertation in Computer Science delving into some issues of loosely coupled computer systems (4). His goal was to develop an arithmetic auxiliary processor to be added to any mini-computer. Both scalar and vector floating point arithmetic hardware math functions and fast-Fourier transforms are done in firmware by this Logic Machine. In fact, by remicroprogramming, nearly any function we can think of can be performed.

Vector arithmetic is extremely fast. For example, when two vectors are operated on to produce a third, the time spent is accounted for just by the arithmetic time; no apparent time is spent in indexing or running the program. Thus a 16-bit mini-computer can obtain performance on vectors formerly restricted to very large machines.

It occurred to us that we might be able to build firmware for operations other than vector arithmetic. Character/stream manipulation is useful for a variety of purposes, notably text editing and certain compilation phases. Both of these tasks consume a great deal of computer time, not only when developed originally, but also during use. Computers that execute high level languages directly have been built (5), but we felt that hardwired compilers were too inflexible. An optimum between inflexible hardwired and flexible, but too detailed, software, is desired.

Designing a compiler giving the desired characteristics is the first task. Push-down stacks were eliminated as both too slow and requiring too elaborate hardware. Something much simpler was needed. In a previous chapter (6) we showed some state logic methods for lexical analysis. Determining a lexeme is the fundamental problem wih lexical analysis, but it is a problem only because multi-character lexemes are allowed. For example, F O R must be recognized as FOR, a single token. If a character/stream functional unit were developed to handle groups of characters as easily as single characters, then nearly all of lexical analysis would be eliminated. The FOR would be recognised immediately. But it is not only in compilation that substring matching of this sort is useful. Text editing and analysis can also use these facilities, as can call-by-name file directories embedded in operating systems. Thus, the hardware is not special purpose, but can be used in a variety of applications (admittedly, applications are limited to character/streams).

ESTABLISHING PRECEDENCE IN ARITHMETIC EXPRESSIONS

Some time after the completion of lexical analysis a compiler must determine the precedence of the operations, i.e., a) exponentiation, b) multiplication and division, and c) addition and subtraction, cover the arithmetic operators. In addition, reserved words, subscripts (parentheses), system subroutines and user subroutines all have precedence relationships, one to another. A portion of compilation is concerned with precedence.

Our firmware approach looks up each operator in a precedence table as it is recognized in the lexical analysis and tags that operation's precedence. Highest precedence goes to left parenthesis, followed by right parenthesis, and so on until all have been established. These are entered in a vector next to the location of the operator's entry (see figure 2). The actual values for precedence are not important; only each value relative to another has meaning.

Next a cumulative precedence is computed by a vector operation, from a scan of precedence. A counter (CUMPRED) starts at zero. Then:

1. If $Precedence_j$= 1000 (a left parenthesis), then
   CUMPRED =CUMPRED + 1000
   CUMPREDCOL = CUMPRED
2. If $Precedence_j$= 990 (a right parenthesis), then
   CUMPREDCOL = CUMPRED - 1000
3. If $Precedence_j$= 100 (an equal sign), then
   CUMPREDCOL = CUMPRED
   CUMPRED = CUMPRED + 10
4. If $Precedence_j$> 0 (some other operator), then
   CUMPREDCOL = CUMPRED + Precedence

A (I,J) = B + C (I + 2, K*3)/ Original line

Lexemes	Precedence	Cumulative Precedence
A		
(	1000	1000
I		
,	10	1010
J		
)	990	1000
=	100	100
B		
+	321	331
C		
(	1000	1010
I		
+	321	1331
2		
,	10	1020
K		
*	322	1332
3		
)	990	1010

Figure 2. Parsing a simple arithmetic expression.

The vector operation which builds the cumulative precedence follows the non-linear rules described above. Finally, a Min-max scan of the cumulative precedence locates the high precedence operator, a generator produces the apropriate code, the table is collapsed and a new Min-max locates the next highest precedence operator until all have been retired.

Details concerning the generated code are not of primary interest, and are not discussed here. What is important is that very little effort is needed to produce a language with this firmware. A side-benefit is a major increase in compilation speed, since most of the operations are vector-like and run at memory bandwidth speeds rather than program speeds (at lest five to ten times faster).

Most important, with such firmware a simple, consistent method of developing a computer language is available. Computer languages are not created exclusively by computer scientists. Any communications with a computer require language, and neurobiologists, sooner or later, must communicate using their own individual terms and expressions. We expect that this firmware will make language development both simple and uniform.

DISCUSSION

DUFFY: You said the operations on this machine were on the order of the 6400. That's a 60-bit machine.

KEHL: Yes, good point. The 6400 has a lot more bandwidth than we have, with a 60-bit word. If your problem requires 60 bits you should be using a 6400. That's not that common, but sometimes you need it.

MCINTYRE: Your semi-automatic wire wrapper looks like a good idea.

KEHL: You have to drive it with a computer. It has a positioning accuracy of a thousandth of an inch. We've wrapped at least a few hundred thousand wires, and we know of only 12 errors, all done very early on. The machine has probably paid for itself ten times over.

MORAFF: What kind of saving is there over doing it manually?

KEHL: At least one hundred fold. I can train an undergraduate to do it in fifteen minutes, and do a different board every day.

OLCH: Is your software exportable?

KEHL: No. It's too complex, and it's the key to the whole system.

RHODE: Have you considered extending the word length?

KEHL: Yes, the bus is expandable, you can make it as wide as you like. The auxiliary processor will almost certainly be 32 bits wide. We can have more than one bus on the same processor, incidentally. We've done it.

RHODE: What's the bus speed?

KEHL: 100 nanoseconds per transfer. Our system now uses BASIL, our macro-level language (Basic System Implementation language), which is itself written in BASIL. The overall objective with BASIL is to develop the concept of a personal, private computer, one you can have in your own laboratory, which most scientists can afford, and provide all the computational power you could ever possibly use.

BROWN: When?

KEHL: BASIL will be up by September for sure.

BROWN: What's the strategy for dissemination?

KEHL: I'll send you the tape. I've considered producing it, and I don't want to. But let me give you a cost analysis. You can build an LM^2 for less money than you can purchase a Honeywell 7600 FM tape transport, and do most things the transport would do, much better. That doesn't say people in our department aren't continuing to buy Honeywell 7600 tape transports, but we can easily beat them out. In fact, we can beat them by a factor of 2, for $10,000. We actually have a program going, trying to get people to use our version. Anything to get them converted.

REFERENCES

1. Clarke, W. A. Macromodular computer systems. In *Proc. 1967 Spring Joint Computer Conference*, (1967).
2. Bell, C. G., Grason, J., and Newell, A. *Designing Computers and Digital Systems*. Digital Press: Maynard, Mass. (1972).
3. Kehl, Moss, and Dunkel. "LM^2 - A Logic Machine Mini-Computer". To appear in "Computer".
4. Burkhardt, Kenneth. A Logic Machine Auxiliary Processor. Ph.D. Thesis, Computer Science Department, University of Wasington.

5. Rice, Rex. The Hardware Implementation of SYMBOL, Compcon, (1972).
6. Kehl, T. H. On uniformity of computer interface design. In *Computer Technology in Neuroscience*, ed. P. B. Brown, Hemisphere Publishing and John Wiley and Sons: Washington, D.C. (1976).

chapter 36

ON UNIFORMITY OF DIGITAL COMPUTER INTERFACE DESIGN

T. H. KEHL

Department of Physiology and Biophysics
University of Washington
Seattle, Washington

Methodologies for computer system design -- both hardware and software -- are a major area of computer science which could have benefit in neurobiology. Software engineering (1), compiler construction (2), logical design of operating systems (3), macromodules (4), RTM's (5) are examples of specific searches for methodologies to simplify each of these respective design tasks. It is gratifying for computer scientists to see a unifying thread emerge from all of these efforts. We shall refer to this thread as State Logic. (Others may call it Sequential State Machines or Graph Theory, etc.) Regardless of what one may call the methodology the point is that a single approach to a wide variety of computer-related design situations is now becoming evident. Computer system development destined for biomedical research could benefit from this methodoloy but, thus far, nearly all biomedical computer systems seem to have been designed ad hoc - i.e., the antitheses of a methodological technique. As this conference is meant to explore the computer requirements of neurobiology, with the view to, perhaps, implementing a system, we urge the use of methodological, rather than ad hoc, design.

In this chapter we shall review a specific methodology (State Logic), outline implementation techniques/problems at several levels of system design and show a consistent thread of design methodology in both hardware and software. But first an overview of State Logic, uncoupled from any specific implementation, will be presented in a tutorial fashion.

STATE LOGIC

State Logic is derived from automata theory and will be familiar to the reader as represented by flowcharts. A state diagram is an abstract representation of a series of time- sequential steps which, when constructed as hardware, results in the execution of an algorithm. Conventionally, hardware is divided into two areas; data part and control part. The data part is simply the data path through the machine and, with MSI and LSI, quite easily implemented. The main concern is always with the sequencing of the device. Consider the pair of state diagrams: Several aspects of state diagrams should be noted. 1) Each state is unique and may cause several unique actions. 2) Only a single state of a state diagram will exist at any instant although two or more state diagrams may be simultaneously active, 3) transitions from state to state may either be conditioned on Boolean variables or may be unconditional, 4) transitions are uniquely defined by the state diagram, i.e. a transition connects one specific state with another specific state and no others, 5) all realized digital systems are represented by closed state diagrams, i.e., each state has at least one transition in and at least one out.

These may be taken as the "rules" for making a state diagram to describe the sequential actions desired of a device. Inevitably the reader will be troubled by some of these statements and so, in anticipation of the difficulities, one should consider the following, corresponding in numerical order to the above statements: 1) If a state were not unique then ambiguity could exist -- and ambiguity is anathema to digital logic; 2) derives from 1) but introduces parallelism by allowing a second state diagram; 3) the data path, in response to data variables, must be able to condition and alter the sequential nature of the state logic; however, all conditioning must be done as Boolean variables -- arithemetic variables must be reduced to Boolean variables before use; item 4) is obvious -- the state logic is sequenced through the steps as indicated and hidden transitions are not allowed; 5) if a state had only an "in" transition then the machine could be permanently trapped in that terminal state -- furthermore, a state which cannot be entered cannot exist.

POWER OF THE METHODOLOGY

It can be shown that any sequential device can be reduced to a sequential state machine (implying a representation by a state diagram.) Furthermore, a minimum sequential state machine is minimal for that operation -- no other technique will permit so few steps of logic. (see Zohavi (6)).

These are very powerful statements: The first says that state logic is universal and the second says that it is best. (Best is very difficult to define -- often we mean minimal but here we are trying to use a broader meaning to imply most powerful, easiest to understand, most consistent with all problems, etc.).

Historically it is the one technique which has stood the test of time. Even though other techniques are actively being explored (Petri Nets, for example, are the inverse of state diagrams) the weight of previous research and the fact that many computer devices have been implemented in state logic means that it is the methodology of choice for all practical situations.

HARDWARE IMPLEMENTATION

A Code-converting Console Printer Interface

Shown in figure 1 are the state diagrams for an interface between a computer and a Texas Instruments Silent 700 thermal printer. In this case the computer uses a character-code other than ASCII and, to avoid reprogramming labor, a ROM table is used as a code converter. In operation, an ASCII character transmitted from the keyboard undergoes a table look-up (by scanning a ROM table, states 4&5). The result of this look-up is the translated character which is then transmitted to the computer. Thus, no alteration to the T.I. thermal printer, the computer or the software is required.

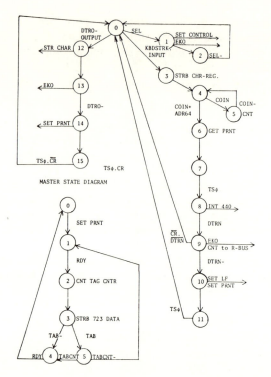

Figure 1. State Diagram of Thermal Printer Interface.

This interface uses two state diagrams; A Master State diagram for overall control and a Procedure State diagram for printing. Printing occurs both as a result of computer output and whenever a keyboard character is struck. Rather than duplicate portions of the Master State diagram, Procedure State logic is used -- just as one would use a software subroutine.

POSITIVE ASSERTION LOGIC VS. DEFAULT LOGIC

Positive Assertion Logic is defined as the logic which allows transitions only as the result of assertions. The antithesis of Positive Assertion Logic is Default Logic. Default Logic works by implication: if not A_1, and not A_2,...and not A_j,..., then it must be (by default) A_n.

Positive Assertion Logic covers all eventualities of a system; nothing is left for the Default case. We strongly urge the use of Positive Assertion Logic because it is impossible to accurately define all of the Default cases -- all error conditions are Default cases and must be included. However, it is difficult to enumerate all error conditions. Consequently, if all transition possibilities except the default possibility are checked then one still cannot conclude it is the default possibility because it may be an error. But which error?

In our experience the use of Default Logic causes so many troublesome system problems that it should never be used. A very simple State Logic scheme to avoid Default Logic is to make State Logic tables (figure 2). A State Logic table is a 2-dimensional array of states by transitions. A ROM of this table, then, represents the next state which is assumed in response to any transition. All entries must be filled, even though some entries have no meaning. The net effect is to force the designer to consider every possible transition, and no Default cases can creep in.

STATE LOGIC IN SOFTWARE

All scientists must be able to communicate with their computers. As disciplines such as neurobiology become increasingly sophisticated in computer usage, and consequently communications become more specialized, it will continue to be necessary to construct specialized computer languages (7,8). State Logic can make uniform methods of producing computer languages as well as hardware.

Two examples of State Logic in software will be given here: 1) Lexical Analysis during compilation and 2) task specification of an operating system.

Lexical Analysis is required to identify a group of contiguous symbols, bounded by delimiters, as a term. For example, FOR must be recognized as FOR opens a FOR-loop. Our scheme is to implement a table-driven analyser (figure 2) examining one character at a time, while keeping track of the state. Whenever a delimited (completed) Lexeme is identified, it is tokenised. Then the Lexeme is replaced with a token so its identity is established. Parsing for code generation consequently only deals with the resulting tokens and operators.

In operation the Lexical Analyser is always in a specific state. The character under consideration is

State	Character				
	Null	Letter	Number	Operator	EOL
1	1	2	3	4	5
2	X	2	2	6	7
3	X	2	3	8	9
4	X	---------Language		Dependent--------	
5	1	2	3	4	5
6	X	2	3	4 (**,<=etc.)	7
7	1	X	X	4	5
8	X	2	X	4	X
9					

State 1: Null State
2: Symbol Building
3: First character after null is number - (could be error)
4: First character after null is operator - (could be error)
5: Null line - get next line
6: Symbol complete - store in symbol table and token-replace
7: Last symbol complete - store in symbol table, token-replace and get next line
8: Number complete - make into valle, store in symbol table and tokenise
9: Same as 8 but also get next line

Figure 2. Table Driven Lexical Analyser.

then examined and a transition to another state is made as specified by the table. One can view the operation as though a specific row of the transition table is activated by the current state, and transitions can occur only to those states listed in the activated row.

OPERATING SYSTEM

Operating systems must be able to start and stop processes which, while running, are quite independent.

Loosely coupled subroutines of this type (usually called coroutines) have the appearance of running in parallel; in fact, coroutines run sequentially as does everything else in a digital computer. Activating coroutines depends on what is already active and whether the resources for activation are available.

To illustrate the use of State Logic in this application consider the following rather common operating system problem for neurobiologists (figure 3): 1) A/D data is to be acquired from an experiment, 2) acquisition is to start in response to a trigger, 3) data is double buffered, 4) while one buffer is filling, the other is to be dumped onto tape, 5) the last completed buffer is to be displayed on the oscilloscope.

MICROPROGRAMMING AND FIRMWARE STATE LOGIC

A microprogrammed computer is controlled by quite fast microinstructions grouped into microroutines. Each microroutine is equivalent to a conventional machine language instruction.

Our definition of firmware is somewhat more narrow than that usually used. Firmware, for us, is the combination of microprogram acting in conjunction with specific (usually special purpose) hardware. Not only is the microprogram specially written for the task, but the hardware is task-specific as well.

Any program (microprogram or macroprogram) can be represented as State Logic -- the flowchart known to all of us. We have, in the LM , extended the use of State Logic to implement transitions in a very direct way. Directness not only makes implementation easy but self-documents and provides maximum speed.

As background, nearly all microprogrammed computers construct, from the macroinstruction, a transition to the microroutine. Usually the mechanism is similiar to indirect addressing where the macroinstruction register is connected directly to the address input of a memory-type (ROM or RAM) device.

In the LM^2 (described in a companion chapter in this book (9)), we extended this addressing mechanism to include microroutines for peripherals as well. Briefly, several addresses representing requests to execute a corresponding number of microroutines are resolved in a priority network. The highest priority peripheral

Input State	A/D Block Change Trigger	Input Buffer Overrun	Output Buffer Complete	D/A Output Buffer Complete	Finish Off Signal	Start
1	2	X	X	X	7	X
2	4	3	X	X	8	X
3	3	3	3	3	9	X
4	4	3	5	6	10	X
5	4	3	5	6	10	X
6	4	3	5	6	10	X
7	2	3	3	3	X?	X
8	8	3	1	3	8	X
9	9	9	9	9	11	X
10	10	9	10	10	X	1
11	11	11	11	11	11	1

State 1: Null State
    2: A/D block but no output
    3: Taking data too fast for size of buffers - error
    4: Switch pointers to continue filling other buffer - start secondary storage output, start D/A display
    5: Continue input and D/A display
    6: Restart D/A display, continue A/D
    7: Wasn't doing anything and told to quit - print condition
    8: Dump first buffer
    9: Taking data too fast for size of buffers and told to quit
   10: Finish off signal present - fill and dump final buffers
   11: Waiting for start command

Figure 3. State Logic to control coroutines: A/D Input, Display, and Tape Output.

generating a request is allowed to place its microroutine address in the microengine's PC, thus causing a branch to that microroutine. Then, simply by defining macroinstruction execution as lowest priority (i.e. if nothing else requires processing then execute macros) microroutines are allowed to run in a quasi-parallel fashion as co-routines.

The point is that by considering the microroutine branching function as several completing transitions to coroutines we were able to build a minicomputer which simply, and yet at very high speed, controls coroutines not only for the CPU but for peripherals as well.

UNIFORMITY OF INTERFACE DESIGN

By now it should be clear that uniformity of interface design means much more than interfacing a peripheral to a computer. In the preceding we have given specific examples of how one might use State Logic in many facets of a system; both hardware and software. The main advantages of State Logic implementation are 1) universality, 2) simplicity, 3) economy of parts, 4) economy of documentation/ understanding, 5) rigor, 6) implementation independence. We would strongly urge anyone undertaking a system design for neurobiology to use State Logic.

Let us examine some of the advantages in detail in order to fully appreciate their potential. Universality means that any facet can be implemented, and as simply as possible. Parts (I.C.'s, program statements, etc.) are quite inexpensive to obtain but quite expensive to keep in operation. Reducing the number of parts makes it easier to identify a failing part.

Most importantly, understanding how a system is supposed to operate is necessary to build it, maintain it and improve it. In our view a formal, rigorous approach such as State Logic allows computer personnel to approach each of these aspects knowing in advance a great deal about the system. That is to say, without even opening a manual one will know a great deal about system organization if he knows it is done in State Logic.

Finally, independence means that one can discuss the operation of a system without mentioning flip-flops, gates, dimension statements, or any of these details. It can be used to describe, from one person to another, the flow of processing wihout stating implementation.

But whether or not State Logic is used in the implementation of a system, some consistent methodology should be used to save us from most of our computer chaoses.

DISCUSSION

D.O. WALTER: What goes wrong with this kind of thing?

KEHL: What goes wrong with anything? An overvoltage can burn out an IC. Usually that kind of failure. There's one sneaky problem, involving hazards in race conditions, but that's another story in itself. By the way, let me make a comment: This approach eliminates most of the electrical engineering courses in logic design. That's what they talk about all quarter: mini-max, all that. You can forget it, because you can't do it more minimally than this. At the gate level, of course you've done all possible decodes, and that's certainly not a minimum. But who cares? The chip costs less than two bucks. In package count, you can't be more minimum than that.

MCINTYRE: That looks like register transfer logic.

KEHL: It's not. I'm talking about register transfer logic in the next paper. First of all, register transfer logic is asynchronous. This is completely synchronous.

REFERENCES

1. See *IEEE Transactions on Software Engineering*.
2. Gries, D. *Compiler Construction for Digital Computers*. John Wiley and Sons, Inc. (1971).
3. Shaw, A. *The Logical Design of Operating Systems*, John Wiley and Sons, Inc. (1974).
4. Clark, W. A. Macromodular computer systems. *1967 Proceedings of the Spring Joint Computer Conference* (1967).
5. Bell, C. G., Grason, J. and Newell, A. *Designing Computers and Digital Systems*. Maynard, Massachusetts: Digital Press (1972).
6. Zohavi. *Switching and Finite Automata*, McGraw-Hill, New York (1970).

7. Kehl, T. H. Interactive real time computation. Computers and Biomed. Res. 1:590-604 (1968).
8. Ransil, B. J. Use of table file structures in a clinical research center. FASEB Proc. 33 (1974).
9. Kehl, T. H. and Dunkel, L. Uses of LM^2 in neurobiology. In *Computer Technology in Neuroscience* ed. P. B. Brown, Washington, D.C.: Hemisphere Publishing and John Wiley and Sons (1976).

chapter 37

HIGH LEVEL LANGUAGE COMPUTERS: POTENTIALS AND LIMITATIONS IN NEUROBIOLOGY

T. H. KEHL

Department of Physiology and Biophysics
University of Washington
Seattle, Washington

INTRODUCTION

It is probably not possible, in terms of the current market place, to restrict our attention to neurobiology. A vendor cannot view neurobiology as a market because far too few computers are sold to neurobiologists. I would estimate that no more than 100 neurobiology computers are sold in the United States on an annual basis. Nearly all are minicomputers and competition for minicomputer sales can only be described as fierce. Because of this fierce competition a vendor cannot build highly specialized systems--either hardware or software. A vendor has to estimate his potential market in terms of many thousands of units. Neurobiology simply is not large enough to be a market. While it is true that most vendors will build specialized equipment for a small market, the cost is beyond anything neurobiology can hope to fund.

Most of us are of the opinion that current computer systems are not adequate to meet the needs of neurobiology. Indeed, a major focus of this meeting is to determine what can be done to alleviate this problem. It seems to me that there are three alternative courses of action: 1. Acquire vendor supplied systems (both hardware and software) and limit neurobiology research to within the capabilities of the system; 2. Modify vendors'

hardware and/or software for more capabilities;
3. Build one's own computer system. Each of these
alternatives has its own virtues and drawbacks. Since
our laboratory favors the third alternative and because
this approach is very controversial, most of this
discussion will focus on it.

STRICTLY VENDOR SUPPLIED SYSTEMS

An axiom of computer design is "Minimize the layers
of logic". Following this axiom, probably the best
advice is to limit a computer system to the application
either in progress or foreseeable. One must remember
that a system larger than is needed includes components
not wanted. If it were the case that the unneeded
components could simply be ignored, then no problem would
exist. However, this is often not the case. Take for
example a time-sharing system purchased with the (vague)
eventual prospect of doing time sharing. If the primary,
first uses are for on-line experiments the extra hardware
and software will almost certainly get in the
experimenter's way. Assuming the experimenter does not
want to modify the system himself he will have to try to
reconfigure from the vendor's stock shelf; he will have
to backtrack. Invariably the extra cost and frustration
will seriously and perhaps permanently thwart the
experimenter.

I have put the case for overpurchasing first because
it is at least as serious as underpurchasing. All of us
are familiar with the problems encountered from too
little hardware. The point is that both underpurchasing
and overpurchasing are dangerous; somehow the
neurobiologist must acquire computer systems within the
range of his immediate needs and relatively short-term
expansion.

MODIFY HARDWARE OR SOFTWARE

Often a neurobiologist, faced with an inadequate
computer system, is tempted to modify it to meet his
requirements. With medium- and large-scale integrated
circuits hardware modifications are continually becoming
easier to accomplish. Unfortunately, the unbundling of

software has resulted in a vendor tendency to give out only the binaries rather than source code; software is harder to adapt.

When a laboratory decides to make hardware or software modifications it is invariably necessary to have one technician for each. Occasionally one can hire a person capable of both, but this is exceptional. Even a single annual salary can burden a grant so it is difficult to justify such a person. Particularly when the sort of person desired is competitively sought by many others (vendors, industry, etc.). Often the neurobiologist must be satisfied with a person of less than full competence. But software or hardware is a full-time job and just keeping abreast of these rapidly developing areas requires a great deal of effort. Seldom can individual neurobiologists support the effort required and conseqently the marginally competent employee becomes, relatively, less competent as time goes on. For example, a neurobiologist might hire someone to make some interfaces. Usually the person assigned the task does an inadequate job and requires more time and money than would have been required if the interfaces had been purchased in the first place. And, by the time the interfaces are completed, new, better devices are available. A vicious cycle then begins with the personnel getting farther and farther behind the state-of-the-art.

In my opinion this state of affairs exists because neurobiologists (and other scientists, as well) have not recognized computer science as the keystone to their work. Physics, mathematics, and some phases of electrical engineering are considered as foundations to neurobiology. Yet I have never seen a neurobiologist take any cognizance of computer science as important to him, even when he spends most of his time doing computer science. Seldom, if ever, do neurobiologists or their graduate students delve into computer science topics; even though most computer problems encountered by neurobiologists have solutions, neurobiologists approach computers as though no one else had ever worked on similar problems. And they reinforce their own, and their colleagues', prejudices by describing to one another their modest, inadequate accomplishments.

With this dismal situation extant I will now advocate "taking the bull by the horns" and describe the simplest way out of the forest. Drastic situations call for drastic remedies. Overall, my proposed remedy is to make computer science an important contributing subject in neurobiology. My solution is drastic; build your own computer systems from the ground up. In my opinion, building one's own system will inject computer science into neurobiology in the most efficient way. Large scale integration makes this possible; high level language computers (i. e., a new approach to computer design) make system development easy enough for neurobiologists to pursue.

SELF-CONSTRUCTED HIGH-LEVEL LANGUAGE COMPUTERS

Two occurences make HLL computers a reality: Large-scale integration (LSI) and compiler generation methodology. LSI allows a designer to quickly and easily put together large functional units -- units which process a significant portion of an algorithm. The "CPU-on-a-chip" LSI has been extended to second generation models containing 16-bit registers, arithmetic logic unit, multiplexer gating and control logic in ways that allow a 16-bit (or 32-bit, or any grouping of 2 or 4 bits) device to be built quite easily. Microprogram control units to control this data part are also available in LSI. LSI memories and peripheral device controllers make it feasible to build computers and I suggest that neurobiologists build their own; not only will they obtain a computer at low cost but they will understand how their computer works, i. e., they will gain a major portion of computer science knowledge.

While it is relatively easy to build the hardware it is much more difficult to build the software. An operating system, for example, will take several man-years to complete, if it is of conventional "do-all-be-all" design. Although less complete operating systems are all that are required by neurobiolobists (and should even be preferred) the neurobiologist will still face the task of a compiler or interpreter for a higher level language. BASIC and FORTRAN, written on a conventional machine, can easily take a couple of man-years. Obviously the software effort far exceeds the hardware effort.

However, an HLL computer executes a high level language including operating systems and compilers/interpreters. An HLL for an HLL computer must include system implementation language (SIL) but these are already available in computer science. The point is: An HLL computer with an HLL-SIL language will come with at least a prototype operating system and a compiler for immediate use. Extensions and improvements for local use can easily be made to suit a system to a particular experimental situation. And these modifications represent the educational injection of computer science in neurobiology which is so necessary.

LIMITATIONS OF HLL COMPUTERS

An obvious limitation of HLL computers is that few have been built. SYMBOL (1) and EULER (2) are the two notable exceptions. SYMBOL is a hardwired machine executing a single language -- SYMBOL. SYMBOL has many interesting features including variable precision arithmetic and internal data structures. EULER was microprogrammed on an IBM 360/30. Reports on both of these machines ecphasize the computation speed due to execution of the HLL directly. But neither is likely to be more than "one-of-a-kind". My guess is that both are too inflexible and, although they execute the designed-in language very efficiently, no language improvements can be made. These machines execute a built-in language and that's all.

We have been developing an HLL computer -- BASIL -- using our own modular computer design system dubbed "The Logic Machine" (3, 4). A Logic Machine consists of a user-microprogrammable control processor, one or more bidirectional busses, several functional units and microprogram all arranged to accomplish a specific digital algorithm--in this case an HLL minicomputer. A variety of digital devices have been constructed including a graphics display terminal (5), a floating-point arithmetic processor (6), an auxiliary processor (7), and a minicomputer (LM^2) (8). The BASIL language was first implemented on the LM^2 computer (9) and is now being embedded in firmware. As reported by the SYMBOL and EULER groups, BASIL is significantly faster when running on hardware tailored to its execution, i.e., running on BASIL rather then LM^2. But the significant point is that BASIL has high level language primitives embedded in firmware. We could have embedded the entire compiler in firmware but BASIL would then become as static as the previous HLL machines. We tried to leave enough control for the systems programmer so that he can still evolve new languages. But we tried to build in powerful primitives so that systems programming is easy. Thus our primitives rapidly and easily perform table driven parsing, but with tables in memory so that operators, reserved words and symbols can be altered to alter the language constructs. It should be possible to evolve entirely new systems from BASIL so that the machine becomes, for instance, a PASCAL or FORTRAN machine. Furthermore, BASIL's primitives are useful in applications such as text editing and vector arithmetic, although we have constructed an auxiliary processor (also a Logic Machine) to handle the latter.

Nowhere in industry does there seem to be a trend in the HLL computer direction. Obviously our small group cannot support a large, nation-wide support function. Our aim is to furnish enough of our design experience to other scientists so that they will be able to construct their own BASIL machines. If we are able to do this, we feel that our most important goal of injecting computer science into other sciences will then, in part, become a reality.

DISCUSSION

D. O. WALTER: What with CAD, and the rise of computer-assisted implementation of design, doesn't that suggest that neurobiologists will be able to buy technical expertise to custom-build their own machines?

KEHL: Yes, that's what I'm saying.

OLCH: But you also say that we can't expect manufacturers to gear their efforts toward our needs.

KEHL: If you expect the vendor to write a software package for you to run on your current systems, you're wrong. We can't do what I've been projecting today. The modules I've been talking about might be available soon, maybe sooner than you think. Computer scientists always have trouble distinguishing the present from the future, and I try to make a very sharp distinction. They're not going to do it in the current technology, they can't afford to. Maybe, if all my projections are right, maybe that other thing-- plug them together like op amps, only on a much greater scale. But only if my projections are right.

BROWN: The standardization problems are staggering.

KEHL: There are lots of problems. But I think that's the direction things are going.

MCCORMICK: The fact that neurobiology is not a market doesn't necessarily imply that single-processor process control isn't a market, because of all the other applications. What strikes me from the past few days' talks is how much neurophysiological data processing is moving back into contact with the rest of the world.

KEHL: Yes, I got the same thing you did, and I haven't digested it yet.

MCCORMICK: This is the old signal-processing world. All these modules people have been talking about could have their applications in industry. It might be very worthwhile for some subgroup to see if we can tag along with industry using a lot of things being developed for them.

KEHL: I must admit I haven't digested what I got out of yesterday. I was very much impressed by the fact that I was hearing the same story that I had heard ten years ago in a whole different area. I don't know what that marketplace is like now.

MCINTYRE: All the transform processors went away before they got cheap enough for us to buy any.

KEHL: You just rang my bell. Comparing the PDP-11 FFT algorithm to the LM^2 auxiliary processor, the speed-up factor is about a factor of 2. This is close on to what the special-purpose processors can do. LM^2 can do most of what any special-purpose processor can do.

MCCORMICK: But I don't see how you can get language standardization if you don't have hardware standardization.

KEHL: That would be like saying "Let's force English as the universal language." What are we going to do, burn the French literature? That would be stupid. You should try for it, where it's useful. My approach with LM^2 BASIL is to say, "I'll let you develop any language you want, but I'll try to make BASIL so good that you won't need anything else." But the hooks are there for any language you want. As long as some of you are going to do it anyway, I'll make it as easy as possible. Hopefully you won't want to if I can show you a better way.

MORAFF: As long as the field is moving fast enough, you can't afford standardization. I'm not convinced that evolution of computer architecture has slowed down enough to allow standardization in the near future.

KEHL: I'm in complete agreement. I hope we don't slow down, the opportunities are too good. There are too many great things that can be done yet.

MCCORMICK: Although I don't want to see things slow down, it would be good to have standard packages that people can use.

KEHL: I agree to a certain extent. I don't think there's anything more to be learned, for example, about how to do lexical analysis. We know how to do it, and it's not very interesting; to spend all kinds of systems programming time doing any more of that is just ridiculous. We know how to parse, at least algebraic languages. And it's not interesting, it's not important, and we can embed that into firmware. It's all right to make standard ways of doing those things. I wouldn't go much beyond that. It's like floating point arithmetic. We used to do it all in software, and the literature was full of articles on efficient algorithms. When the hardware came in, there was no way you could beat it. So that was the end of all that software. So embed it in the hardware, and who cares? But there are still a lot of things that can't be.

MORAFF: Don't you also standardize so you can share things, communicate them to other people? What do you want to communicate besides your neurophysiology results?

BROWN: You want to communicate tools as well as results.

REFERENCES

1. Rice, R., and Smith, W. P. SYMBOL-- a major departure from classic software dominated von Neumann computing systems. Proc. SJCC 38:575 (1971).
2. Weber, H. A microprogrammed implementation of EULER on the IBM system 360/30. CACM, Sept., 1967:549.
3. Torode, J. Q., and Kehl, T. H. The logic machine: a modular computer design system. IEEE Trans. Comp. C-35,11 (1974).
4. Kehl, T. H. The logic machine: a modular design system for digital hardware experimentation. COMPCON 71: 305 (1975).

5. Torode, J. Q., and Kehl, T. H. A graphics display terminal logic machine. COMPCON 75: 313 (1975).
6. Torode, J. Q. *A Microprogrammable Logic Machine* Ph.D. dissertation, University of Washington, (1972).
7. Burkhardt, K., and Kehl, T. H. A logic machine auxiliary processor. COMPCON 75:309 (1975).
8. Kehl, T. H., Dunkel, L., and Moss, C. LM -- a logic machine micomputer. IEEE Computer 8:11 (1975).

Societies of computing professionals also provide a clearinghouse function in their publications. But is it reasonable to expect a neurophysiologist to join any or all of these groups? And how much of the available software is well enough documented so that he can extend and modify it as his research needs grow and change? Computer manufacturers and commercial software firms advertise their software products in computer trade journals (such as *Datamation*, *Modern Data*) and these journals publish listings and surveys of commercially available software. But data on evaluation of these software products is not easy to come by. The question of whether someone else's program will work in his environment represents the issue of program transportability. It is not an uncommon experience to find that a program of some complexity, transplanted from one machine to another of the same model and design specifications, will fail to run without modifications.

At the 1974 MUMPS users' group meeting in Denver, Dr. G. Octo Barnett, one of the originators of the MUMPS (Massachusetts General Hospital Utility Multi-Programming System) language and a member of the MUMPS users' group task force on program transportability spoke (1) on factors in transportability of MUMPS application programs. Dr. Barnett emphasized that a MUMPS program, in order to be successfully transported, must clearly satisfy a recognized deficiency in medical information processing; that it must be operating and in routine use; and that it must be supported by extensive documentation, not only of how it works but also of its objectives, specific benefits of its use, and cost of implementation. It helps if the program to be transported is simple in structure and aimed at a straightforward application not dependent on the characteristics of the local environment. Then, he noted, the actual transport will best be accomplished by having the originators of the program produce and support a standard implementation. He observed that the need for demonstrations of such program transportability is urgent, and offered suggestions for getting started.

Numerous research projects are under way in which investigators are seeking ways to improve the transportability of software. For instance, International Mathematical and Statistical Libraries, Inc. (IMSL) are currently studying portability in the

chapter 38

LABORATORY PROGRAMMING: HOW CAN WE REALLY GET IT DONE?

H. MORAFF
New York Veterinary College Computer Facility
Cornell University
Ithaca, New York

Recently one of our colleagues needed a signal averaging package, so he hired a programmer to write it for him. Soon after, the same programmer wrote a spike histogram package, then left for another position. Our colleague has since learned that an improved disk operating system is likely to become available for his computer. He has just hired another programmer, a student, to adapt the averaging and histogram packages, combining them in one package, and possibly arranging for compatibility with the new operating system. He's getting what he needs, but it's slow, and costly. He's learned lessons about documentation and contract writing.

It seems that there should be a better way. Let's look at some alternatives: 1. Get software from others; 2. Ship your data; 3. Do your own software; 4. Share local concentrated resources.

GET SOFTWARE FROM OTHERS

Computers have been used in laboratories long enough that someone, somewhere, has probably developed the programs our friend needs. But how can he find them, and what are the prospects of getting them to work on his computer, in his laboratory? The computer user groups (SHARE, GUIDE, COMMON, DECUS, Data General Users Group, etc.) serve as clearing houses for user written software.

specific areas of mathematical and statistical subroutines (2). They are preparing a Basis Library of programs, which runs on one computer type and contains all the information needed to convert it automatically to other computer types. Three computer types are involved: the IBM 370, CDC 6000 series and UNIVAC 1100 series. There are currently several hundred subscribers, about 300 subroutines and two volumes of documentation. This project is supported by the National Science Foundation. The NSF has also joined with the Atomic Energy Commission in support of the NATS project (National Activity to Test Software), in which the Argonne National Laboratories and a number of other institutions have been collaborating in studies of the problems of creating, testing, certifying, disseminating and maintaining mathematical software for a variety of computers. Two software packages have been created. The first is EISPACK (3), a widely distributed collection of matrix eigensystem subroutines which are nearly machine independent. The second package, FUNPACK (4), is a collection of special function subroutines certified for use on three different lines of computers and not intended to be easily transportable to other machines. These efforts and others (such as the BMD statistical package created by the UCLA Health Sciences Computing Facility) have led to some understanding of the problems of transportability, including the major realization that making usable, disseminable software is very expensive. The work goes on. Recent activities include the NSF-supported planning project for a multi-institutional Mathematical Software Alliance, centered at the University of Colorado (Wayne R. Cowell, principal investigator). The alliance is to be designed for the creation, evaluation and dissemination of mathematical software.

If and when the major problems which currently inhibit the transport of programs are solved, a good part of the needs of our laboratory researcher could be met by software made by experts and transported to him.

SHIP YOUR DATA

An interesting solution to the problem of transporting programs is to keep the programs fixed at their locations of origin and move the data instead.

The growth of computer networking has enabled serious exploration of this approach. As a result, some of the health science computing resources have offered access to their computing power, primarily for specialized needs, to outside investigators on a regional or even a national basis. Such remote access has so far been most successful in relatively low data rate situations, such as sophisticated modeling applications, but a few significant projects have involved rather high data rates. For instance, the Stanford University Medical Experimental Computing Facility (SUMEX) is a resource which specializes in applications of analytical methodologies to biomolecular characterization and in applications of artificial intelligence in medicine. They are well known for their development of the heuristic DENDRAL program, which employs artificial intelligence techniques for the analysis of mass spectra. Their goals include development of effective ways to accomplish "intelligent" real-time management of instrumentation involving high data volumes and rates, to serve medical applications both locally and remotely over a communications network. Whether such goals can be realized economically with today's data communications technology is not clear. This project has substantial grant funding. The experience of other projects is that the high cost of long distance communications makes high data rate transmission uneconomical, and that the best current uses of networking are for the sharing of large data bases stored centrally, such as the protein data bank, stored at the Brookhaven National Laboratories and accessed remotely via the experimental CRYSNET network. Of course, successful implementation of such projects as Stanford's SUMEX/AIM resource will provide a strong impetus toward solution of the communications problems.

DO YOUR OWN SOFTWARE

The third alternative involves development of a high level laboratory-oriented language, which should be sufficiently general and easy to use, that our neurophysiologist, after a reasonable training effort, could write his own software. This is an ideal solution in the sense that the investigator could really get what he needs, since the serious difficulty of communication between biologist and computer specialist would be eliminated, and the programs could be custom tailored.

I believe that such a language has a place in the inventory of of research tools, and that eventually a substantial portion of the software used in the neurophysiology laboratory may be based on this approach. But I feel that this is still a long way off. Languages and modular program packages for laboratory applications have been around for a long time. For instance, the FOCAL language has found use in low data rate laboratory environments (5,6). More specialized approaches such as DEC's Laboratory Applications Package have satisfied some research needs. But these approaches have suffered either from limited scope of application for the specific modular program packages, or from extremely limited speed of operation for the more general languages. Differences among designs of hardware and systems software pose another serious problem for any implementation which is intended to be really general. Finally, neurophysiologists have always managed to design experiments which call for computer performance beyond the state-of-art capabilities, and then they have proceeded to program in assembly language to push the computer to its limits. This will likely continue, but recent computer technology predictions such as those of Rein Turn at the Rand Corporation offer some hopes of real progress toward practical, efficient high level languages and application-matched hardware. For instance, Turn predicts that hardware developments should simplify the programmer-hardware interface, enable production of cost-effective, problem-oriented systems, and generate systems which process higher-order languages directly. We can already design economical systems which have random access memory sizes many times larger than those of a few years ago, so memory space is no longer a real problem. Now is the time to be doing the planning and developmental groundwork for the high-level language approach, and the next two workshops in this series will certainly address the issue directly. But what can we offer right now?

SHARE LOCAL CONCENTRATED RESOURCES

When computers began to find widespread use, universities created computation centers, to enable sharing of these very expensive resources and of the associated expertise to develop and operate them. This centralization served its purpose. It also created problems, including failures in establishing fair and workable priorities among users, the lack of administrative control over center operations by the user community, disappointments stemming from remarkably poor and inefficient performnce of the resource-sharing mechanisms in the system software, and a general inadequacy of availability of individual technical assistance for the users who need it. This situation, coupled with the appearance on the market of medium and small scale systems with impressive capabilities has led over the past decade to a proliferation of computers and resources at the level of the small organizational unit in the university, a movement in which the university computing center has generally failed to take part, and which in fact has become competitive in many cases with services provided by the university centers.

Laboratories, departments and divisions have individually and in some cases cooperatively gotten into the computer research resource game, often with support from the NIH, NSF, and other funding agencies. Some important examples are the Health Science Computing Facility at UCLA, the Chemistry-Physics Tri-level Computing Network at the University of Kansas and the MIRACLE (Multidisciplinary Integrated Research Activities in Complex Laboratory Environments) Resource at Purdue. At the University of Chicago, an interesting thing happened. There, the University has provided a centralized supporting resource for minicomputers, under the leadership of Dr. Robert Ashenhurst, who has set up one of the earliest campus networks of minicomputers. The system, called Minicomputer Interface Support System, or MISS, provides a central minicomputer which supports higher level programming and use of fancier peripherals and more powerful computing technology than can be afforded in most laboratories. This approach has had considerable success, and is being repeated in many places.

Last summer, at a conference on the computer as a research tool in the Life Sciences, Dr. Jerry Cox gave a talk (8) on the economics of computing. In that talk, he discussed three main ways to achieve economical computing. The first is to tailor the problem to existing shared facilities. Utilization must be high to realize savings, and this requires compatible tasks: dissimilar jobs make full utilization difficult to manage. The second way is to tailor the equipment to the problem: this is what our prototypical researcher has done in acquiring his own computer. Utilization is then typically so low that the potential economies may not be realized. The third way combines economies of scale and of equipment specialization by identifying computing needs which are common to many users, and providing efficient secialized (tailored) equipment, accessed directly or via data communication lines, thereby achieving the high volume of use needed for economic operation. The key point is that a computer resource can more economically serve a user community which is large enough to share the costs widely and yet small enough or specialized enough to have a reasonably limited scope of service needs.

So, while individual researchers are getting into the computer game in increasing numbers as hardware prices fall, I feel that the interesting action still lies with the multi-user resource. This is the level at which expert staff can be supported, and computer system research and development can be afforded. This is the training ground for the growing population of research-computer specialists, who offer competence in on-line, real-time and interactive program development, often coupled with a good to excellent appreciation for the application areas they serve. With this kind of concentration of resources and expertise, the staff can become very proficient and professional. Those individual investigators whose departments have not yet gotten involved in acquisition of computer resources, often have access to excess services and system capacity at local resources of other departments: such excess capacity is essential to the healthy operation of these resources. My own resource at Cornell has dispensed its surplus capacity in this way for almost ten years. We at Cornell are in fact beginning to see the emergence of a marketplace for research computation programming, hardware design and implementation, and other forms of service. So far, as in most universities, the situation has developed on a more or less ad hoc basis. However, I feel that as the market for specialized computing grows and becomes more visible to university administrations, we will see more central control exerted over its development. I can see benefits to the researcher from the formalization of the division of services by resources, in terms of advertisement, coordination, provision of fiscal help for stability, and effective representation of user needs.

While we may eventually produce truly usable high-level programming tools which satisfy essentially all the needs of the researcher, it's encouraging to know that in the interim, there is a viable source of the help he needs.

In summary, I feel we have a long way to go in achieving any significant degree of generality in laboratory software. Accomplishments in that direction will be impeded by the variety of machine types which must be accommodated, and aided by improvements and innovations in hardware. While efforts such as the ones to which these workshops may lead are necessary, we should recognize that the interim is going to be a substantial period, and I suggest that sharing of people resources -- the most expensive aspect of computing activities -- is a viable solution.

DISCUSSION

RANSIL: Having been at Chicago, now at Harvard, and watched this kind of development of multi-user resources, I've tried to get some kind of handle on how these things evolved. You've seen it, and I think PROPHET is an example of the same sort of thing, but with a national user group rather than a local user group. Could you speak to some of the organizational things that bring this kind of local computer resource into existence?

MORAFF: It's mostly political. If the Board of Trustees likes it, they'll go after it in a big way. I had the remarkable experience myself of going to Albany for support for my own facility. I experienced a tremendous reception and I had no idea why. As soon as they realized this was the only Veterinary college in the country with an online medical information system, they snapped that up immediately and said, "What do you need"? I learned only later that historically, the State University of New York got into computing late, with no large-scale operation, so there was no way they could achieve a leadership role in large-scale university computing. So they're looking for something else which is unique about their operations, which they can point to and say, "We may not be a leader in large computer systems, but we are damned good at this, and this, and this". They saw an opportunity in my resource to make one of these really good small-scale operations. Now that's a political decision. We never had a chance to discuss our needs in any kind of technical or medical terms. They only wanted to hear that we had something juicy, it didn't matter what it was. I'm afraid that all too often this is how things are decided: somebody says, "I want this big machine because it'll make us look better than somebody else." Frequently what determines which university in a university system gets a particular machine is which asks for it first, not which one needs it the most. This happens all the time. In Texas, the University of Texas and Texas A&M were competing for a regional center. Now why should they even want to? Don't they understand that what they really want to do is serve their users? Why would anyone in his right mind want to have that kind of facility in his own backyard? I should think it would be the other way around. If Texas wants the computer, A&M ought to buy their service and be free and clear. Harvard did it that way: they buy it from everybody else.

KEHL: I think particularly in the last few years, we've had too damned much politics in science. It's time to get back to science.

MORAFF: Absolutely.

KEHL: I think that anything that will release the scientist from politics is all to the good.

MORAFF: A real thorn in the side for many institutions is the existence of a group which must pass on all computer acquisitions. Now, you can get around that sometimes, by not calling a computer a computer. All kinds of instruments have them built in now, so you can in effect smuggle one in, in a mass spectrometer for example. In fact, now when a computer costs less than $10,000, the administrators can't be bothered with it anymore. So we could bring in a computer and its peripherals on separate orders, and not have to fuss with the administration.

KEHL: In Washington, it's against the law!

MCINTYRE: It's not the hardware, though, it's the people who run these things that cost the money.

MORAFF: Definitely. One of the things I've tried to do is convince Cornell that they should have their systems programmers start doing application programming for users. They refused, because they only had enough people to keep the system going. Now my own staff currently has as many people, for one college, as the Cornell central facility has for the whole university. The notion of a central facility is fine, but they don't realize that their facility is worthless to a lot of people because it's understaffed.

RANSIL: While politics may have gotten your operation going, you have to provide services to keep functioning. In neuroscience, we have to define the commonalities of needs, and how they can be met. Are there any parallels between the neuroscience thing we're trying to do now and the pharmacologists' experience? They got together and defined a multi-center resource.

MORAFF: These are all subsets. What we have here are people who need to do modelling, online data acquisition, signal analysis, analysis of data bases, and all kinds of other computing. We're not really in the information business as much as we are in the signal business. The needs are somewhat different. But we do have a lot to learn from people like the chemists and physicists that have preceded us by a few years in tackling these problems.

MCCORMICK: We've had a lot of experience on the AIM network, for the use of artificial intelligence in medicine, and I'd like to report on one aspect of that: we have a number of young assistant professors trying to get going on various applications, and the fact that they're in direct communication with people at MIT, Stanford, Rutgers, and Pittsburgh, is an enormous aid. There's mail transferred all the time, an interchange of software, and we can get setups going much faster. It strikes me the difference in neurophysiology is just that the terminal has to be a minicomputer with a bunch of local peripherals.

MORAFF: That's not really a very big difference, these days.

MCCORMICK: Right. But I must say this aspect of trying out people's codes on one machine is a great advantage, where you don't have to worry about portability.

MORAFF: What you have is a hard-wired user's group.

MCCORMICK: Yes, and a national one, on a very minor level of specialization.

MORAFF: And it's an extremely valuable resource.

D.O.WALTER: I'd like to comment on the concept of a

hard-wired users' group. The uniformity of configuration is really enforced, and really pays off. You're getting paid in released funds, released time, and increased productivity.

VIDAL: Someone made the point earlier that there is no common data base for neurobiologists, and I think we should examine this. I'm not so sure that there isn't a common data base or data bases. Some other investigators and I have been exchanging huge amounts of evoked response data for a year and a half, using facilities at MIT's MULTICS for data processing. These exchanges have been going on routinely now for some time. There's no reason why spike train data couldn't be handled the same way. There were hidden costs in that operation that were transparently covered, by the way.

MORAFF: Absolutely, and that's a point I'd like to make to you, Dr. McCormick. While your multi-user group is beautiful, it's lavishly funded. I don't mean they're getting too much money, but simply that we're really dealing with two sets of people: the haves and the have nots. And so far we're hearing from the haves. Every once in a while, someone talks about A/D on a shoestring, and he's a have not, but he's really a have in terms of having people who can get it going for him. We really need to get at the have nots, and I asked Paul at lunch just what he's trying to do: sell computers to the natives, or advance the state of the art, or what? It's an issue that needs to be discussed. How are we going to get this out to the troops? If we just get together and talk and come back with elevated opportunities for doing our own research, we haven't accomplished all that much. There are people out there who are going to be buying computers, and they may even get turned off about computers when they discover they can't hack it. How do they get the help they need so they can hack it? That's the issue.

MCINTYRE: Networking will do that. One of the tricks is to figure out where to put all the data and then just let everybody come in. The terminal costs are trivial.

BROWN: We still have the problem of real-time processing.

MORAFF: For modelling and offline analysis, a central data base and central facility is fine, but what about the guy who has a cat on the table?

MCINTYRE: Then you have to have your own mini.

HARTLINE: I'm completely turned off by that, because

there isn't anybody else's data base that could possibly be relevant to me. And my stuff would be useless for anyone else.

MORAFF: Thank you. That's just what I wanted to hear. So what you need is a very efficient language to write your programs in.

RANSIL: For a person to operate independently, we have to cut his overhead development costs. That's what we're after.

MCCORMICK: In the AIM network, the hotshots aren't the users, they're in the network itself. I think its extremely important that the users are developing their own applications programs. We don't want the central shop to do that for us.

MORAFF: A network could fill a real need. There are a lot of people out there, off by themselves, who have a lot to contribute. You could look at a network as a single shop, and people who can afford to get on would clearly benefit. But it won't solve the problems of today in anything like today's terms. That's a good five years off.

BROWN: Let me emphasize one very important point, however. Inevitably the scientist who depends on a facility that he doesn't direct becomes very bitter at one point or another, from the fact that over and over again he has an animal open on the table, and he can't sign on the machine, or he discovers there's been a system change since he was on last. People at the central facility at various institutions have asked me, "Why don't you take more advantage of our facilities"? The last four times I tried to get on our central university facility here, it was down. Dr. Moraff can recall my experience at Cornell, when after a year of designing and implementing hardware and software, I experimented for half a year and then his computer facility changed machines. It took another nine months before I could experiment online again. There were absolutely no measures to provide continuity of service, and indeed any such measures would probably have been economically unfeasible. Multi-user facilities do this all the time, to avoid obsolescence. A personally owned facility doesn't have to.

HARTLINE: The other thing about centralization of computer power is the fact that they just aren't fast enough. A central shop where I could get programming done would be great, but you have to take the hardware to the user.

MORAFF: That depends on the application. A star network may be most appropriate for some groups, but a central shop should do the programming and even the maintenance, for economy of scale.

CLARKE: The computing power's going out to the peripheral minis, clearly. That means we need to develop distributed as well as local networks.

O'LEARY: How typical is your mythical friend who has a computer and no one to help him make it work for him? It seems to me that younger people in neuroscience who realize that they will need computing are learning to write personnel into their budgets along with the hardware they need.

MORAFF: He has that money. He's used it to hire several programmers in succession. He has transient labor, he can't hire the right people because he doesn't know enough about the task to pick them. He can't afford the cost of supporting the kind of stability and high quality he needs -- he really has to share it with someone else.

ROBERTS: Another question is, if you are a neuroscientist, should you be spending your time computing or doing neuroscience?

MORAFF: What's the difference if he needs the one to do the other?

ROBERTS: The difference is that your time is not just money, it's productivity, and you need to delegate tasks that are a waste of your time.

MORAFF: If you can delegate it, I can envision a computer you program by twiddling knobs, like an oscilloscope. Even that would be pretty complicated. The problem is, he can't always delegate tasks which at least superficially you would think are scut-work. He doesn't hire out the surgery or the performance of experiments, if he can't teach someone else to do it for him reliably. Often it would take too long to teach someone. He's not like the microbiologist, who can often train a couple of technicians, and from then on only needs to give them a schedule of experiments and peruse the results, without going into the lab for weeks at a time. We can't realistically tell the neurobiologist that he can completely forget about the programming. Even if he can delegate it, he has to be able to train people he can delegate it to.

REFERENCES

1. Barnett, G. Octo, Program transportability - controversy and challenges. Proc. 1974 MUMPS User's Group Meeting, MUMPS Users' Group, Biomedical Computer Lab., St. Louis, (1974).
2. Battiste, E. L. Mathematical Software Patterns. ACM SIGNUM newsletter, 10:17-20, (Jan. 1975).
3. Garbow, B. S. EISPACK - a package of matrix eigensystem routines. Computer Phys. Comm. 7:179-184 (1974).
4. Cody, W. J. The FUNPACK package of special function subroutines. ACM Trans. Math. Software 1:13-25 (1975).
5. English, J. C. FOCAL used in a data acquisition and control system - advantages and disadvantages. DECUS Proc. 191-193 (spring 1972).
6. Mullen, S. FOCLAB - a computer controlled experimentation system. DECUS Proc. 125-126 (spring 1974).
7. Turn, R. Computers in the 1980's - trends in hardware technology. Info. Process. 74 North Holland, Amsterdam (1974).
8. Cox, J. R. Jr. Economy of scale and specialization revisited. Fed. Proc. 33:2357-2360 (Dec. 1974).

chapter 39

PANEL DISCUSSION

SHANTZ: I'd like to address a few comments to the subject of computer-aided neuroanatomical reconstruction, my own area. First of all, I think there's a great danger of generating more work for less results by using the computer in structural analysis. You have to be very careful to choose the right things to do because you can spend a lot of time trying to solve fundamental problems in pattern recognition in order to make the computer do things that the human can do better to start with. A lot of neuroanatomists will argue that they can look at a Golgi- stained neuron, focus up and down, and put together a pretty good picture in their heads, without spending years programming a computer to generate a movie that rotates a cell in technicolor. The question arises, "How much do you gain by going through the whole song and dance of computer processing?".

One possible application where the computer can be helpful is in what I consider to be a necessary approach now, namely the analysis of large numbers of cells, since they sometimes don't fall into the nice stereotyped morphological classes that many of us thought they might. The computer should be good at doing statistical analyses of morphological properties. So you have to spend as little time processing individual neurons as possible, and the computer should be used to maximize your efficiency in the reconstruction task. In our development of the picture processing system, we tried not to get into pattern recognition problems. We do automated reconstruction in the cheapest possible way, letting the computer do things it can do without extensive software development.

Also, I think there's probably a limited return from the generation of displays which don't give good structural descriptors; the computer seems to be useful in extracting such descriptors, and in correlating a descriptor with other descriptors, both structural and functional.

The computer's very good at generating certain types of analyses. In particular, if you have a nerve cell described in a multi-dimensional space, the computer can plot projections of the space very easily, letting you do cluster analysis.

Finally, we're starting to use the picture processing system to do two-dimensional modelling, viewing the retina as a two-dimensional transducer. A picture-producing system with Fourier transform capabilities provides a natural tool for modelling two-dimensional transfer functions in two-dimensional spatial white noise analysis of retinal function.

RHODE: In retrospect, after developing our own system, we've reached a few conclusions about how to do it. I agree with Ted Kehl, that we should keep the system as simple as possible. If you look at our system, it's not particularly simple, but I think it's the best we could do for our needs, with the technology available to us.

We're very much concerned with online feedback. We want as much feedback about the process we're investigating as possible, as quickly as possible, implying a fair amount of computer power within your own laboratory. There's good reason for that. I'm sure a good number of you who do physiological experiments realize that you can collect a lot of data and go back maybe a week or sometimes, I admit, months later, and find out the interpretation of the data is not clear-cut, and would have wanted to repeat particular paradigms, but now the animal's dead and it's too late.

The small computers usually used in this mode are not powerful enough to spit back transforms, statistics, or simulations during experiments, even though that may be the kind of feedback you really ought to have.

Thirdly, we have a strong commitment to flexible graphics. Probably everyone has some type of graphic feedback, at least offline.

We'd like to see some sort of experimental control language which could possibly look like a state diagram implementation language. We want to be able to easily and quickly modify experimental paradigms during an experiment, to allow us to try new things on neurons, particularly when we want to investigate some peculiarity of a neuron which we couldn't anticipate. Presently we can't modify our LINC programs during an experiment in any substantial way. Typically it could take a month or two to make any significant modifications to an experimental paradigm.

We certainly want to work with a higher-level language. We re-programmed our LINC software on the new system in FORTRAN, in a month or two. That's significant. When I first got the system in I sat down and in a matter of two or three hours I had an interval histogram program running on the Datacraft. You just can't do that in machine language.

Another aspect of our philosophy is to put anything we can in hardware. It relieves your computer of mundane tasks, freeing it to do things you wouldn't have time for otherwise. That's really the same philosophy as the classic LINC: in that machine, there are single instructions which perform powerful I/O operations. That's a very nice concept, eliminating a lot of software development. I suppose in terms of current technology that means we'll be able to use microprocessors to handle interfacing to the lab.

One other point: We rely heavily on autonomy, and therefore have no problems with other people making decisions about our computer system which we don't like. We have full control over the system, and would fight strongly to keep that control.

It seemed to me in the course of our discussions, that there are four principle areas we might consider working on: (1) experimental control; (2) data acquisition; (3) simulation; and (4) data management.

KEHL: I think I can talk more about what not to do. We've been talking about scars, and I've got all of mine. First, minimize the layers of logic between you and your problem, and work on your problem. That doesn't mean you can reduce them to zero and it doesn't mean your problems will always be simple. But don't make it more complicated than it already is: that's just Occam's Razor.

To invent problems, particularly political problems in a university environment, is a real error. We have to go back to science. I think a lot of the computer organization problems, computer politics, are just distractions from science. I'm not too worried about neurobiologists falling into that trap: there's a long and strong tradition of neurophysiologists paying attention to physiology. The senior people in the field will slap the wrists of any young upstarts who don't. That's really good. That doesn't mean they aren't open to introduction of new ideas and techniques. They are, and we can see it happening.

My own position contradicts this advice, because I've built it all up from scratch. You can argue that that's not minimizing the work. I think it is in the long run, although I can't prove it. I hope someday to run a session in which I take some young neurophysiologists and train them to know everything they need to know about an LM^2-BASIL computer in three months. Each person would build one and take it with him to his laboratory. If that works, and I expect it to, because that's exactly what happened with the LINC evaluation program, then I don't think anything else will be needed. Certainly I don't think any more power is needed than is provided by LM^2-BASIL. The big problem is the software. I can't prove my approach is the best way to go, at least not yet. Maybe next year at this time.

MORAFF: It's been said, "Give a physicist a computer, and you lose a physicist," and I worry that this is happening in neurophysiology. I've seen it happen to some neurophysiologists, and I'm afraid it may happen to many more if we don't do something about it. When Paul came to me with the proposal for this workshop, my first reaction as a former vendor of computer services was, "It's crazy. You'll never write a language that a neurobiologist will be able to use and which will enable him to do everything locally." But then I thought about it some more and realized that we really have to try. The reviewers agreed, so here we are.

Neurobiologists are rugged individualists; they never want exactly what you give them. They even change their oscilloscopes -- I've seen things hanging off the front of a Tektronix scope that you wouldn't believe. Even something that universal doesn't always do the job they want to do. That's the way research is.

Neurobiologists need our support in the implementation of two kinds of applications: those they've thought of, and those they haven't. We can't completely define ahead of time what kind of software a neurobiologist is going to need. So whatever we do, it has to be pretty general. It has to be extensible. As soon as we tell him he can do 100,000 A/D conversions per second, he'll ask, "What would it take to get 200,000?"

We can't possibly predict all the needs, because the needs will always grow out of reach of the technology.

The neurobiologist needs very good I/O for experimental data, and intimate control of all the computer functions. If he's not in full control, his research will be inadequate. He can't tell his programmer what he wants done and stand back and watch the computer run his lab. So you have to give him a language. You can't just trade software. He's still going to go home and want to do something new, with as little outside help as he can get away with.

Finally, all of the laboratory hardware I've seen has tended toward individual possession. You could see this with electron microscopes. Institutions now realize that each large user of electron microscopes should have his own or his research will suffer. That would have been unheard of 15 years ago. Computers are going this way. Now individuals have them. I think the neurobiologist ultimately should have his own machine, which he alone can use. Until that's happened, we haven't finished the job.

BROWN: I'd like to restrict my comments to an extended reply to each of you. First, Mr. Shantz: We all do use microscopes, oscilloscopes, and calculators. If we view the computer as a research tool in this context, we can arrive at a reasonable sense of the sorts of things we can accomplish in the sort of program we have proposed. You've raised the very real philosophical problem of where you draw the boundary between man and machine in performing difficult integrative analyses. I think we've seen from the work presented here that the boundary can quite legitimately be drawn at any of a number of different places. As long as we put it in a reasonable place, then we're all right. I think that as much as possible, we should try to specify to the machine what we want in terms of an analysis. My reason for this stance is that, to paraphrase Wittgenstein, "If you can't define what you want to do in the form of an algorithm, you can't do it." I know this runs against the grain with people who maintain that a lot of science is art, and I recognize their view as having value -- I only differ on where to place the boundary between art and engineering. I've seen too many cases where, on re-examination of data using quantitative methods, apparently firm conclusion based on eyeball examination had to be thrown out.

I was pleased to hear you hint at one possible development, and that is using the computer to go from structural and physiological data to realistic simulation of networks. That's very hard to do - we don't have good software for plugging experimental data into simulations. We need it badly.

With regard to Dr. Rhode's comments, I couldn't agree more with the notion of simplicity of hardware and software systems, and minimization of development effort. But as you pointed out, online analysis requires a certain minimum of performance which in turn requires a corresponding minimum configuration. I think, as you do, that you have to have good peripherals and a good command language. The advantages of these are manifold. For example, people used to count iterations of a program loop to time intervals. That is probably offensive to all of us today. We all use other peripherals to make life easier: We use disks to store programs instead of feeding them in through the switches. We don't use cards or paper tape to store data, we use tape or disk. To do otherwise would be crazy. Well, it's just as crazy to make do with inadequate peripherals for data acquisition. Ignorance is no excuse -- it's easy to get the design you need from someone else now, especially after this symposium is published. I personally have never had trouble getting help implementing other people's hardware and software. Also, if a given operating system can have device handlers for system hardware, it can have handlers for laboratory interfaces. You can always get software -- at least flow diagrams -- from the same people you get the hardware designs from. When we complain about the lack of uniformity in these designs, we're raising a bogus issue. There is no more uniformity in system hardware than there is in laboratory hardware, and most laboratory peripherals are a lot easier to handle than are system devices like disks and tape drives.

A few other comments on peripherals: Display processors are clearly going to have an enormous impact on the sorts of things we can do, especially as they get cheaper. It's not unreasonable to expect verbal communication between man and machine in the next decade, which will have tremendous consequences. We've already touched on the concept of privately-owned computers, and I believe we have only scratched the surface in considering the consequences, good and bad, of such a trend. Surely we will soon have universal programmable interfaces, programmable black boxes that will allow you to plug laboratory instruments in at one end and computers at the other end, achieving compatibility with software or firmware, in ROM, RAM, or PROM. The technology already exists, in the form of microprocessors.

The whole problem of peripherals reminds me of the rule of thumb for a hi fi system: spend at least half your money on the transducers. We've all learned, sometimes the hard way, that peripherals always cost much more than the central processor. You have to plan for that. And probably a very large fraction of your programming time will be spent writing handlers, although that may someday come to less than 25% of the total, with cheap hardware options like existing arithmetic elements, but of a much higher order of sophistication.

With regard to Dr. Kehl's comments, let me say only that in order to keep things simple for neurobiologists, those of us with the scars will have to put out that extra bit of effort to make sure that others don't have to go through the same mistakes we did. It would be criminal not to share our experiences.

Dr. Moraff, I think your pessimism is quite well warranted. But let me put the problem in a more optimistic perspective: if we could meet the needs of 30% of the newcomers, neurobiologists who are setting up new laboratories, we would have saved so much money and manpower that we would have had a significant effect on the productivity of neurobiologists. I don't think that's an unreasonable goal. I think this can be accomplished in just a few years, because we can do it by soliciting help from people who have already developed their own systems. All the information needed for an excellent neurobiology system exists right now. The only reason it would take that long is that you have to pull it all together and get it all to work on the same machine. In the process, we can implement a simple, consistent command language. That's my strategy, at any rate.

In conclusion, I'd like to point out that there are a number of applications that were not represented here. To name a few, they include source-sink (current source density) analysis, several other types of stimulus generation, handling of multi-dimensional attribute spaces, quantification of other types of histological data besides those discussed here, and contingency programming of stimulation sequences where the computer will sequence stimulus variations as a function of the data from ongoing stimulus runs. We can't expect to do all of these things, but they should be kept in mind during Workshops II and III.

VIDAL: The idea of a command language is no different from just specifying the tasks. If we can do that, we can have a command language. I say that that is just a definition of science: you can't perform a task until you have defined it. Once the task is defined, the command language can be devised.

RANSIL: I'll second that, because just from the PROPHET experience, we found it possible to structure a command language only because there was a consensus as to precisely what the users wanted. The definition of what you want to do is where your major problem lies.

BROWN: Of course those are the objectives of the next two workshops.

FOOTE: Neuroscience has reached a point where the quantitative techniques for data manipulation have recently become in and of themselves an important part of the developing concepts of the science. The programs that have been discussed here are widely varied. I'd like to see the Society for Neuroscience set up a program library and a user's group. This could be a subscription service, with a catalogue of available software.

BROWN: That solution might prove disastrous. The people implementing new systems might be seduced into using this library, rather than the better one we're trying to produce.

FOOTE: I'm not interested in something better that I can't get my hands on today, I have work I want to get done, and my only interest in computers is to use them to get answers I can't get any other way. I haven't seen too many systems that will really do that.

BROWN: I disagree. Every system you've seen in the past three days has been used to do neuroscience research, which couldn't have been done without them.

BROWNELL: That's an experience we've all had. We've all seen that neurobiologists with no computer experience are terrified of computers. Let's talk about them for a minute. These people are living in a very limited world -- they are constantly running against needs that can only be met by computers, and so they have to back away. They are limiting the things they can do.

KEHL: I was waiting for that argument. That's the great unwashed argument: "Let's worry about the have-nots and make them haves." That's not science.

BROWN: Then what was the LINC development program?

KEHL: There's an interesting answer to that, a surprising answer. Wes Clark was interested in neurophysiology and he wanted to design something that would enable him to do experiments the way he thought they should be done. I'm saying, concentrate on what you want to do, and don't worry about the great unwashed. They'll want the same things you do.

HARTLINE: I think that's wrong. Those of us who've risen above a certain threshold do get down to doing it and they use computers. They waste a lot of their time programming. There are a lot of other people who, given a system designed for neurobiologists, wouldn't have as high a threshold. Such a system would make life an awful lot easier for us too.

KEHL: I agree with all that. I was saying, make it on the basis of your own experience.

MCINTYRE: There's a serious flaw in the library approach, and that's one Dr. Moraff alluded to earlier: software maintenance. When somebody puts a program in a user's library, and somebody else abstracts it, and somebody else distributes it... nobody's responsible for it. If you could get something like that to run on your machine, I would be very surprised. It always falls flat.

HARTLINE: I'm currently using three programs I got from other neurobiologists, with no problem.

FOOTE: Paul's been complaining about having to go through much the same implementation every time he sets up a new system, and I think he's right, that's a waste of time. You don't want to waste programming time on software that someone else has written elsewhere, either.

MORAFF: It would be interesting to know just how many histogram packages have been funded by NIH. One of the things that should go in the questionnaire you're sending out is a list of applications programs in each lab. Everyone writes an analog logging program, an averaging program, a histogram program -- I'm impressed with the idea of a command language and a package of general-purpose algorithms. I think the neurobiologists of the world have enough energy to implement it. And it will let the computerniks go on to more creative things.

Improved efficiency in algorithms, if centrally distributed, would then be a worthwhile effort, because of the massive savings of time and money resulting from their widespread use. We're not talking about fancy programs, but rather the everyday programs that everybody needs. If we can provide those, we've accomplished a lot.

CLARKE: There are some needs which are unique to neurobiologists, but not all of them are. One reason there have been so many histogram packages developed is that there has not been developed a unified means of making modules that will fit into any syntactical system. There's no structured programming.

MORAFF: The main reason everybody writes his own is that the alternative involves making a literature search, getting the code, making it compatible with his configuration, and then discovering there's a glitch in the thing that the person he got it from is used to working around. Writing your own, although a horrible way to do it, is still better than that. A good language would let the neurobiologist write his own without wasting a mountain of effort doing it.

DUNNING: As one of the great unwashed, perhaps the only one here, I'd like to reply to some of the comments I've heard. I've been coming to hear most of the papers, and I have learned a little, not as much as I had hoped. Mainly I think you've all been talking to each other and you're all committed, as has been said before. I think computers are important, but you've all been talking kilobucks and kilobucks and kilobucks, and most of us haven't got it. I think the idea of a language is also important, even if I do have to wait a few years before I can run it on the 360, if there were only some way to do it.

KEHL: The language doesn't solve the kilobuck problem; as a matter of fact it generates new kilobuck problems.

MORAFF: We all have the money problem. I dealt with it from the other side for a few years, and I know painfully well, even for a few people in this room, just how serious a problem it is. Perhaps if we can come up with a more uniform product Washington will buy more of it.

BROWN: One of the things we have to do is convince review committees that a computer in a proper configuration is just as routine a research tool as an oscilloscope.

KEHL: The committees are already convinced. I've been on these committees, and if you come in with a good biological proposal to do something worthwhile, to find out something people need to know, then they'll go for whatever you say. If you come in with a proposal to acquire a computer and there's no neurobiology, then you can forget it.

BROWN: Curiously enough, my own research absolutely requires computer support, and has been funded with the computer items deleted. This has happened regardless of how painfully detailed my explanation to the effect that I can't do cross correlograms on my fingers.

MORAFF: One of the biggest problems we'll face in Workshops II and III will be in coming up with a package that will not only be useful initially, but will remain useful and expandable.

OLCH: It seems to me that we're talking about developing an implementation-free command language. This means developing pieces of code for a variety of machines, constituting the library from which the higher-level language can assemble routines for different jobs. The interface between these levels should be fuzzy, so the user can write the exact piece of code needed for his hardware, and attach it easily. The command language should be supplemented by a library of user- written code for specific hardware.

MCINTYRE: When somebody in the field gets this package and it doesn't work, who will he be able to gripe to? We're not a computer manufacturer or a software house.

MORAFF: That's something this group had better work out.

HARTLINE: I agree with Dr. Moraff that extensibility is very important. Certainly neurobiologists have to have this feature, and that's one of the reasons why I decided to write my own. It's very gratifying to be able to go in there and change it as I need.

RHODE: Let me comment on our experience with the Harris computer, trying to write a program library. It was something of a fiasco, really, but there was at least one good idea: anyone who submitted a program should not only provide documentation, but also a data set that you could run through on your machine, to verify that you get the same answer he did. Also, maybe we should go in the direction of ACM's *Collected Algorithms*. We've used Knuth's book extensively, and one of the reasons our Versatec plotting package runs so fast is because someone looked in his book and found a great algorithm for sorting.

MORAFF: If we succeed on a couple of machines, other groups will copy it for other machines. That is the MUMPS experience.

BROWN: We plan to do this on two or three machines. We would be responsible for its working on individual machines in the field for some period of time.

FOOTE: This group should do two things: draw up a timetable for the language and figure out how to get help before the language is ready.

BROWN: I suppose this is a good time to discuss our plans. Workshops II and III will define the major needs of neurobiologists in computer aided research, and the outline of what the operating system should be like. We will implement the system here, under my direction, with consultation from other laboratories. We will get the system running on two or three machines here, so they'd all do the same jobs and get the same answers. We would use existing algorithms from your laboratories wherever possible. The packages would be tested in real experiments in real laboratories. Then they would be field-tested in a number of other laboratories. In a final phase, we would set up a facility to service and install the system, provide modifications, and implement on other machines, as the market requires.

INDEX

ALGOL, 21
Aliasing, 447-458
Alignment of serial sections, 131-133
Amplitude analysis, 169-175, 215-218, 337-369
Artifact rejection, 188-189, 418-419
ASPRIN, 535-540
Autoregression, 177-209

Bandwidth limiting, 447-458
BASIL, 599, 617-618
Behavioral applications, 211-220, 411-438, 475-541
Brain-computer interface laboratory, 411-438

CARTOS, 97-112
CELLSIM, 85-96
CLINFO, 32
Compiler generators, 19, 23, 24
Contour mapping, 321-335
Counting high-contrast objects, 135-137

Data selection, 298-304
DDL, 264-267
Describing functions, 459-473
DESOLV, 402-404, 408-410
Distributed processing, 4-15, 411-438

EEG, 177-209, 411-438
EVE, 530-532

FOCAL SB, 532-535
Frequency-domain analysis, 447-473

Graphics, 97-167, 560-561

High level language computers, 613-621

Image analysis, 113-133, 135-137
Information system design, 29-39
Interface design, uniformity of, 601-611
Interval analysis, 169-175, 215-383, 553-555, 569-586
Intracellular recordings, 169-175
ISL, 67-83

Lexical analysis, 595-596
LM-SQUARE, 591-600
Logic machines, 591-600

Macromodules, 592
Modelling, 41-83, 85-96, 124-127, 153-167, 395-410
Multiple laboratory facilities, 1-15, 17-27

NVAR 337-369
Networking, 623-626, 633
Neuroanatomical applications, 97-151, 321-335
Neuron population analysis, 337-369
Neuronal reconstruction, 97-139
Neurophysiological applications, 1-15, 41-65, 169-175, 211-383, 439-508, 543-600

ODAC, 12
OLERT, 213
OMNITAB, 30, 37, 38
Optical sectioning, 119-122

Parameter-flagged data storage, 439-446
Parzen estimation, 273-291
Perspective generation, 143-146

PL/I, 21-24
Plasma display, 166
PLATO, 153-167
Politics, 630-632
Postsynaptic potentials, 169-175
Precedence in arithmetic expressions, 596-598
Preprocessing, 3-15
PROPHET, 30-32, 37, 38

Real time, types of, 2-4
Recognition of nerve impulses, 215, 221-251
Register transfer modules, 592-593
Rotation and translation of lines, 140-143
RPL. 6, 17-27

Satellite processors, 8-15
Schema, 264-267
Simulation, 41-83, 85-96, 124-127, 153-167, 395-410
Sinc function, 447-458
SNAX, 41-65, 81, 82

Spectral analysis, 459-473, 558-560
Standardization, 619-620
Star networks, 1-15
State logic design, 601-611
Stationarity, analysis of, 271-291
Stimulus space specification, 255-261

Table, as basis for data analysis, 385-394
Time-shared systems, 1-15, 153-167, 211-220
Transportability of programs, 36, 37, 81, 82
TUTOR, 155-167

Vector generation, 146-149

Waveform analysis 160-162, 169-175, 215, 221-251, 555-558, 586-587, 594-595
White noise, 371-383, 550

XPL/S 19, 23, 24